Advances in Intelligent and Soft Computing

154

Editor-in-Chief

Prof. Janusz Kacprzyk
Systems Research Institute
Polish Academy of Sciences
ul. Newelska 6
01-447 Warsaw
Poland
E-mail: kacprzyk@ibspan.waw.pl

For further volumes:
http://www.springer.com/series/4240

Miguel P. Rocha, Nicholas Luscombe,
Florentino Fdez-Riverola, and
Juan M. Corchado Rodríguez (Eds.)

6th International Conference on Practical Applications of Computational Biology & Bioinformatics

Editors
Miguel P. Rocha
Dep. Informática / CCTC
Universidade do Minho
Braga
Portugal

Nicholas Luscombe
EMBL Outstation - Hinxton
European Bioinformatics Institute
Wellcome Trust Genome Campus
Hinxton
Cambridge
United Kingdom

Florentino Fdez-Riverola
Department of Informatics
ESEI: Escuela Superior de Ingeniería Informática
University of Vigo
Edificio Politécnico. Campus Universitario
As Lagoas s/n
Ourense
Spain

Juan M. Corchado Rodríguez
Department of Computing Science and Control
Faculty of Science
University of Salamanca
Salamanca
Spain

ISSN 1867-5662
e-ISSN 1867-5670
ISBN 978-3-642-28838-8
e-ISBN 978-3-642-28839-5
DOI 10.1007/978-3-642-28839-5
Springer Heidelberg New York Dordrecht London

Library of Congress Control Number: 2012933123

Printed on acid-free paper

Springer is part of Springer Science+Business Media (www.springer.com)

Preface

In the last few years, the exponential growth of Bioinformatics and Computational Biology fields has increased the need for computational algorithms able to efficiently handle the huge amounts of data produced by new experimental techniques. In this context, the fruitful interaction of researchers from different scientific areas is, more than ever, of foremost importance for boosting the research efforts in the field and contributing to the education of a new generation of Bioinformatics scientists.

The International Conference on Practical Applications of Computational Biology & Bioinformatics (PACBB) is an annual international meeting dedicated to emerging and challenging applied research in Bioinformatics and Computational Biology. After successful recent events, the 6th edition of PACBB Conference will be held on 28-30th March 2012 in the University of Salamanca, Spain. In this occasion, a special issue published by the Journal of Integrative Bioinformatics will cover extended versions of best refereed works.

This volume presents the contributions that have been finally accepted for the 6th edition of the PACBB Conference after being reviewed by, at least, three different reviewers, from an international committee composed of 74 members from 12 countries. PACBB'11 technical program included 32 papers from a submission pool of 61 papers spanning many different sub-fields in Bioinformatics and Computational Biology. Therefore, the conference will certainly have promoted the interaction of scientists from diverse research groups and with a distinct background (computer scientists, mathematicians, biologists, etc.). The scientific content will definitely be challenging and will promote the improvement of the work that is being developed by each of the participants.

We would like to thank all the contributing authors and sponsors: IEEE Systems, Man & Cybernetics Society, CNRS (Centre national de la recherche scientifique), Plate-forme AFIA 2001, AEPIA (Asociación Española de Inteligencia Artificial), APPIA (Associação Portuguesa Para a Inteligência Artificial), as well as the members of the Program Committee and the Organizing Committee for their hard and highly valuable work and support. Their effort has helped to contribute to the success of the PACBB'12 event. PACBB'12 wouldn't exist without your assistance. This symposium is organized by the Bioinformatics, Intelligent System and Educational Technology Research Group (http://bisite.usal.es/) of the University of Salamanca and the Next Generation Computer System Group (http://sing.ei.uvigo.es/) of the University of Vigo.

Miguel P. Rocha
Nicholas Luscombe
PACBB'12 Programme Co-chairs

Juan M. Corchado Rodríguez
Florentino Fdez-Riverola
PACBB'12 Organizing Co-chairs

Organization

General Co-chairs

Miguel P. Rocha	CCTC, Univ. Minho, Portugal
Nicholas Luscombe	European Bioinformatics Institute, United Kingdom
Florentino Fernández	University of Vigo, Spain
Juan M. Corchado	University of Salamanca, Spain

Program Committee

Miguel P. Rocha (Chairman)	CCTC, Univ. Minho, Portugal
Nicholas Luscombe (Chairman)	European Bioinformatics Institute, United Kingdom
Alamgir Hossain	Bradford University, UK
Alfonso Valencia	Structural Biology and BioComputing Programme (CNIO), Spain
Alicia Troncoso	University Pablo de Olavide, Spain
Ana Rojas	IMPPC, Barcelona, Spain
Anália Lourenço	IBB/CEB, University of Minho, Portugal
Arlo Randall	University of California Irvine, USA
Armando Pinho	University of Aveiro, Portugal
Caludine Chaouiya	Gulbenkian Institute, Portugal
Christopher Henry	Argonne National Labs, USA
Daniel Glez-Peña	University of Vigo, Spain
David Posada	University of Vigo, Spain
Eva Lorenzo	University of Vigo, Spain
Fernando Diaz-Gómez	University of Valladolid, Spain
Florencio Pazos	CNB/CSIC, Madrid, Spain
Frank Klawonn	Ostafilia University of Applied Sciences, Wolfenbuettel, Germany
Giovani Librelotto	Universidade Federal de Santa Maria, Brasil
Gonzalo Gómez-López	UBio/CNIO, Spanish National Cancer Research Centre, Spain
Hagit Shatkay	University of Delaware, USA

Heri Ramampiaro	Norwegian University of Science and Technology, Trondheim, Norway
Isabel C. Rocha	IBB/CEB, University of Minho, Portugal
Jorge Vieira	IBMC, Porto, Portugal
José Luis Oliveira	University of Aveiro, Portugal
José-Jesús Fernández	CNB/CSIC, Madrid, Spain
Juan Antonio García Ranea	University of Malaga, Spain
Juan F. de Paz	University of Salamanca, Spain
Juanma Vacquerizas	European Bioinformatics Institute, EMBL-EBI, UK
Julio R. Banga	IIM/CSIC, Vigo, Spain
Julio Saez-Rodriguez	European Bioinformatics Institute, EMBL-EBI, UK
Katarzyna Stapor	Silesian University, Institute of Computer Science, Poland
Karthi Sivaraman	European Bioinformatics Institute, EMBL-EBI, UK
Kaustubh Raosaheb Patil	Max Planck Insittute for Informatics, Germany
Kiran R. Patil	EMBL, Heidelberg, Germany
Loris Nanni	University of Bologna, Italy
Lourdes Borrajo	University of Vigo, Spain
Luis Figueiredo	European Bioinformatics Institute, EMBL-EBI, UK
Luis M. Rocha	Indiana University, USA
Manuel J. Maña López	University of Huelva, Spain
Manuel Rodriguez	University of Salamanca, Spain
Mª Dolores Muñoz Vicente	University of Salamanca, Spain
Marie-france Sagot	University of Lion, France
Miguel Reboiro	University of Vigo, Spain
Monica Borda	University of Cluj-Napoca, Romania
Nuno Fonseca	CRACS/INESC, Porto, Portugal
Reyes Pavón	University of Vigo, Spain
Rita Ascenso	Polytecnic Institute of Leiria, Portugal
Rosalía Laza	University of Vigo, Spain
Rui Brito	University of Coimbra, Portugal
Rui C. Mendes	CCTC, University of Minho, Portugal
Rui Camacho	LIAAD/FEUP, University of Porto, Portugal
Rui Rijo	Polytecnic Institute of Leiria, Portugal
Sara C. Madeira	IST/INESC ID, Lisbon, Portugal
Sérgio Deusdado	Polytecnic Institute of Bragança, Portugal
Silas Vilas Boias	Univerity of Auckland, New Zealand
Thierry Lecroq	Univeristy of Rouen, France
Wolfgang Wenzel	Karlsruhe Institute of Technology, Germany

Organising Committee

Juan M. Corchado (Chairman)	University of Salamanca, Spain
Florentino Fdez-Riverola (Chairman)	University of Vigo, Spain
Javier Bajo	Pontifical University of Salamanca, Spain
Juan F. De Paz	University of Salamanca, Spain
Sara Rodríguez	University of Salamanca, Spain
Dante I. Tapia	University of Salamanca, Spain
Fernando de la Prieta Pintado	University of Salamanca, Spain
Davinia Carolina Zato Domínguez	University of Salamanca, Spain
Cristian I. Pinzón	University of Salamanca, Spain
Rosa Cano	University of Salamanca, Spain
Belén Pérez Lancho	University of Salamanca, Spain
Angélica González Arrieta	University of Salamanca, Spain
Vivian F. López	University of Salamanca, Spain
Ana de Luís	University of Salamanca, Spain
Ana B. Gil	University of Salamanca, Spain
Jesús García Herrero	Universidad Carlos III de Madrid, Spain
Hugo López-Fernández	University of Vigo, Spain

Contents

Data and Text Mining I

Parallel Spectral Clustering for the Segmentation of cDNA Microarray Images 1
Sandrine Mouysset, Ronan Guivarch, Joseph Noailles, Daniel Ruiz

Prognostic Prediction Using Clinical Expression Time Series: Towards a Supervised Learning Approach Based on Meta-biclusters . . . 11
André V. Carreiro, Artur J. Ferreira, Mário A.T. Figueiredo, Sara C. Madeira

Parallel *e*-CCC-Biclustering: Mining Approximate Temporal Patterns in Gene Expression Time Series Using Parallel Biclustering 21
Filipe Cristóvão, Sara C. Madeira

Identification of Regulatory Binding Sites on mRNA Using in Vivo Derived Informations and SVMs 33
Carmen Maria Livi, Luc Paillard, Enrico Blanzieri, Yann Audic

Data and Text Mining II

Parameter Influence in Genetic Algorithm Optimization of Support Vector Machines 43
Paulo Gaspar, Jaime Carbonell, José Luís Oliveira

Biological Knowledge Integration in DNA Microarray Gene Expression Classification Based on Rough Set Theory 53
D. Calvo-Dmgz, J.F. Galvez, Daniel Glez-Peña, Florentino Fdez-Riverola

Quantitative Assessment of Estimation Approaches for Mining over Incomplete Data in Complex Biomedical Spaces: A Case Study on Cerebral Aneurysms 63
Jesus Bisbal, Gerhard Engelbrecht, Alejandro F. Frangi

ASAP: An Automated System for Scientific Literature Search in PubMed Using Web Agents 73
Carlos Carvalhal, Sérgio Deusdado, Leonel Deusdado

Case-Based Reasoning to Classify Endodontic Retreatments 79
Livia Campo, Vicente Vera, Enrique Garcia, Juan F. De Paz, Juan M. Corchado

A Comparative Analysis of Balancing Techniques and Attribute Reduction Algorithms 87
R. Romero, E.L. Iglesias, L. Borrajo

Phylogenetics and Other Applications

Sliced Model Checking for Phylogenetic Analysis 95
José Ignacio Requeno, Roberto Blanco, Gregorio de Miguel Casado, José Manuel Colom

PHYSER: An Algorithm to Detect Sequencing Errors from Phylogenetic Information 105
Jorge Álvarez-Jarreta, Elvira Mayordomo, Eduardo Ruiz-Pesini

A Systematic Approach to the Interrogation and Sharing of Standardised Biofilm Signatures 113
Anália Lourenço, Andreia Ferreira, Maria Olivia Pereira, Nuno F. Azevedo

Visual Analysis Tool in Comparative Genomics 121
Juan F. De Paz, Carolina Zato, María Abáigar, Ana Rodríguez-Vicente, Rocío Benito, Jesús M. Hernández

From Networks to Trees 129
Marco Alves, Joãd Alves, Rui Camacho, Pedro Soares, Luísa Pereira

Biomedical Applications

Procedure for Detection of Membranes in Three-Dimensional Subcellular Density Maps 137
A. Martinez-Sanchez, I. Garcia, J.J. Fernandez

A Cellular Automaton Model for Tumor Growth Simulation 147
Ángel Monteagudo, José Santos

Ectopic Foci Study on the Crest Terminalis in 3D Computer Model of Human Atrial 157
Carlos A. Ruiz-Villa, Andrés P. Castaño, Andrés Castillo, Elvio Heidenreich

SAD_BaSe: A Blood Bank Data Analysis Software 165
Augusto Ramoa, Salomé Maia, Anália Lourenço

A Rare Disease Patient Manager 173
Pedro Lopes, Rafael Mendonça, Hugo Rocha, Jorge Oliveira, Laura Vilarinho, Rosário Santos, José Luís Oliveira

MorphoCol: A Powerful Tool for the Clinical Profiling of Pathogenic Bacteria 181
Ana Margarida Sous, Anália Lourenço, Maria Olívia Pereira

Sequence Analysis and Next Generation Sequencing

Applying AIBench Framework to Develop Rich User Interfaces in NGS Studies 189
Hugo López-Fernández, Daniel Glez-Peña, Miguel Reboiro-Jato, Gonzalo Gómez-López, David G. Pisano, Florentino Fdez-Riverola

Comparing Bowtie and BWA to Align Short Reads from a RNA-Seq Experiment 197
N. Medina-Medina, A. Broka, S. Lacey, H. Lin, E.S. Klings, C.T. Baldwin, M.H. Steinberg, P. Sebastiani

SAMasGC: Sequencing Analysis with a Multiagent System and Grid Computing 209
Roberto González, Carolina Zato, Rocío Benito, María Hernández, Jesús M. Hernández, Juan F. De Paz

Exon: A Web-Based Software Toolkit for DNA Sequence Analysis 217
Diogo Pratas, Armando J. Pinho, Sara P. Garcia

On the Development of a Pipeline for the Automatic Detection of Positively Selected Sites 225
David Reboiro-Jato, Miguel Reboiro-Jato, Florentino Fdez-Riverola, Nuno A. Fonseca, Jorge Vieira

Compact Representation of Biological Sequences Using Set Decision Diagrams 231
José Ignacio Requeno, José Manuel Colom

Systems Biology and Omics

Computational Tools for Strain Optimization by Adding Reactions 241
Sara Correia, Miguel Rocha

Computational Tools for Strain Optimization by Tuning the Optimal Level of Gene Expression 251
Emanuel Gonçalves, Isabel Rocha, Miguel Rocha

Efficient Verification for Logical Models of Regulatory Networks 259
Pedro T. Monteiro, Claudine Chaouiya

Tackling Misleading Peptide Regulation Fold Changes in Quantitative Proteomics 269
Christoph Gernert, Evelin Berger, Frank Klawonn, Lothar Jänsch

Coffee Transcriptome Visualization Based on Functional Relationships among Gene Annotations 277
Luis F. Castillo, Oscar Gómez-Ramírez, Narmer Galeano-Vanegas, Luis Bertel-Paternina, Gustavo Isaza, Álvaro Gaitán-Bustamante

Author Index 285

Parallel Spectral Clustering for the Segmentation of cDNA Microarray Images

Sandrine Mouysset, Ronan Guivarch, Joseph Noailles, and Daniel Ruiz

Abstract. Microarray technology generates large amounts of expression level of genes to be analyzed simultaneously. This analysis implies microarray image segmentation to extract the quantitative information from spots. Spectral clustering is one of the most relevant unsupervised method able to gather data without a priori information on shapes or locality. We propose and test on microarray images a parallel strategy for the Spectral Clustering method based on domain decomposition and with a criterion to determine the number of clusters.

1 Introduction

Image segmentation in microarray analysis is a crucial step to extract quantitative information from the spots [7], [9], [3]. Clustering methods are used to separate the pixels that belong to the spot from the pixels of the background and noise. Among these, some methods imply some restrictive assumptions on the shapes of the spots [10], [6]. Due to the fact that the most of spots in a microarray image have irregular-shapes, the clustering based-method should be adaptive to arbitrary shape of spots and should not depend on many input parameters. Spectral methods, and in particular the spectral clustering algorithm introduced by Ng-Jordan-Weiss [5], are useful when considering no a priori shaped subsets of data. Spectral clustering exploits eigenvectors of a Gaussian affinity matrix in order to define a low-dimensional space in which data points can be easily clustered. But when very large data sets are considered, the extraction of the dominant eigenvectors becomes the most computational task in the algorithm. To address this bottleneck, several approaches about parallel Spectral Clustering [8], [2] were recently suggested, mainly

Sandrine Mouysset
University of Toulouse - UPS - IRIT, 118 Route de Narbonne, 31062 Toulouse, France
e-mail: sandrine.mouysset@irit.fr

Ronan Guivarch, Joseph Noailles and Daniel Ruiz
University of Toulouse - INPT(ENSEEIHT) - IRIT, 2 rue Camichel, 31071 Toulouse, France
e-mail: {ronan.guivarch,joseph.noailles,daniel.ruiz}@enseeiht.fr

M.P. Rocha et al. (Eds.): 6th International Conference on PACBB, AISC 154, pp. 1–9.
springerlink.com

focused on linear algebra techniques to reduce computational costs. In this paper, by exploiting the geometrical structure of microarray images, a parallel strategy based on domain decomposition is investigated. Moreover, we propose solutions to overcome the two main problems from the divide and conquer strategy: the difficulty to choose a Gaussian affinity parameter and the number of clusters k which remains unknown and may drastically vary from one subdomain to the other.

2 Parallel Spectral Clustering: Justifications

2.1 *Spectral Clustering*

Let's first give some notations and recall the Ng-Jordan-Weiss algorithm [5]. Let's consider a microarray image I of size $l \times m$. Assume that the number of targeted clusters k is known. The algorithm contains few steps which are described in Algorithm 1.

Algorithm 1. Spectral Clustering Algorithm

Input: Microarray image I, number of clusters k

1. Form the affinity matrix $A \in \mathbb{R}^{n \times n}$ with $n = l \times m$ defined by equation (1).
2. Construct the normalized matrix: $L = D^{-1/2} A D^{-1/2}$ with $D_{i,i} = \sum_{r=1}^{n} A_{ir}$,
3. Assemble the matrix $X = [X_1 X_2 .. X_k] \in \mathbb{R}^{n \times k}$ by stacking the eigenvectors associated with the k largest eigenvalues of L,
4. Form the matrix Y by normalizing each row in the $n \times k$ matrix X,
5. Treat each row of Y as a point in $\mathbb{R}^k$, and group them in k clusters via the *K-means* method,
6. Assign the original point I_{ij} to cluster t when row i of matrix Y belongs to cluster t.

First, the method consists in constructing the affinity matrix based on the Gaussian affinity measure between I_{ij} and I_{rs} the intensities of the pixel of coordinates (i, j) and (r, s) for $i, r \in \{1, .., l\}$ and $j, s \in \{1, .., m\}$. After a normalization step, the k largest eigenvectors are extracted. So every data point I_{ij} is plotted in a spectral embedding space of $\mathbb{R}^k$ and the clustering is made in this space by applying K-means method. Finally, thanks to an equivalence relation, the final partition of data set is defined from the clustering in the embedded space.

2.2 *Affinity Measure*

For image segmentation, the microarray image data can be considered as isotropic enough in the sense that there does not exist some privileged directions with very different magnitudes in the distances between points along theses directions. The step between pixels and brightness are about the same magnitude. So, we can include both 2D geometrical information and 1D brightness information in the spectral clustering method. We identify the microarray image as a 3-dimensional rectangular set in which both geometrical coordinates and brightness information are normalized. It is equivalent to setting a new distance, noted d, between pixels by equation (2). So by considering the size of the microarray image, the Gaussian affinity A_{ir} is defined as follows:

$$A_{ir} = \begin{cases} \exp\left(-\frac{d(I_{ij},I_{rs})^2}{(\sigma/2)^2}\right) & \text{if } (ij) \neq (rs), \\ 0 \text{ otherwise}, \end{cases} \tag{1}$$

where σ is the affinity parameter and the distance d between the pixel (ij) and (rs) is defined by:

$$d\left(I_{ij},I_{rs}\right) = \sqrt{\left(\frac{i-r}{l}\right)^2 + \left(\frac{j-s}{m}\right)^2 + \left(\frac{I_{ij}-I_{rs}}{256}\right)^2} \tag{2}$$

This definition (2) permits a segmentation which takes into account the geometrical shapes of the spots and the brightness information among them. In the same way, for colored microarray images with Cy3 and Cy5 hybridizations, we can consider 5D data with 2D geometrical coordinates and 3D color levels.

3 Method

By exploiting the block structure of microarrays, clustering can be made on subdomains by breaking up the data set into data subsets with respect to their geometrical coordinates in a straightforward way. With an appropriate Gaussian affinity parameter and a method to determine the number of clusters, each processor applies independently the spectral clustering (Algorithm 1) on a subset of data points and provides a local partition on this data subset. Based on these local partitions, a grouping step ensures the connection between subsets of data and determines a global partition thanks to the following transitive relation: $\forall I_{i_1 j_1}, I_{i_2 j_2}, I_{i_3 j_3} \in I$,

$$\text{If } I_{i_1 j_1}, I_{i_2 j_2} \in C^1 \text{ and } I_{i_2 j_2}, I_{i_3 j_3} \in C^2 \text{ then } C^1 \cup C^2 = P \text{ and } I_{i_1 j_1}, I_{i_2 j_2}, I_{i_3 j_3} \in P \tag{3}$$

where I is the microarray image, C^1 and C^2 two distinct clusters and P a larger cluster which includes both C^1 and C^2. We experiment this strategy whose principle is represented in Fig.1(a) on several microarray images of the Saccharomyces cerevisiae database from the Stanford Microarray database[1] like the one in Fig.1(b).

It is important to see how the parallel approach can take advantage of the specificities of this particular application. Indeed, when splitting the original image into overlapping sub-pieces of images, the local spectral clustering analysis of each sub-piece involves the creation of many affinity matrices of smaller size. The total amount of memory needs for all these local matrices is much less than the memory needed for the affinity matrix covering the global image. Additionally, the analysis of each

[1] http://smd.stanford.edu/index.shtml

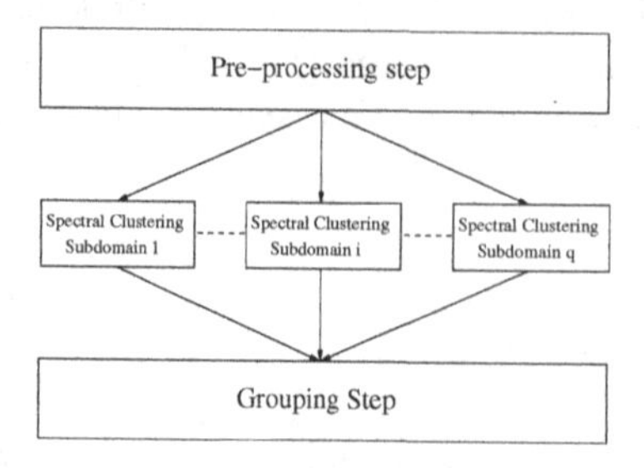

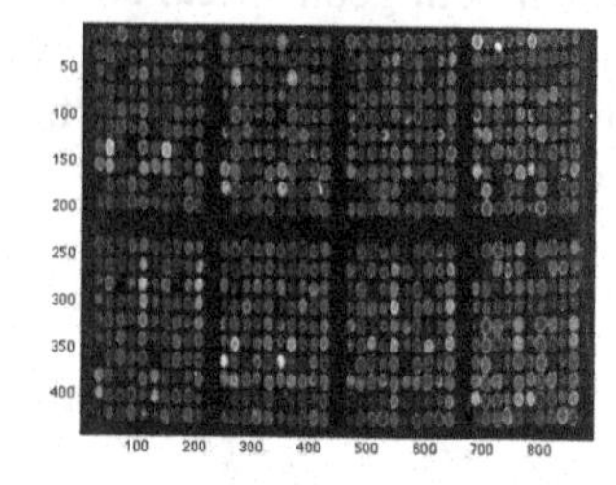

Fig. 1 Principle of the parallel strategy for microarray image : (a) Principle of parallel spectral clustering and (b) Block structure of microarray image.

subproblem is made from the extraction of eigenvectors in the scaled affinity submatrices, keeping in mind that one eigenvector is needed for each identified cluster in the corresponding sub-image. In that respect, the parallel approach enables us to decrease drastically the cost of this eigenvector computation.

3.1 Choice of the Affinity Parameter

The Gaussian affinity matrix is widely used and depends on a free parameter which is the affinity parameter, noted σ, in equation (1). It is known that this parameter conditions the separability between clusters in spectral embedding space and affects the results. A global heuristics for this parameter was proposed in [1] in which both the dimension of the problem as well as the density of points in the given p-th dimensional data set are integrated. With an assumption that the data set is isotropic enough, the image data set I is included in a p-dimensional box bounded by D_{max} the largest distance d (defined by (2)) between pairs of points in I:

$$D_{\max} = \max_{\substack{1 \leq i,r \leq l \\ 1 \leq j,s \leq m}} d(I_{ij}, I_{rs}).$$

A reference distance which represents the distance in the case of an uniform distribution is defined as follows:

$$\sigma = \frac{D_{\max}}{n^{\frac{1}{p}}}, \tag{4}$$

in which $n = l \times m$ is the size of the microarray image and $p = 3$ (resp. $p = 5$) with 2D geometrical coordinates and 1D brightness (resp. 3D color). From this definition, clusters may exist if there are points that are at a distance no more than a fraction of this reference distance σ. This global parameter is defined with the whole image data set I and gives a threshold for all spectral clustering applied independently on the several subdomains.

3.2 Choice of the Number of Clusters

The problem of the right choice of the number of clusters k is crucial. We therefore consider in each subdomain a quality measure based on ratios of Frobenius norms, see for instance [1]. After indexing data points per cluster for a value of k, we define the indexed affinity matrix whose diagonal affinity blocks represent the affinity within a cluster and the off-diagonal ones the affinity between clusters Fig.2(a). The ratios, noted r_{ij}, between the Frobenius norm of the off-diagonal blocks (ij) and that of the diagonal ones (ii) could be evaluated. Among various values for k, the final number of cluster is defined so that the affinity between clusters is the lowest and the affinity within cluster is the highest:

$$\hat{k} = argmin \sum_{i \neq j} r_{ij}. \tag{5}$$

Numerically, the corresponding loop to test several values of k until satisfying (5) is not extremely costly but only requires to concatenate eigenvectors, apply K-means, and a reordering step on the affinity matrix to compute the ratios. Furthermore, this loop becomes less and less costly when the number of processors increases. This is due to the fact that eigenvectors become much smaller with affinity matrices of smaller size. Also, subdividing the whole data set implicitly reduces the Gaussian affinity to diagonal subblocks (after permutations). For the 4×2 greyscaled spotted microarray image which corresponds to one subdomain, the original data set and its clustering result are plotted in Fig.2(b) for $k = 8$.

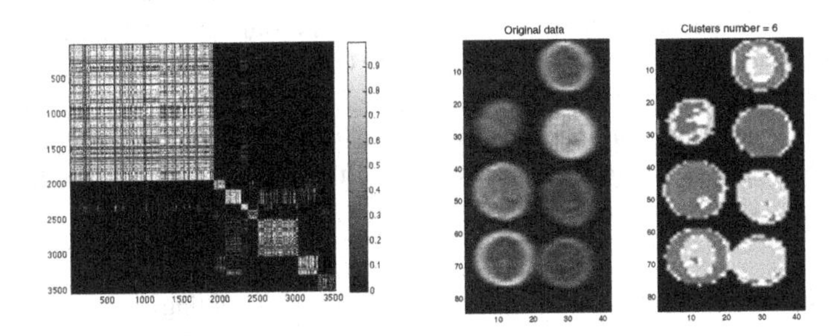

Fig. 2 Clustering on one sub-domain made by 4×2 greyscaled spotted microarray image (3500 pixels): (a) Block structure of the indexed affinity matrix for $k = 8$ and (b) Original data and its clustering result.

3.3 Parallel Implementation of the Spectral Clustering Algorithm

The FORTRAN 90 implementation of the parallel Spectral Clustering Algorithm follows the Master-Slave paradigm with the MPI library to perform the communications between processors (algorithms 2 and 3).

Algorithm 2. Parallel Algorithm: Slave

1: Receive the sigma value and its data subset from the Master (MPI_RECV)
2: Perform the Spectral Clustering Algorithm on its subset
3: Send the local partition and its number of clusters to the Master (MPI_SEND)

Algorithm 3. Parallel Algorithm: Master

1: Pre-processing step
    1.1 Read the global data and the parameters
    1.2 Split the data into subsets regarding the geometry
    1.3 Compute the affinity parameter σ
2: Send the sigma value and the data subsets to the other processors (MPI CALL)
3: Perform the Spectral Clustering Algorithm on subset 1
4: Receive the local partitions and the number of clusters from each processor (MPI CALL)

5: Grouping Step
    5.1 Gather the local partitions in a global partition thanks to the transitive relation
    5.2 Give as output a partition of the whole data set S and the final number of clusters k

4 Numerical Experiments

The numerical experiments were carried out on the Hyperion supercomputer[2] of the CICT. For our tests, the domain is successively divided in $q = \{18, 32, 45, 60, 64\}$ subboxes. The timings for each step of parallel algorithm are measured. We test this Parallel Spectral Clustering on one microarray image from the Stanford Microarray Database. For a decomposition in 64 subboxes, the clustering result is plotted in Fig.3. The original microarray image of 392931 pixels which represents 8 blocks of 100 spots is plotted on the left of the figure. After the grouping step, the parallel spectral clustering result has determined 11193 clusters which are plotted on the right of Fig.3. Compared to the original data set, the shapes of the various hybridization spots are well described.

Table 1 Microrarray image segmentation results for different splittings.

Number of proc.	Number of points	Time σ	Time parallel SC	Time Grouping	Total Time	Memory Cons.
18	22000	1413	36616	892	38927	7.75
32	12500	1371	7243	794	9415	2.50
45	9000	1357	2808	953	5127	1.30
60	6800	1360	1153	972	3495	0.74
64	6300	1372	1030	744	3157	0.64

[2] `http://www.calmip.cict.fr/spip/spip.php?rubrique90`

We give in Table 1, for each distribution, the number of points on each processor, the time in seconds to compute σ defined by (4), the time in the parallel Spectral Clustering step, the time of the grouping phase, the total time and the memory consumption in GigaOctets.

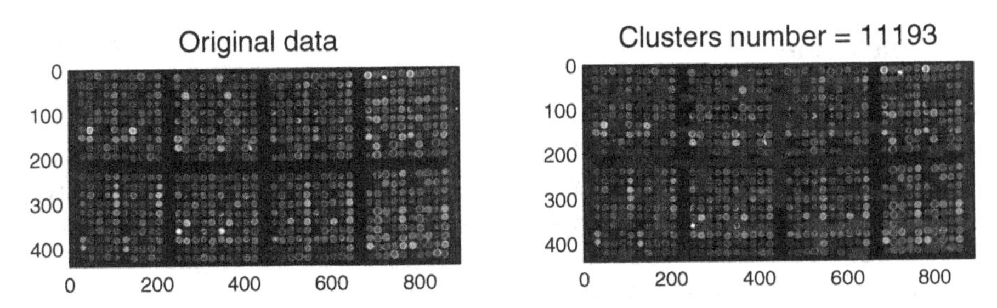

Fig. 3 Original microarray image and its clustering result.

The first remark is that the total time decreases drastically when we increase the number of processors. Logically, this is time of the parallel part of the algorithm (step 3.) that decreases while the two other steps (1 and 5), that are sequential, remain practically constant.

To study the performance of our parallel algorithm, we compute the speedup. Because we cannot have a result with only one processor in order to have a sequential reference (lack of memory), we take the time with the 18 processors, the minimum number of processors in order to have enough memory by processor. The speedup for q processors will then be defined as $S_q = \frac{T_{18}}{T_q}$.

We can notice in Fig.4(a) that the speedups increase faster than the number of processors: for instance, from 18 to 64 processors, the speedup is 12 although the number of processors grows only with a ratio 3.55. This good performance is confirmed if we draw the mean computational costs per point of the image.We define, for a given number of processors, the parallel computational cost (resp. total computational cost) the time spent in the parallel part (parallel Spectral Clustering part) (resp. total time) divided by the average number of points on each subdomain. We give in Fig.4 (b), these parallel (plain line) and total (dashed line) computational costs.

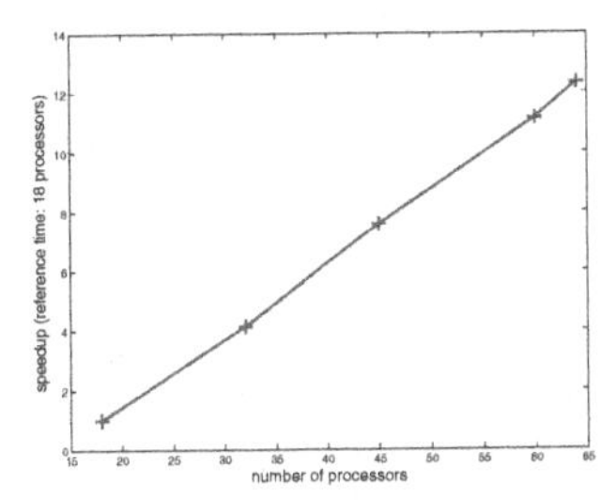

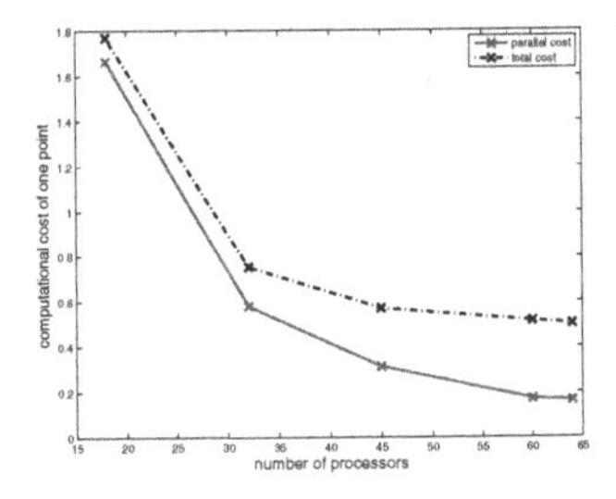

Fig. 4 Performances of the parallel part : (a) Speedup with the 18 processors time as reference and (b) Parallel and total computational costs.

We can observe from Table 1 that the less points we have by subset, the faster we go and the decreasing is better than linear. This can be explained by the non-linearity of our problem which is the computation of eigenvectors from the Gaussian affinity matrix. There are much better gains in general when smaller subsets are considered.

5 Conclusion

With the domain decomposition strategy and heuristics to determine the choice of the Gaussian affinity parameter and the number of clusters, the parallel spectral clustering becomes robust for microarray image segmentation and combines intensity and shape features. The numerical experiments show the good behaviour of our parallel strategy when increasing the number of processors and confirm the suitability of our method to treat microarray images.

However, we find two limitations: the lack of memory when the subset given to a processor is large and the time spent in the sequential parts which stays roughly constant and tends to exceed the parallel time with large number of processors. To reduce the problem of memory but also to reduce the spectral clustering time, we can study some techniques for sparsifying the Gaussian affinity matrix: some sparsification techniques, such as thresholding the affinity between data points, could also be introduced to speed up the algorithm when the subdomains are still large enough. With sparse structures to store the matrix, we will also gain a lot of memory. However, we may have to adapt our eigenvalues solver and use for example ARPACK library [4]. To reduce the time of the sequential parts, we could investigate parallelization of the computation of the σ parameter and ability to separate the spotted microarray image in sub-images.

References

1. Mouysset, S., Noailles, J., Ruiz, D.: Using a Global Parameter for Gaussian Affinity Matrix in Spectral Clustering. In: Palma, J.M.L.M., Amestoy, P.R., Daydé, M., Mattoso, M., Lopes, J.C. (eds.) VECPAR 2008. LNCS, vol. 5336, pp. 378–390. Springer, Heidelberg (2008)
2. Chen, W.-Y., Yangqiu, S., Bai, H., Lin, C.-J., Chang, E.Y.: Parallel Spectral Clustering in Distributed Systems. Preprint of IEEE Transactions on Pattern Analysis and Machine Intelligence (2010)
3. Giannakeas, N., Fotiadis, D.: Image Processing and Machine Learning Techniques for the Segmentation of cDNA Microarray Images (2008)
4. Lehoucq, R., Sorensen, D., Yang, C.: ARPACK users' guide: solution of large-scale eigenvalue problems with implicitly restarted Arnoldi methods. SIAM (1998)
5. Ng, A.Y., Jordan, M.I., Weiss, Y.: On spectral clustering: analysis and an algorithm. In: Proceedings in Advance Neural Information Processing Systems (2002)
6. Rueda, L., Qin, L.: A New Method for DNA Microarray Image Segmentation. In: Kamel, M.S., Campilho, A.C. (eds.) ICIAR 2005. LNCS, vol. 3656, pp. 886–893. Springer, Heidelberg (2005)
7. Rueda, L., Rojas, J.C.: A Pattern Classification Approach to DNA Microarray Image Segmentation. In: Kadirkamanathan, V., Sanguinetti, G., Girolami, M., Niranjan, M., Noirel, J. (eds.) PRIB 2009. LNCS, vol. 5780, pp. 319–330. Springer, Heidelberg (2009)

8. Song, Y., Chen, W., Bai, H., Lin, C., Chang, E.: Parallel spectral clustering. In: Proceedings of European Conference on Machine Learning and Pattern Knowledge Discovery, Springer, Heidelberg (2008)
9. Uslan, V., Bucak, O., Cekmece, B.: Microarray image segmentation using clustering methods. Mathematical and Computational Applications 15(2), 240–247 (2010)
10. Yang, Y., Buckley, M., Speed, T.: Analysis of cDNA microarray images. Briefings in Bioinformatics 2(4), 341 (2001)

Prognostic Prediction Using Clinical Expression Time Series: Towards a Supervised Learning Approach Based on Meta-biclusters

André V. Carreiro, Artur J. Ferreira, Mário A.T. Figueiredo, and Sara C. Madeira

Abstract. Biclustering has been recognized as a remarkably effective method for discovering local temporal expression patterns and unraveling potential regulatory mechanisms, critical to understand complex biomedical processes, such as disease progression and drug response. In this work, we propose a classification approach based on meta-biclusters (a set of similar biclusters) applied to prognostic prediction. We use real clinical expression time series to predict the response of patients with multiple sclerosis to treatment with Interferon-β. The main advantages of this strategy are the interpretability of the results and the reduction of data dimensionality, due to biclustering. Preliminary results anticipate the possibility of recognizing the most promising genes and time points explaining different types of response profiles, according to clinical knowledge. The impact on the classification accuracy of different techniques for unsupervised discretization of the data is studied.

1 Introduction

Over the past years we have witnessed an increase in time-course gene expression experiments and analysis. Until then, gene expression experiments were limited to static analysis. The inclusion of temporal dynamics on gene expression is now enabling the study of complex biomedical problems, such as disease progression and

André V. Carreiro · Sara C. Madeira
KDBIO group, INESC-ID, Lisbon, and Instituto Superior Técnico,
Technical University of Lisbon, Portugal
e-mail: acarreiro@kdbio.inesc-id.pt, sara.madeira@ist.utl.pt

Artur J. Ferreira
Instituto de Telecomunicações, Lisbon, and Instituto Superior de Engenharia de Lisboa,
Lisbon, Portugal
e-mail: arturj@isel.pt

Mário A.T. Figueiredo
Instituto de Telecomunicações, Lisbon, and Instituto Superior Técnico,
Technical University of Lisbon, Portugal
e-mail: mtf@lx.it.pt

M.P. Rocha et al. (Eds.): 6th International Conference on PACBB, AISC 154, pp. 11–20.
springerlink.com

drug response, from a different perspective. However, studying this type of data is challenging, both from a computational and biomedical point of view [1].

In this context, recent biclustering algorithms, such as CCC-Biclustering [10], used in this work, have effectively addressed the discovery of local expression patterns. In the specific case of expression time series, the relevant biclusters are those with contiguous time-points.

In this work, we propose a supervised learning approach based on meta-biclusters for prognostic prediction. In this scenario, each patient is characterized by gene expression time series and each meta-bicluster represents a set of similar biclusters. These biclusters thus represent temporal expression profiles potentially involved in the transcriptomic response of a set of patients to a given disease or treatment. The advantage of this approach, when compared to previous ones, is both the interpretability of the results and the data dimensionality reduce data dimensionality. The first is crucial in medical problems whereas the latter results from biclustering.

We present results obtained when analyzing real clinical expression time series with the goal of predicting the response of multiple sclerosis (MS) patients to treatment with Interferon (IFN)-β. The interestingness of these results relies on the possibility to analyze class-discriminant biclusters and find promising genes to explain different expression profiles found for different types of treatment response, and not only on the classifier's accuracy. We also assess the impact of unsupervised discretization on this type of data and its effects on classification accuracy.

1.1 Classification of Clinical Expression Time-Series

Regarding the case study, there are three main works which focused on it in recent years. Baranzini et al. [2] collected the dataset and proposed a quadratic analysis-based scheme, named *integrated Bayesian inference system* (IBIS). Lin et al. [7] proposed a new classification method based on *hidden Markov models* (HMMs) with discriminative learning. Costa et al. [4] introduced the concept of constrained mixture estimation of HMMs. A summary of their results can be found in [3].

Following those works, Carreiro et al. [3] have recently introduced biclustering-based classification in gene expression time series. The authors proposed different strategies revealing important potentialities, especially regarding discretized data. The developed methods included a biclustering-based kNN algorithm, based on different similarity measures: between biclusters, expression profiles, or between whole discretized expression matrices (per patient), and also a *meta-profiles* strategy, where they searched for the biclusters with similar expression profiles, computed the respective class proportions, using these as a classifying threshold. In the work reported in this paper, compared with [3] we note that the main advantages of meta-biclusters is the easier interpretation of the results, as we get, from the most class-discriminant meta-biclusters, the most promising sets of genes and time points (biclusters) involved in patient classification. In the meta-profiles method [3], we first have to compute the biclusters which present the respective expression profiles.

Hanczar and Nadif [6] adapted bagging to biclustering problems. The idea is to compute biclusters from bootstrapped datasets and aggregate the results. The authors perform hierarchical clustering upon the collection of computed biclusters and select K clusters of biclusters, defined as *meta-clusters*. Finally, they compute the probabilities of an element *(Example, Gene)* belonging to each meta-cluster by assigning the elements to the most probable one. The sets of *Example* and *Genes* associated to each meta-cluster define the final biclusters. This technique has shown to reduce the biclustering error and mean squared residue (MSR) in both simulated and real datasets. However, gene expression time series or classification problems, as we introduce here, were not considered.

1.2 Feature Discretization

Unsupervised feature discretization (UFD) has been shown to reduce the memory requirements and improve classification accuracy [11]. The first discretization scheme used in this work is based on *variations between time points* (VBTP), as performed by Madeira et al. [10]. This yields patterns of gene expression temporal evolution between consecutive time-points, with three symbols: D, N, and U (denoting *decrease*, *no change*, and *increase*, respectively). Still in this context, two other efficient techniques are commonly used: *equal-interval binning* (EIB), which corresponds to uniform quantization with a given number of bits per feature; *equal-frequency binning* (EFB), that is, non-uniform quantization yielding intervals such that, for each feature, the number of occurrences in each interval is the same.

Recently, two scalar UFD methods, based on the Linde-Buzo-Gray (LBG) algorithm [8], have been proposed [5]. The first method, named U-LBG1, applies the LBG algorithm individually to each feature, and stops when the mean square error (MSE) distortion falls below a threshold Δ or when a maximum number, q, of bits per feature is reached. Thus, a pair of input parameters (Δ, q) needs to be specified; using Δ equal to 5% of the range of each feature and $q \in \{4, \dots, 10\}$ has been shown to be adequate [5]. Naturally, U-LBG1 may discretize features using a variable number of bits. The second method, named U-LBG2, results from a minor modification of U-LBG1 by using a fixed number of bits per feature, q.

Both these UFD methods exploit the same key idea that a discretization with a low MSE will provide an accurate representation of each feature, being suited for learning purposes. Previous work [5] has shown that this discretization method leads to better classification results than EFB on different kinds of (sparse and dense) data.

2 Methods

In this section we present the proposed supervised learning approach based on meta-biclusters, outlined in Fig. 1 with its three main steps. The first step is the biclustering of the multiple expression time series. In the second step, a distance matrix is built for all the computed biclusters, on which a hierarchical clustering is performed. Cutting the resulting dendrogram at a given level returns a set of meta-biclusters. A **meta-bicluster** is thus a cluster of biclusters returned by a cut in a dendrogram of biclusters, that is, a set of similar biclusters. The third step starts by building a binary matrix representing, for each patient, which meta-biclusters contain biclusters from that patient. An example of such a matrix is also represented in Fig. 1. Finally, in order to classify the instances, this binary matrix is used as input in a classifier.

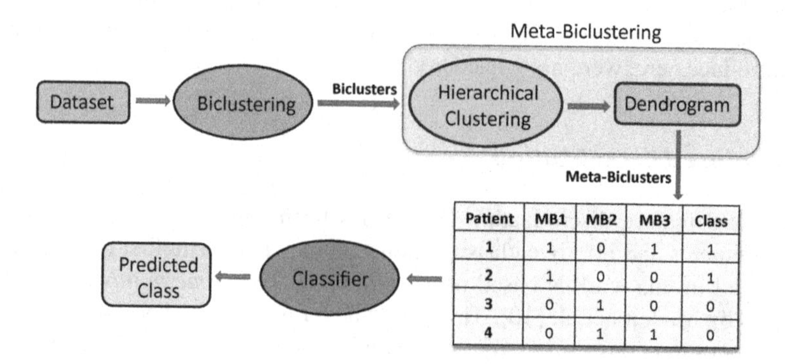

Fig. 1 Workflow of a classifier based on meta-biclusters.

2.1 *Biclustering*

Each patient is represented by a matrix with N_G (number of genes) rows and N_T (number of time points) columns. Since we are using gene expression time series, the biclustering algorithm used is CCC-Biclustering [10]. We end up with a set of CCC-Biclusters (named biclusters for simplicity) for each patient. Two examples of biclusters are represented in Fig. 2.

2.1.1 CCC-Biclustering

The values of an expression matrix A can be discretized to a set of symbols of interest, Σ, that represent distinctive activation levels. After discretization, matrix A' is transformed into matrix A, where $A_{ij} \in \Sigma$ represents the discretized value of the expression level of gene i in time point j. In Fig. 2 a three symbol alphabet $\Sigma = \{D,N,U\}$ was used, corresponding to the case of VBTP discretization.

The goal of biclustering algorithms is to identify a set of biclusters $B_k = (I_k,J_k)$, where each bicluster is defined by a subset of genes and a subset of conditions, such that each bicluster satisfies specific characteristics of homogeneity [9]. For time series gene expression data analysis, Madeira et al. [10] defined the concept of CCC-Bicluster as follows: A *contiguous column coherent bicluster* (CCC-Bicluster) A_{IJ} is a subset of rows $I = \{i_1,\ldots,i_k\}$ and a subset of **contiguous** columns $J = \{r,r+1,\ldots,s-1,s\}$ such that $A_{ij} = A_{lj}$, $\forall i,l \in I$ and $\forall j \in J$. A CCC-Bicluster defines a string (a symbolic pattern) common to every gene in I for the time points in J. A CCC-Bicluster A_{IJ} is **maximal** if no other CCC-Bicluster exists that properly contains A_{IJ}: for all other CCC-Biclusters A_{LM}, $I \subseteq L \wedge J \subseteq M \Rightarrow I = L \wedge J = M$.

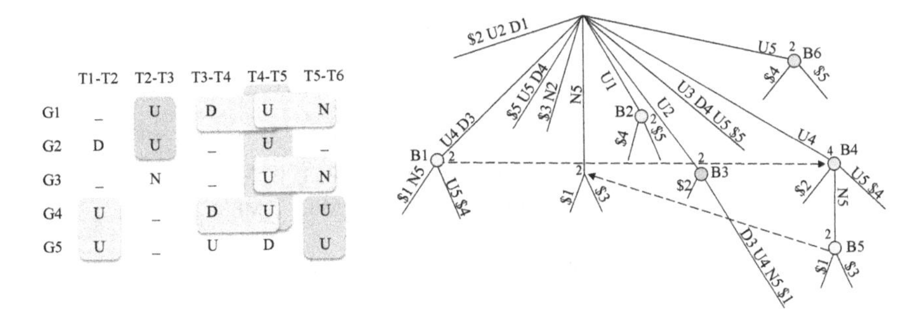

Fig. 2 Maximal CCC-Biclusters in the discretized matrix and related nodes in the suffix tree, for the case of VBTP (Variation Between Time-Points) discretization.

Consider now the matrix obtained by preprocessing matrix A using a simple alphabet transformation, that appends the column number to each symbol in the matrix and the generalized suffix tree built for the set of strings corresponding to each row in A. CCC-Biclustering is a linear time biclustering algorithm that finds and reports all maximal CCC-Biclusters based on their relationship with the nodes in the generalized suffix tree (see Fig. 2 and [10] for more details).

2.2 *Meta-biclustering*

¿From the set of computed biclusters for all the patients, we compute the similarity matrix, S, where S_{ij} is the similarity between biclusters B_i and B_j. This similarity is computed with an adapted version of the Jaccard Index given by

$$S_{ij} = J(B_i, B_j) = \frac{|B_{11P}|}{|B_{01}| + |B_{10}| + |B_{11}|}, \tag{1}$$

where $|B_{11P}|$ is the number of elements common to the two biclusters that have the same symbol. $|B_{10}|$ and $|B_{01}|$ represent the number of elements belonging exclusively to bicluster B_i and B_j, respectively. Finally, $|B_{11}|$ represents the number of elements in common to both biclusters, regardless of the symbol. Note that it is important to consider the discretized symbols, since we are also comparing biclusters from different patients, and biclusters sharing the same genes and time points might not represent similar expression patterns. The similarity matrix S ($0 \leq S_{ij} \leq 1$) is then turned into a distance matrix D, where $D_{ij} = 1 - S_{ij}$.

Using D, we perform a hierarchical clustering of the biclusters, building a dendrogram representing their similarity relationship. An example of such a dendrogram is shown in Fig. 3. Using the dendrogram and a cutting-level, we obtain K meta-biclusters (clusters of similar biclusters).

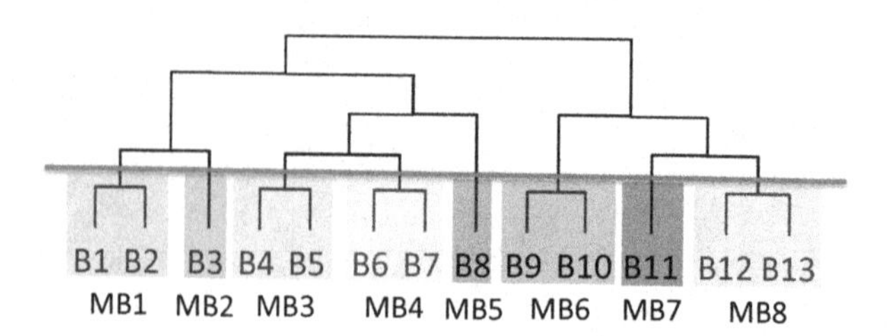

Fig. 3 Meta-biclusters represented as clusters of biclusters in the dendogram.

2.3 Classification

The final step is the inference of the patients' response class. For this purpose, we build a binary matrix, C, with N_P rows (number of patients) and N_{MB} columns (number of meta-biclusters). C_{ij} equals 1 if Patient$_i$ has at least one bicluster represented by Meta-Bicluster$_j$, and equals 0 otherwise. This binary matrix C is then used as input to supervised learning classifiers. In this work, we use decision trees (DT), k-nearest neighbors (kNN), support vector machines (SVM), and radial basis function network (RBFN) classifiers, available in the Weka toolbox (`www.cs.waikato.ac.nz/ml/weka`).

3 Results and Discussion

In this section, we present and discuss the specificities of the MS case study, including the dataset description and preprocessing. The main results obtained with the proposed classification approach are also shown and discussed.

3.1 Dataset and Preprocessing

The dataset used as case study in this work was collected by Baranzini et al. [2]. Fifty two patients with RR-MS were followed for a minimum of two years after the treatment initiation. Then, patients were classified according to their response to the treatment, as good or bad responders. Thirty two patients were considered good responders, while the remaining twenty were classified as bad responders to IFN-β. Seventy genes were pre-selected by the authors based on biological criteria, and their expression profile was measured in seven time points (initial point and three, six, nine, twelve, eighteen and twenty-four months after treatment initiation), using one-step kinetic reverse transcription PCR [2].

In order to apply CCC-Biclustering [10], as part of the proposed meta-biclusters classifier, we normalized and discretized the expression data, using the techniques in Section 1.2. In the case of standard classifiers, not able to deal with missing values directly, these were filled with the average of the closest neighboring values, after data normalization. Although CCC-Biclustering is able to handle missing values, for comparison purposes, the results reported in this paper were obtained with filled missing values, also for the meta-biclusters classifier.

3.2 Performance Evaluation

3.2.1 Meta-biclusters Classifier

Fig. 4 summarizes the mean prediction accuracies obtained by the method based on meta-biclusters with different discretization approaches (Section 1.2). The results for a differential version of these techniques (except for VBTP) are shown below.

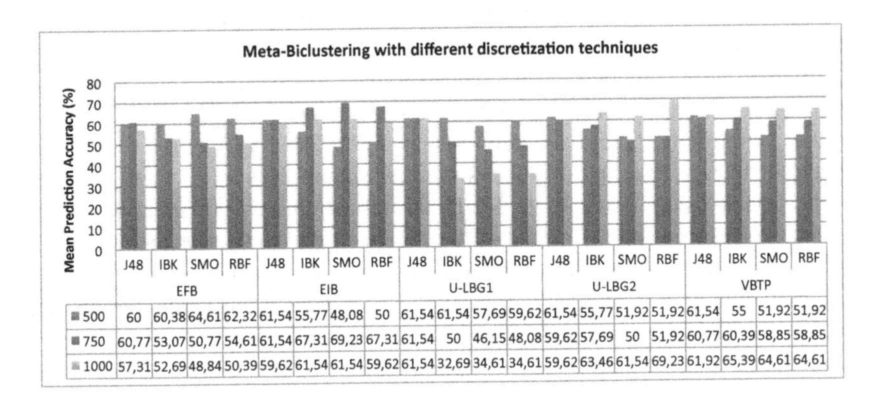

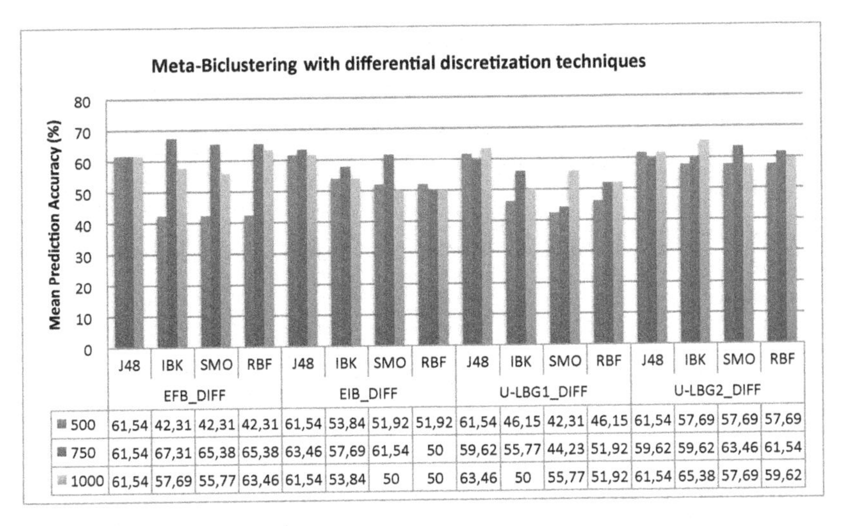

Fig. 4 Meta-Biclustering with the new discretization techniques EFB, EIB, U-LBG1 and U-LBG2 (top) and their differential versions (except for VBTP) (bottom). Classification is performed by Weka classifiers: J48 (DT), IBK (kNN), SMO (SVM), and RBF. 5 x 4-fold CV is used.

3.2.2 Tests with Real-Valued and Discretized Versions of the Original Dataset

Fig. 5 shows the mean prediction accuracy values obtained for the different state-of-the-art classifiers in the real-valued expression data, together with different discretized versions of the expression data (Section 1.2). We aimed to assess the influence of the discretization process used in the meta-biclusters classifier. The results show that the standard classifiers performed better when tested on the real-valued dataset. The use of any discretized version of the data lowers the classifiers' performance significantly (p-value < 0.05 for all classifiers except J48, paired t-test). This suggests that these classifiers cannot deal well with this discretized dataset. Note, however, that these discretization approaches were already used in other expression data, improving the classification results when faced against the real-valued dataset [5]. In fact, the meta-biclusters method outperforms all of these standard classifiers when using, for example, VBTP discretization and 1000 meta-biclusters. The main conclusion is that, instead of the choice of a discretization technique, the problem lies in the use of discretization itself. It does not seem possible to discretize the dataset without losing critical information. Nevertheless, this preprocessing step is essential to CCC-Biclustering [10].

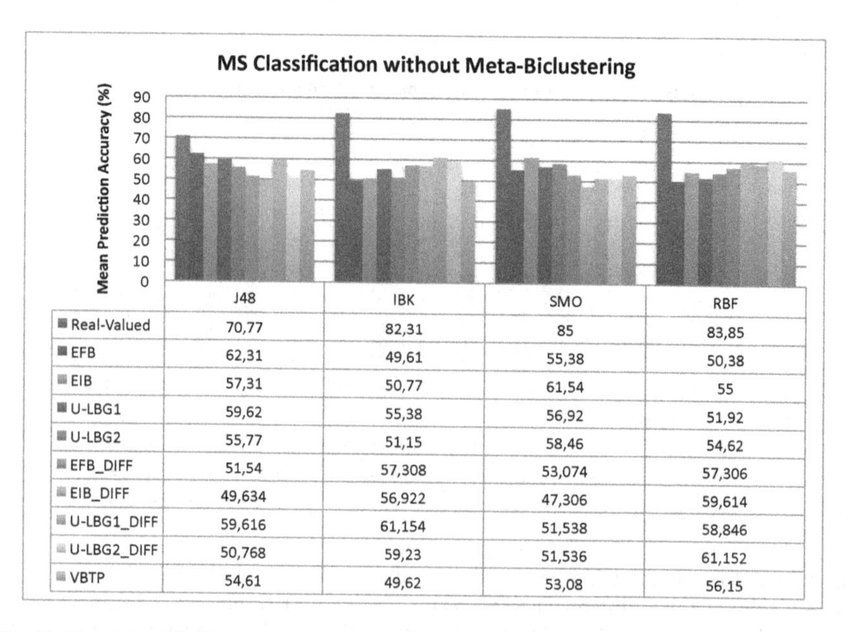

	J48	IBK	SMO	RBF
Real-Valued	70,77	82,31	85	83,85
EFB	62,31	49,61	55,38	50,38
EIB	57,31	50,77	61,54	55
U-LBG1	59,62	55,38	56,92	51,92
U-LBG2	55,77	51,15	58,46	54,62
EFB_DIFF	51,54	57,308	53,074	57,306
EIB_DIFF	49,634	56,922	47,306	59,614
U-LBG1_DIFF	59,616	61,154	51,538	58,846
U-LBG2_DIFF	50,768	59,23	51,536	61,152
VBTP	54,61	49,62	53,08	56,15

Fig. 5 Classification without Meta-Biclustering: real-valued and discretized data (EFB, EIB, U-LBG1, U-LBG2, VBTP and their differential (DIFF) versions (except for VBTP). Classification is done by Weka classifiers: J48 (DT), IBK (kNNs), SMO (SVM), and RBF. 5 x 4-fold CV is used.

4 Conclusions and Future Work

The dataset herein considered presents a singular characteristic that might justify the difficulty in the classification: good responders have many similar biclusters in common with other good responders, but also with the bad responders. Bad responders, however, have few similar biclusters in common between the class. This suggests that there are different expression signatures associated to a poor response to IFN-β treatment or an absence of signature present in good responders [3].

Other features of the data can also explain some of the challenges faced by the biclustering-based classifiers. These include class imbalance, biasing the prediction towards the good responders, and the reduced number of time points, in comparison with the number of genes, possibly resulting in overfitting. Additionally, a common problem to clinical time series analysis is the reduced number of patients, also introducing important inconsistencies. Concerning interpretability, since the results are not very satisfactory, the analysis of the most class-discriminant biclusters will be addressed in future work.

The results presented in this work show that the required discretization step is the most limiting aspect of the proposed method, carrying the risk of loss of important information. Thus, we also studied the impact of the discretization on this dataset without Meta-Biclustering. In this context, since the results obtained with state-of-the-art classifiers were comparable to the ones obtained with Meta-Biclustering, we conclude that the discretization is not a method-specific problem. Instead, it is related to the dataset itself, possibly due to its time-series nature. We note, however, that the discretization step is key to guaranteeing the completeness and efficiency of the biclustering algorithm, which would otherwise have to rely on heuristics. In this context, as future work, we would like to validate the Meta-Biclustering method on different datasets (even if they are not clinical expression time-series, where we would use other biclustering approaches). Preferably, the first experiment should be carried out with a dataset for which discretization has been shown not to be an issue.

Acknowledgements. This work was partially supported by FCT (INESC-ID multiannual funding) through the PIDDAC Program funds and NEUROCLINOMICS - Understanding NEUROdegenerative diseases through CLINical and OMICS data integration (PTDC/EIA-EIA/111239/2009). A preliminary version of this work was presented at INForum2011, a Portuguese conference with non-indexed local proceedings. The method was not changed but new extensive results studying the impact of discretization are presented.

References

1. Androulakis, I.P., Yang, E., Almon, R.R.: Analysis of time-series gene expression data: methods, challenges, and opportunities. Annual Review of Biomedical Engineering 9, 205–228 (2007)
2. Baranzini, S.E., Mousavi, P., Rio, J., Stillman, S.J.C.A., Villoslada, P., Wyatt, M.M., Comabella, M., Greller, L.D., Somogyi, R., Montalban, X., Oksenberg, J.R.: Transcription-based prediction of response to IFN-beta using supervised computational methods. PLoS Biology 3(1) (2005)

3. Carreiro, A.V., Anunciaçao, O., Carriço, J.A., Madeira, S.C.: Prognostic prediction through biclustering-based classification of clinical gene expression time series. Journal of Integrative Bioinformatics 8(3), 175 (2011)
4. Costa, I.G., Schönhuth, A., Hafemeister, C., Schliep, A.: Constrained mixture estimation for analysis and robust classification of clinical time series. Bioinformatics 25(12), i6–i14 (2009)
5. Ferreira, A., Figueiredo, M.: Unsupervised joint feature discretization and selection. In: Vitrià, J., Sanches, J.M., Hernández, M. (eds.) IbPRIA 2011. LNCS, vol. 6669, pp. 200–207. Springer, Heidelberg (2011)
6. Hanczar, B., Nadif, M.: Using the bagging approach for biclustering of gene expression data. Neurocomputing 74(10), 1595–1605 (2011)
7. Lin, T.H., Kaminski, N., Bar-Joseph, Z.: Alignment and classification of time series gene expression in clinical studies. Bioinformatics 24(13), i147–i155 (2008)
8. Linde, Y., Buzo, A., Gray, R.: An algorithm for vector quantizer design. IEEE Trans. on Communications 28, 84–94 (1980)
9. Madeira, S.C., Oliveira, A.L.: Biclustering algorithms for biological data analysis: a survey. IEEE/ACM Transactions on Computational Biology and Bioinformatics 1(1), 24–45 (2004)
10. Madeira, S.C., Teixeira, M.C., Sá-Correia, I., Oliveira, A.L.: Identification of regulatory modules in time series gene expression data using a linear time biclustering algorithm. IEEE/ACM Transactions on Computational Biology and Bioinformatics 7(1), 153–165 (2010)
11. Witten, I., Frank, E.: Data Mining: Practical Machine Learning Tools and Techniques, 2nd edn. Elsevier, Morgan Kauffmann (2005)

Parallel *e*-CCC-Biclustering: Mining Approximate Temporal Patterns in Gene Expression Time Series Using Parallel Biclustering

Filipe Cristóvão and Sara C. Madeira

Abstract. The ability to monitor the change in expression patterns over time, and to observe the emergence of coherent temporal responses using gene expression time series, obtained from either microarray or RNAseq technologies, is critical to advance our understanding of complex biomedical processes such as growth, development, response to stimulus, disease progression and drug responses. In this paper, we propose parallel *e*-CCC-Biclustering, a parallel version of the state of the art *e*-CCC-Biclustering algorithm, an efficient exhaustive search biclustering algorithm to mine approximate temporal expression patterns. Parallel *e*-CCC-Biclustering implemented using functional programming and achieved a super-linear speed-up when compared to the original sequential algorithm in test cases using synthetic data.

1 Introduction

e-CCC-Biclustering [3] is a state of the art biclustering algorithm specifically developed for time series expression data analysis that finds and reports all maximal contiguous column coherent biclusters with approximate expression patterns in time polynomial in the size of the expression matrix. The polynomial time complexity is obtained by manipulating a discretized version of the original expression matrix and by using efficient string processing techniques based on suffix trees. These approximate patterns allow a given number of errors, per gene, relatively to an expression profile representing the expression pattern in the bicluster. However, although efficient, this algorithm is inherently sequential.

Filipe Cristóvão
Software Engineering Group (ESW), INESC-ID, Lisbon, and Instituto Superior Técnico, Technical University of Lisbon, Lisbon, Portugal
e-mail: filipe.cristovao@ist.utl.pt

Sara C. Madeira
Knowledge Discovery and Bioinformatics (KDBIO) group, INESC-ID, Lisbon, and Instituto Superior Técnico, Technical University of Lisbon, Lisbon, Portugal
e-mail: sara.madeira@ist.utl.pt

M.P. Rocha et al. (Eds.): 6th International Conference on PACBB, AISC 154, pp. 21–31.
springerlink.com

In this paper we propose parallel *e*-CCC-Biclustering, a parallel version of the original *e*-CCC-Biclustering algorithm. A key feature of *e*-CCC-Biclustering is its ability to efficiently perform an exhaustive search to find all contiguous columns biclusters with approximate patterns. As such, a speedup in its runtime is key to guarantee a reasonable runtime in expression data with longer time series now becoming available, such as microarray experiments containing a large number of time points, RNA-seq data and potentially other temporal (biomedical) data, while enabling a larger number of errors, which might be crucial to find interesting expression patterns in noisy data. We note that guaranteeing the efficiency of the algorithm implies not only an efficient mining of all approximate patterns representing maximal *e*-CCC-Biclusters but also an efficient postprocessing (usually based on sorting and filtering steps using similarity and statistical criteria) of the huge amount of biclusters discovered. To our knowledge this is the first proposal of parallel biclustering in exhaustive mining of (temporal expression) patterns. Moreover, only few approaches explored the use of parallel biclustering algorithms when biclustering gene expression data using heuristic approaches [7, 6, 4].

Using functional programming, we were able to harness the already easily available multicore architectures. We could have chosen to implement specialized mutable concurrent data structures, but these are known to be error prone. Instead, we used the higher order operations of the Scala[1] collections [5] to implement parallel *e*-CCC-Biclustering. The performed tests, varying the set of genes and number of time points, achieved interesting results: although in a single-core the algorithm can have slowdowns (which were to be expected) due to the increased overhead of functional programming, we were able to achieve speedups of about 26 times faster with only 8 cores available. This super-linear speed-up is due to a problem in a mutable data-structure that the parallel algorithm does not suffer from. We show that, as the input size increases, the parallel algorithm performs better as the overhead costs are diluted and all the cores are used to do efficient work.

2 Methods

2.1 Sequential e-CCC-Biclustering

Let A' be an $|R|$ row by $|C|$ column time series gene expression matrix defined by its set of rows (genes), R, and its set of columns (time points), C. In this context, A'_{ij} represents the expression level of gene i in time point j. Matrix A is a discretized version of matrix A', where $A_{ij} \in \Sigma$ represents the discretized value of the expression level of gene i in time point j. Following the definition of CCC-Bicluster, an *e*-CCC-Bicluster and a maximal *e*-CCC-Bicluster are defined as follows [3]:

Definition 1 (CCC-Bicluster). A *contiguous column coherent bicluster* A_{IJ} is a subset of rows $I = \{i_1, \ldots, i_k\}$ and a subset of **contiguous** columns $J = \{r, r+1, \ldots, s-1, s\}$ such that $A_{ij} = A_{lj}$, for all rows $i, l \in I$ and columns $j \in J$. Each CCC-Bicluster defines a string S common to every row in I for the columns in J.

[1] http://www.scala-lang.org/

Definition 2 (e-CCC-Bicluster). A contiguous column coherent bicluster with e errors **per gene**, e-CCC-Bicluster, is a CCC-Bicluster A_{IJ} where all the strings S_i that define the expression pattern of each of the genes in I are in the e-Neighborhood of an expression pattern S that defines the e-CCC-Bicluster: $S_i \in N(e,S), \forall i \in I$. The e-Neighborhood of a string S of length $|S|$, defined over the alphabet Σ with $|\Sigma|$ symbols, $N(e,S)$, is the set of strings S_i, such that: $|S| = |S_i|$ and $Hamming(S,S_i) \leq e$, where e is an integer such that $e \geq 0$.

Definition 3 (maximal e-CCC-Bicluster). An e-CCC-Bicluster A_{IJ} is maximal if it is row-maximal, left-maximal and right-maximal. This means that no more rows or **contiguous** columns can be added to I or J, respectively, maintaining the coherence property in Definition 2.

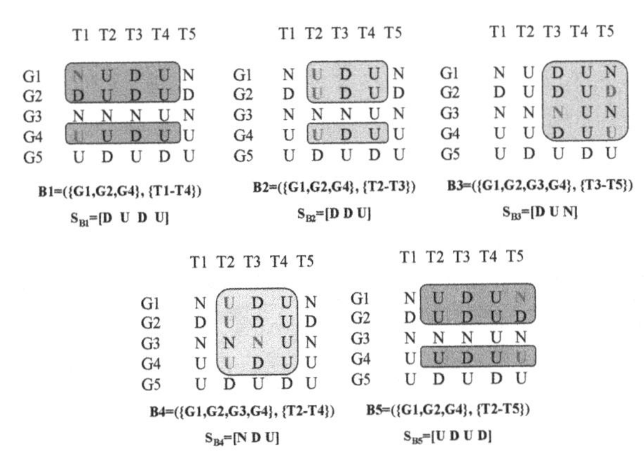

Fig. 1 Maximal e-CCC-Biclusters with row/column quorum constraints in a discretized matrix.

Under these settings, sequential e-CCC-Biclustering [3] identifies and reports all maximal e-CCC-Biclusters $B_k = A_{I_k J_k}$ such that I_k and J_k have at least q_r rows and q_c columns, respectively. Fig. 1 shows the five maximal 1-CCC-Biclusters with at least 3 rows/columns ($q_r = q_c = 3$) that can be identified in the discretized matrix presented in the example. These 1-CCC-Biclusters are defined, respectively, by the following patterns: S_{B1} = [D U D U], S_{B2} = [D D U], S_{B3} = [D U N], S_{B4} = [N D U] and S_{B5} = [U D U D]. Also clear from this figure is the fact that the same e-CCC-Bicluster can be defined by several patterns. For example, 1-CCC-Bicluster B1 can also be identified by the patterns [N U D U] and [U U D U]. An interesting example is the case of 1-CCC-Bicluster B2, which can also be defined by the patterns [N D U], [U N U], [U U U], [U D D] and [U D N]. Note however, that B2 cannot be identified by the pattern [U D U]. If this was the case, B2 would not be right maximal, since the pattern [U D N] can be extended to the right by allowing one error at time point 5. In fact, this leads to the discovery of the maximal 1-CCC-Bicluster B5. Moreover, e-CCC-Biclusters can be defined by expression patterns not occurring in the discretized matrix. This is the case of 1-CCC-Biclusters B2 and

B4, defined respectively by the patterns S_{B2} = [D D U] and S_{B4} = [N D U], which do not occur in the matrix in the contiguous time points defining B2 and B4.

Figure 2 presents the workflow of *e*-CCC-Biclustering and highlights its four main steps (see [3], for details). The asymptotic complexity of this exhaustive search biclustering algorithm is $O(\max(|R|^2|C|^{1+e}|\Sigma|^e, |R||C|^{2+e}|\Sigma|^e))$: computing all valid models corresponding to right-maximal *e*-CCC-Biclusters using procedure `computeRightMaximalBiclusters` is $O(|R|^2|C|^{1+e}|\Sigma|^e)$; deleting from *modelsOcc* all valid models that are not left-maximal using procedure `deleteNonLeftMaximalBiclusters` is $O(|R||C|^{2+e}|\Sigma|^e)$; deleting from *modelsOcc* all models representing the same *e*-CCC-Biclusters with procedure `deleteRepeatedBiclusters` takes $O(|R|^2|C|^{1+e}|\Sigma|^e)$; and reporting all maximal *e*-CCC-Biclusters using procedure `reportMaximalBiclusters` is $O(|R|^2|C|^{1+e}|\Sigma|^e)$.

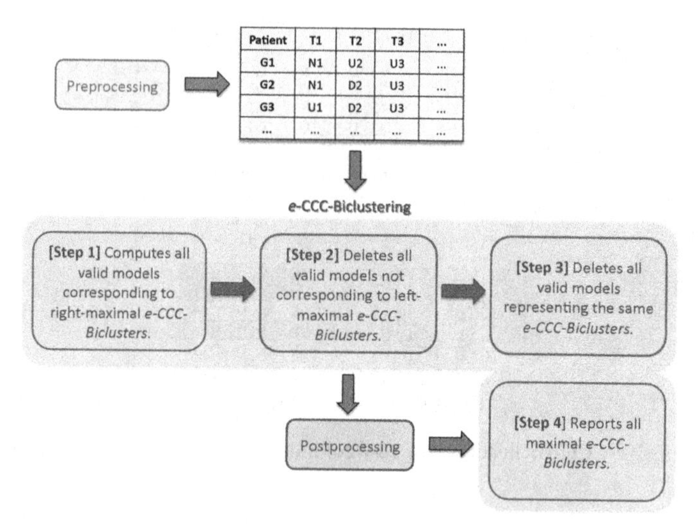

Fig. 2 Basic workflow of the (sequential) *e*-CCC-Biclustering algorithm.

2.2 Parallel *e*-CCC-Biclustering

The parallel version of *e*-CCC-Biclustering follows the four key steps in the sequential version (Fig. 2). Algorithm 1 presents its pseudocode.

We followed a functional approach to the problem, that is, an approach using immutable objects and functions that operate on them. Immutable objects are usually easier to reason about since they do not have states that change over time. This is particularly important in concurrent programming since no threads concurrently accessing an object may corrupt its state from the moment it was constructed. However, this approach does have its drawbacks: the main disadvantage of immutable objects is the fact that they can require a large object graph to be copied where otherwise an update could be done *in place*. An example is the insertion of an element at the end of a simply-linked list. In this case, using a simple pointer update in the last element would solve the problem in $O(1)$, while a functional approach would

Algorithm 1. Parallel *e*-CCC-Biclustering

Input : A, Σ, e, q_r, q_c
Output: Maximal *e*-CCC-Biclusters.

```
1 {S_1,...,S_|R|} ← alphabetTransformation(A, Σ)
2 modelsOcc ← computeRightMaximalBiclusters(Σ, e, q_r, q_c, {S_1,...,S_|R|})
3 modelsOcc ←deleteNonLeftMaximalBiclusters(modelsOcc)
4 if e > 0 then
5     modelsOcc ←deleteRepeatedBiclusters(modelsOcc)
6 reportMaximalBiclusters(modelsOcc)
```

have to copy the whole list, taking $O(n)$ time, and doubling the space needed. The results section shows the impact of these issues in the results.

2.2.1 Computing Valid Right-Maximal Biclusters

The first step of *e*-CCC-Biclustering, corresponding to the execution of procedure `computeRightMaximalBiclusters` (pseudocode in [3]), computes all right-maximal biclusters and is divided in two distinct phases: building a generalized suffix tree for a set of strings corresponding to the set of genes in A, followed by the "spelling" of valid models based on the suffix tree (without modifying it). Since building the suffix tree is $O(|R||C|)$ and spelling the models using `spellModels` is takes $O(|R|^2|C|^{1+e}|\Sigma|^e)$ time, we parallelized `spellModels`.

Parallel `spellModels` (Algorithm 2) follows the ideas of its sequential counterpart: uses the transformed matrix A as input and recursively constructs a set *modelsOcc*, storing triples with the following information (m, $genesOcc_m$, $number OfGenesOcc_m$), where m is the model, $genesOcc_m$ contains the distinct genes in the node-occurrences of m, Occ_m, and $numberOfGenesOcc_m$ is the number of genes where the model occurs. However, parallel `spellModels` not only returns the set of model occurrences, as the sequential version, but it also returns a boolean indicating whether or not should the model from the previous recursion step be included in the set of model occurrences. This is so since, when recursively extending the models, one might find an extended model that indicates that the father model was not the rightest-maximal bicluster possible. Since the algorithm is based on the information provided by extending models, this boolean is required to ensure that a model that would not be right-maximal after being extended is not stored.

Parallel `spellModels` starts by checking if the current model should be added to the set of triples (`keepModel` in [3]). It then checks whether the father model should also be kept and returns the result of this verification to the caller.

Parallel `spellModels` takes advantage of the fact that there is no data dependency in the extension of a model. As such, instead of extending a model one symbol α at a time, we compute the extension with all the symbols in parallel. This parallel extension can itself create a new model to be extended, creating the recursion step (line 7 in Algorithm 2). If, instead, the model is not extendable, then the result of the algorithm is the tuple $(true, \emptyset)$, indicating that the father model may be kept, and that no model was created extending the father with the symbol α.

The reduce step in line 10 condensates the results of all the extensions performed (a set of tuples $(keepModel, result)$): $keepModel$ will be `false` if any of the extensions indicated that this model should not be kept (and `true` otherwise), and $result$ will contain the union of all the sets of extended models. Finally, the result of `spellModels` will be only the result of its child if any of them indicated that this model was not to be kept (line 13) or the union of their results with itself (line 16), but only if its a valid model, as checked before (line 18).

Algorithm 2. spellModels()

```
   /* Called recursively. Returns right-maximal e-CCC-Biclusters
      */
   Input  : Σ, e, q_r, q_c, T_right, m, length_m, Occ_m, Ext_m, father_m, numberOfGenesOcc_father_m
   Output: (keepFatherModel, modelOccurrencesSet)
 1 (keepModel, modelOcc)←—shouldModelBeKept (q_r, q_c, T_right, m, length_m, Occ_m,
      father_m, numberOfGenesOcc_father_m) /* returns true if this model is
      to be added to the result set, and a new ModelOccurrence in
      that case, false otherwise                                        */
 2 keepParentModel←—shouldParentModelBeKept(q_r, q_c, modelOcc, father_m,
   numberOfGenesOcc_father_m)
 3 if length_m ≤ |C| then
      /* |C| is the length of the longest model                         */
 4    childResults ←— forall symbols α in Ext_m such that α is not a string terminator do
 5       (Occ_mα, Colors_mα, Ext_mα, maxGenes, minGenes) ←— extendModel (Σ, e,
         T_right, Occ_m, m, α)
 6       if modelHasQuorum(maxGenes, minGenes, Colors_mα, q_r) then
 7          spellModels(Σ, e, q_r, q_c, T_right, mα, length_m + 1, Occ_mα, Ext_mα,
            father_mα, numberOfGenesOcc_m)
 8       else
 9          (true, ∅)
10    reduceResult ←—reduce childResults
11    switch reduceResult do
12       case (false, result)
            /* somewhere along the way, some extended model said
               that this model it's not to be kept                       */
13          (keepFatherModel, result)
14       case (true, result)
15          if keepModel then
16             (keepFatherModel, result ∪ modelOcc)
17          else
18             (keepFatherModel, result)
19 else
20    if keepModel then
21       (keepFatherModel, modelOcc)
22    else
23       (true, ∅)
```

2.2.2 Deleting Non Left-Maximal Biclusters

The step of deleting non left-maximal biclusters (Algorithm 3) is also divided in two distinct phases: a trie is first built based on the models and occurrences computed in Step 1, followed by the exploration of the trie (in a Depth-First Search style) to find the occurrences that are not left-maximal. Again, we explored the absence of data dependencies in the exploration of the tree and parallelized the search in it.

Algorithm 3. deleteNonLeftMaximalBiclusters()

```
Input  : modelsOcc
Output: modelsOcc
/* Returns a new modelsOcc without the non left maximal
   biclusters                                                  */
1 T_left ⟵ createTrie ()
2 foreach model and occurrences (m, genesOcc_m, numberOfGenesOcc_m) in modelsOcc do
3     m_r ⟵ ReverseModel(m)
4     addReverseModelToTrie(T_left, m_r)
5 nonLeftMaximalBiclusters ⟵getNonLeftMaximalBiclusters(T_left)
6 return modelsOcc\ nonLeftMaximalBiclusters
```

In the recursive function `getNonLeftMaximalBiclusters`, each node is inspected in parallel, exploring the subtries until there is a leaf (in which case a model is certainly a left-maximal bicluster). After reaching a leaf, the recursion traces back accumulating the maximal number of genes seen in each subtrie along with all non left-maximal biclusters found (line 10).

Function `getNonLeftMaximalBiclusters`

```
Input  : T_left
Output: (maxNumberOfGenes, modelOcc)
/* Returns the max number of genes found in a trie along with
   the modelsOcc that are not left maximal                     */
1  nodesResults⟵ foreach nodes v in T_left do
2      if node is leaf then
           /* It's certainly left maximal, so we return an empty
              set                                               */
3          (-∞, ∅)
4      else
5          (maxGenesOfSubtree,
           nonLeftMaximalBiclusters)⟵getNonLeftMaximalBiclusters(v)
6          if genes_v = maxGenesOfSubtree then
               /* It's not left maximal                         */
7              (max(genes_v, maxGenesOfSubtree), nonLeftMaximalBiclusters ∪ v)
8          else
9              (max(genes_v, maxGenesOfSubtree), nonLeftMaximalBiclusters)
10 return reduce nodesResults
```

2.2.3 Deleting Valid Models Representing the Same *e*-CCC-Biclusters

When errors are allowed, different valid models may identify the same *e*-CCC-Bicluster. Step 3, described in procedure `deleteRepeatedBiclusters` (Algorithm 5), uses an identity function to identify from *modelsOcc* all the valid models that, although maximal (left and right), represent repeated *e*-CCC-Biclusters.

In this case, there is no need for any other data-structure to be built and all the algorithm can be performed in parallel. All model occurrences are, in parallel, attributed a key (based on the model *m* and its first and last column). After that, a set is built based on these keys and consequently eliminate all the repeated biclusters.

Algorithm 5. deleteRepeatedBiclusters()

Input : *modelsOcc*
Output: *modelsOcc*
`/* Returns a new` *modelsOcc* `without the repeated biclusters */`
1 *allKeys*⟵ **foreach** *model and occurrences* $(m, genesOcc_m, numberOfGenesOcc_m)$ *in* *modelsOcc* **do**
2 $firstColumn_m = C(m[1])$
3 $lastColumn_m = C(m[length_m])$
4 $key \longleftarrow$ `createKey(`$firstColumn, lastColumn, genesOcc_m$`)`
5 **return** *allKeys* **toSet**

3 Results and Discussion

Parallel *e*-CCC-Biclustering, as most algorithms, has an inherently sequential portion. In this case, Steps 1 and 2 were divided in two distinct phases, the first of which was entirely sequential and the second was parallelized. Step 3 was entirely performed in parallel. Amdahl's Law [1] states that, due to these sequential portions of the program, there is a maximum theoretical speedup possible, depending on the percentage of the time that is run sequentially. This is discussed below.

In order to evaluate the proposed parallel *e*-CCC-Biclustering algorithm we used synthetic test cases with the following structure: each test had as input *R* genes expressed in *C* columns, where gene expression is randomly generated using an uniform distribution in the interval $[-1, 1]$. We ran the test cases in two distinct machines: Phobos, with two Intel Xeon E5520 (8 real cores + 8 hyperthreading), 24Gb of RAM, running Ubuntu 10.04; and Deimos, with four AMD Opteron 6168 (totaling 48 cores), 128Gb of RAM, running Red Hat Enterprise Linux 6.1.

Each test case was run 10 times in each machine varying, the number of cores available. In case of the sequential algorithm it was allowed to run in one core, while the parallel algorithm was allowed to run with a different number of cores for each machine: in the case of Phobos, it was run with 1, 2, 4, 8 and 16 cores, while in the case of Deimos it was run with 1 core and then with multiples of 6 until the 48 cores. We took measures of all the phases of both algorithms although we decided to present only the averaged total of the runs due to space considerations.

We present the total time of the algorithms and the speedups associated to the parallel algorithm. Speedup measures how faster (or slower) the parallel algorithm when compared to the sequential version. As such, values under 100% indicate that the parallel algorithm was actually slower, and values greater than 100% represent how much faster the parallel version was (200% was two times faster, and so on). The Y-axis in the graphics are in logarithmic scale.

Table 1 shows that in the case of 1500 genes and 20 columns and when restricted to 1 or 2 cores in Phobos, the parallel version has a worse performance than its sequential counterpart. This was already expected since, as we stated earlier, we decided to follow a functional approach to the problem, which have some performance penalties. However, with 4 cores, the algorithm already surpasses the sequential version.

Table 1 Phobos (top) and Deimos (bottom) total execution times.

Total Time (s)	L1500-C20	L1500-C40	L3000-C20	L3000-C40	L6000-C20	L6000-C40
Sequential	16	116	50	495	178	2518
Parallel (x1)	51	86	72	142	114	305
Parallel (x2)	27	48	36	84	65	199
Parallel (x4)	13	27	22	54	45	142
Parallel (x8)	9	18	16	40	35	108
Parallel (x16)	8	17	14	37	34	103
Sequential	26	199	82	770	260	3367
Parallel (x1)	155	212	184	300	254	493
Parallel (x6)	26	41	36	84	65	199
Parallel (x12)	18	27	25	55	47	138
Parallel (x18)	15	24	21	53	44	140
Parallel (x24)	13	21	19	47	40	130
Parallel (x30)	11	20	17	46	38	128
Parallel (x36)	10	18	16	45	37	125
Parallel (x42)	10	18	16	43	36	123
Parallel (x48)	10	18	16	43	36	127

Another achievement of the proposed parallel *e*-CCC-Biclustering algorithm can be seen in Table 1, comparing the sequential algorithm with the parallel with only 1 core when only is used: in the case when there are more than 3000 genes and 40 columns, the sequential algorithm performs much worse than the parallel. This was unexpected. Further inspection showed that, since the sequential algorithm stores all the results in one structure (and the operations in the structure are not $O(1)$, nor could be), operations in this structure get slower as the program progresses. In the parallel algorithm, this is not the case: the final result is constructed based on the results of recursion steps. As such, only at the end of the algorithm there is all the results in one structure (built by joining several other smaller structures).

Furthermore, observing Fig. 3 also shows that the parallelism is evident with increasing speedups as more cores are available to the algorithm. From the figures, we can also see that, eventually, no matter how many cores we use, the performance goes no further. This happens since, eventually, there is no work that can be done in parallel. One such example is the reduce operation in line 10 of Algorithm 2: the operation of joining the several result sets from the model extension is done sequentially, even though the extension themselves were done in parallel.

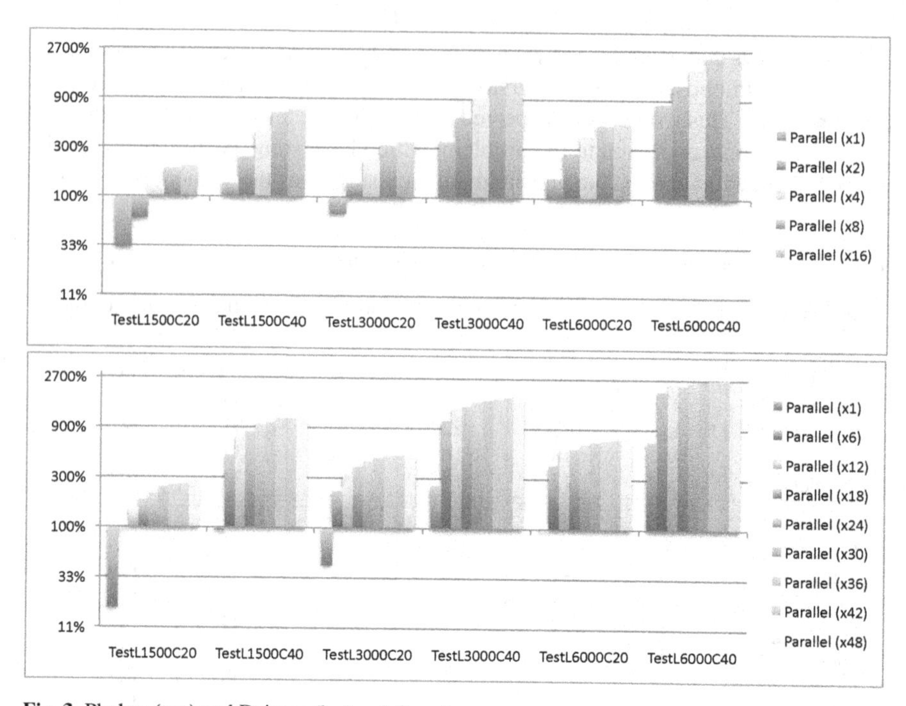

Fig. 3 Phobos (top) and Deimos (bottom) Speedups.

It is also clear from the results that, when one doubles the number of columns in the input, the sequential algorithm performs about an order of magnitude worse while the parallel algorithm only doubles its total time. Even in this case, the parallel algorithm makes use of the cores that are available.

We have estimated that the inherently sequential portion of the parallel algorithm varied between 7% and 13%. According to Amdahl's Law [1], given these percentages and infinite cores, one could only expect 1400% to 769% of speedup. We have actually performed better than this because of what we explained earlier, not due to the parallelization. If we consider Gustafson's law [2], one should focus in the amount of processing that can be done with parallelism: instead of looking how faster the program becomes, we focus on how much bigger of a problem one can solve. From Table 1 it is also clear that problems that could easily take almost an hour now take only 2 minutes. In this case, bigger inputs will make better use of the number of cores available, making these inputs tractable.

4 Conclusions and Future Work

We proposed parallel *e*-CCC-Biclustering, a parallel version of the original *e*-CCC-Biclustering algorithm, a polynomial time exhaustive search algorithm to find all maximal contiguous column coherent biclusters with approximate patterns. Parallel *e*-CCC-Biclustering was implemented using a functional programming approach and proved its effectiveness by taking advantage of multicore architectures. The tests performed with synthetic data, increasing both the number of genes and the number of time points, showed that, as the input size increases, the parallel algorithm performs better than the original sequential algorithm as the overhead costs are diluted and all the cores are used to do efficient work.

As future work we intend to test the performance parallel *e*-CCC-Biclustering using different (temporal) biomedical datasets, such as microarray experiments containing a large number of time points, RNA-seq data, etc. Since the efficiency of *e*-CCC-Biclustering relies not only on the efficient mining of all approximate patterns representing maximal *e*-CCC-Biclusters, which we are already able to perform, but also on an efficient postprocessing of the huge amount of biclusters discovered, the postprocessing step is, in itself, another challenge that can greatly benefit from parallel computing. We are already working on it.

Acknowledgements. This work was partially supported by FCT (INESC-ID multiannual funding) through the PIDDAC Program funds and NEUROCLINOMICS - Understanding NEUROdegenerative diseases throught CLINical and OMICS data integration (PTDC/EIA-EIA/111239/2009).

References

1. Amdahl, G.M.: Validity of the single processor approach to achieving large scale computing capabilities. In: Proceedings of the Spring Joint Computer Conference, pp. 483–485 (1967)
2. Gustafson, J.L.: Reevaluating amdahl's law. Communications of the ACM 31, 532–533 (1988)
3. Madeira, S., Oliveira, A.: A polynomial time biclustering algorithm for finding approximate expression patterns in gene expression time series. Algorithms for Molecular Biology 4(1), 8 (2009)
4. Mejia-Roa, E., Garcia, C., Gomez, J., Prieto, M., Nogales-Cadenas, R., Tirado, F., Pascual-Montano, A.: Biclustering and classification analysis in gene expression using non-negative matrix factorization on multi-gpu systems. In: 11th International Conference on Intelligent Systems Design and Applications, ISDA (2011)
5. Odersky, M.: The Scala Language Specification, version 2.7 (2009)
6. Tewfik, A., Tchagang, A., Vertatschitsch, L.: Parallel identification of gene biclusters with coherent evolutions. IEEE Transactions on Signal Processing 54(6), 2408–2417 (2006)
7. Zhou, J., Khokhar, A.: Parrescue: Scalable parallel algorithm and implementation for biclustering over large distributed datasets. In: 26th IEEE International Conference on Distributed Computing Systems (2006)

Identification of Regulatory Binding Sites on mRNA Using in Vivo Derived Informations and SVMs

Carmen Maria Livi, Luc Paillard, Enrico Blanzieri, and Yann Audic

Abstract. Proteins able to interact with ribonucleic acids (RNA) are involved in many cellular processes. A detailed knowledge about the binding pairs is necessary to construct computational models which can avoid time consuming biological experiments. This paper addresses the creation of a model based on support vector machines and trained on experimentally validated data. The goal is the identification of RNA molecules binding specifically to a regulatory protein, called CELF1.

Keywords: Support vector machines, bioinformatics, machine learning, classification, RNA binding site prediction.

1 Introduction

Gene expression is a complex process by which genes are transcribed into mRNAs that are in turn translated into proteins. RNA-binding proteins (RBP) bind specific RNAs to regulate post-transcriptional processes like alternative splicing, mRNA stability and mRNA translation. The interaction specificity between RBPs and RNA is crucial for the cellular physiology and therefore of special interest to identify the RNA targets for a given RBP, in order to understand its function. Several molecular approaches detect the specificity of RBPs [13, 9], but computational models and binding-predictions would greatly reduce the "hands-on" experimental time. Some proteins use structural patterns to associate an RNA [2, 6], whereas others bind specific motifs on the target RNA. The CELF1 protein (also known as CUGBP1, or EDENBP) is a human RNA-binding protein which binds mainly to single stranded

Carmen Maria Livi · Enrico Blanzieri
Department of computer science DISI, University of Trento, Italy
e-mail: {livi,blanzier}@disi.unitn.it

Yann Audic · Luc Paillard
CNRS, Institut genetique et developpement de Rennes, University of Rennes 1, France
e-mail: {yann.audic,luc.paillard}@univ-rennes1.fr

M.P. Rocha et al. (Eds.): 6th International Conference on PACBB, AISC 154, pp. 33–41.
springerlink.com

UG rich segments of RNA [20, 16]. Being present in the nucleus and cytoplasm of the cell, CELF1 controls post-transcriptional processes at many levels.

In the past few years several studies have focused on the prediction of binding sites on RBPs using machine learning techniques like artificial neural networks [11], support vector machines (SVMs) [22, 5], Naïve Bayes [15, 19] and random forest method [14]. The common idea is the prediction of amino acid residues involved in the binding with an RNA strand. On the RNA side, the literature addresses amongst others the detection of translation initiation sites and of splicing enhancers using neural networks and SVMs [10, 23, 17]. Other papers dissect the structure and the binding sites in RNA-protein complexes to get more information about the particular binding [8, 12]. Interestingly no study focus on the prediction of the binding to an RNA strand. To date *in vitro* and *in vivo* experimentation like "systematic evolution of ligand by exponential enrichment procedures" (SELEX) [13] and "crosslinking and immunoprecipitation experiments and sequencing" (CLIP-seq) [9] are the state of the art procedures for the identification of RBP binding sites on RNA molecules. To extend the use of such approach to RNAs of other cell-types or other species, identified binding sites are modeled as position-specific scoring matrices (PSSMs) which score binding motifs in RNAs. Unfortunately this method fails to take into account the sequence context of the motif. Therefore we propose in this work the use of a SVM trained with 287 features to discriminate CELF1 binding sites *in silico* and describe two experiments to verify the prediction ability of the proposed model.

The paper is organized as follows. First we briefly describe the approach, specify the methods, the datasets and the features. The fourth section details the experiments and the results. In the last section we discuss them and outline future steps.

2 Approach

The approach applies a SVM to discriminate CELF1-binding RNA sites where every RNA sequence is described with a 287 dimensional vector: 30 features are motif scores obtained by PSSMs, 256 features obtained by codifying the RNA sequence in words and one feature by detecting a UGU-rich motif in the sequence.

PSSMs are used in biology to describe motifs [7]. We individuate two binding-motifs for CELF1 based on previously available binding data [16], search a third motif in the CLip dataset and calculate the motif-scores along every RNA strand. The ten highest scores per sequence and motif are used as features for the SVM. In other words we use 10×3 features to describe the binding sites. The binding of an RBP to an RNA is not only characterized by the presence of a specific binding motif but depends also on the environmental sequence. To incorporate the individual characteristics of every sequence we codify the RNA strand in "words". The appearance of every word in the sequence sets the corresponding vector field. After this procedure the feature vector contains in total 286 features. Based on structural information a particular UGU-rich binding motif (see Section 3) is known to be bound by CELF1. We describe the UGU-rich binding motif by a binary feature which is set to 1 if the motif is present in the sequence, otherwise it is set to 0.

A 10-fold-cross validation is used to select the features and two experiments are performed to validate the model. The first experiment implements the SVM and localSVM first on the total dataset of 21662 sequences and second on the balanced dataset by applying a balancing algorithm. The second experiment divides the training data in subsets, each set contains chains of a limited length l, and validates the prediction performance on subsequences of the test data.

3 Materials and Methods

Support vector machine (SVM). The method we use to classify binding and non-binding RNA sequences is the SVM approach [21]. The SVM classifier tries to discriminate linearly between input vectors $x_i \in \mathbb{R}^p$ with p number of features, $i = 1,2,...,n$ which belong to different classes y_i with $y_i \in \{+1,\text{-}1\}$. An input vector x_i belongs to the positive class with label $+1$ (e.g. RNA bound by CELF1) or to the negative class with label -1 (e.g. RNA not bound by CELF1). The goal of a SVM is to find a discrimination function $f(x)$, which divides the two classes in such way that the label for new vectors can be predicted: $f(x) = sign(< w,x > +b)$ where w is the weight vector, the scalar b the bias and $sign$ returns the sign of the argument. $f(x) = 1$ assigns the positive class label, otherwise the negative one. In this work, we use the freely available SVM package LibSVM [3] and apply the linear kernel.

Feature description

PSSM. The identification of a motif in a sequence $s = s_1,...,s_n$ is based on the PSSM matrix-value $pssm(g(\hat{s}_k),k)$ which calculates a motif-score $score_{\hat{s}} = \sum_{k=1}^{m} pssm(g(\hat{s}_k),k)$ for each subsequence $\hat{s} = s_{1+i}\ldots s_{m+i}$. m indicates the motif length $m \leq n$, k the position in the subsequence $\hat{s}$, $g(\hat{s}_k)$ the symbol emitted at position $\hat{s}_k$ and $pssm(g(\hat{s}_k),k)$ the value of the matrix at column k and and row $g(\hat{s}_k)$. Several motif based sequence analysis tools, such as MEME Suite [1] implement this algorithm, discover motifs in sequences and calculate the PSSMs. MEME suite identifies two motifs (motif11 and motif15 shown in Figure 1) in RNA sequences obtained by an independent SELEX experiment with CELF1 [16]. Additionally the tool searches a third motif in the CLIPData. To avoid a circularity problem we extract this third motif in each training fold separately. The motif is not shown as it changes slightly for every fold.

Words. To incorporate the sequence specific environment we codify the RNA strand in "words" of length 4. Words are all possible combinations of nucleotides $\Omega = \{A,U,C,G\}$: $word_x = w_1w_2w_3w_4$ with $w_i \in \Omega$ and $i = 1,2,3,4$. For example $AAAA, AAAU, AAUC, \ldots$ The frequency of every word is extracted.

Binding motif. CELF1 contains 3 RNA recognition motifs (RRMs) with similar specificity. Based on structural information of the CELF1 protein, each RRM may recognizes one UGU trinucleotide. Therefore a known binding pattern for CELF1

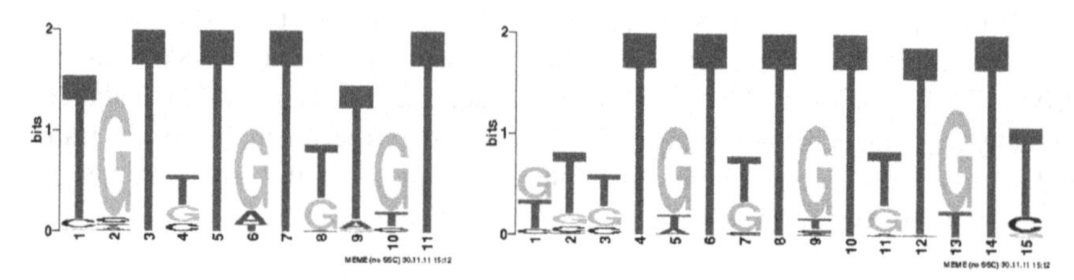

Fig. 1 The two binding motifs, PSSM11 and PSSM15, found in the SELEX experiment with CELF1 [16]. The corresponding PSSMs produce scores which are used as features in our model.

is a UGU-motif with the following structure: $[UGU]N_1[UGU]N_2[UGU]$, where N_i i=1,2 are two nucleotide sequences $n_{i0}...n_{is_i}$ with length $s_i \in [0...20]$.

Datasets

CLIPData. This dataset originates from a CLIP-seq experiment realized in Hela-kyoto cells (Olivier le Tonqueze, unpublished data) and represents the positive data points (+) for the SVM. It consists of 1932 RNA sequences identified to be bound by CELF1 *in vivo*.

NegData. This dataset is constituted by 3'UTRs from genes expressed in Hela-kyoto cells and not bound by CELF1. It represents the negative data points (-) for the SVM. Originally the dataset consists of 36701 transcripts deriving from 9849 genes. In this paper a subset of 24680 transcripts are used as negative dataset in order to provide a negative dataset with the same sequence length distribution as the positive dataset. Table 1 shows a summary of the datasets.

Table 1 Summary of the datasets

dataset	No. of chains	shortest/longest	average length
CLIPData	1932	29/1401	170
original NegData	36701	11/16193	952
NegData	24680	19/999	379

Balancing Dataset. A frequent problem with biological datasets is that they are unbalanced. Usually the number of negative data-points is much higher than the number of positive ones [19, 5, 22]. To overcome the unbalanced training of our model, with 1932(+) binding and 24680(-) non-binding sequences, we attempt different approaches:

1. The application of a synthetic minority over-sampling technique (SMOTE) [4]. The approach amplifies the positive dataset by creating new synthetic instances and forces the classifier to become more general.

2. The application of local support vector machines (localSVM) which use a set of SVMs to create an appropriate local model on each training point. We use the freely available software package FaLKM-lib [18] with the linear kernel.

Classifier evaluation

Statistical measures. The performance of the model is evaluated by calculating Matthews correlation coefficient, sensitivity and specificity. The Matthews correlation coefficient $MCC = \frac{TP \times TN - FP \times FN}{\sqrt{(TP+FP) \times (TP+FN) \times (TN+FP) \times (TN+FN)}}$ is a balanced measure independent from the class sizes and measures the correlation between the observed and predicted classifications. The sensitivity $Sens = \frac{TP}{TP+FN}$ measures the ability of the classifier to identify positive elements whereas the specificity $Spec = \frac{TN}{TN+FP}$ measures the proportion of correctly classified negative elements. TP are the true positive predicted elements, TN the true negative, FP and FN the false positive and false negative, respectively.

Training and testing. A 10-fold cross validation is used to evaluate the performance of the model. We optimize the kernel parameters for the SVM and the localSVM via a grid-search on the highest MCC.

4 Experiments and Results

Several combinations of the previously described features have been tested using a 10-fold-cross validation on a randomly selected subset of 1932 binding and 1989 non-binding sequences (Table 2). The combination of words, 10×3 motif-scores and TGT-motif achieves the best performance values with an MCC=0.63, Sens=0.81 and Spec=0.82. Therefore we build the SVM on these features and describe two experiments to test its classification ability.

First experiment. We first test the performance of the SVM and the localSVM on the total dataset. The results, reported in Table 3, provide important information regarding the influence of the unbalanced dataset to the performance. With a high specificity of 0.9 and a low sensitivity of around 0.4 is it evident that both methods are not discriminative. This is likely induced by the unbalance of the dataset. Therefore we decided to apply the two classification methods after balancing the dataset using the SMOTE algorithm. Sensitivity and specificity increase to 0.92 and 0.77 with the SMOTE+localSVM harboring a MCC of 0.40. Much better performs the SMOTE+SVM approach which increases both, sensitivity and specificity, to 0.86 to reach an MCC of 0.48 (see Table 3). The positive and the negative sequences differ in the composition of mRNA regions known to bind CELF1 and in the full length 3'UTR sequences, respectively. To overcome this difference we apply the model, in a second experiment, on subsequences of fixed length.

Second experiment. From an RNA sequence to be tested we extract subsequences of length x, with an overlap of $x/2$, and apply the classifier on the subsequences.If a subsequence is predicted as binding, the tested sequence is classified as a bind-

Table 2 Different features and their combinations tested on a randomly selected dataset (1932(+)/1989(-)). The combination with the best classification values (last row) is used for the final model.

PSSM11	PSSM15	PSSMClip	words	TGT	Sens	Spec	MCC
3	3	3	0	0	0.63	0.63	0.27
5	5	5	0	0	0.63	0.65	0.28
7	7	7	0	0	0.63	0.66	0.29
10	10	10	0	0	0.62	0.68	0.30
10	10	10	1	1	0.81	0.82	0.63

Table 3 Results of the first experiment. Applying SVM and localSVM on the total dataset (1932(+)/24680(-)) shows that both methods are not discriminative on the unbalanced dataset. After the balancing with SMOTE, SVM and localSVM classify the sequences with high accuracy.

method	Sens	Spec	MCC
SVM	0.39	0.99	0.52
SMOTE+SVM	0.86	0.86	0.48
localSVM	0.41	0.98	0.52
SMOTE+localSVM	0.92	0.77	0.40

Table 4 Results of the second experiment. Different models are trained on different sequence subsets, called set_x. Every set contains sequences of a limited length x. The models are used to classify RNA subsequences of length $l \leq x$ extracted from longer sequences. The table shows the percentages of true-positive (TP) and true-negative (TN) predicted RNAs and the MCC.

set_x	TP prediction (%)	TN prediction (%)	MCC
set_{100}	96.5	14.8	0.20
set_{200}	90.5	49.4	0.44
set_{300}	87.8	62.6	0.46
set_{400}	78.9	75.8	0.45
set_{500}	68.6	80.2	0.36

ing sequence. To bypass the unbalanced data problem described above, the total dataset is divided in several training sets set_x. Every training set contains sequences of a limited length $l \leq x$, x=100, 200, 300...500. For instance: $x = 200$ indicates that the model is trained on sequences with a maximum length of 200 (set_{200}). The corresponding model classifies subsequences of the same length, extracted from a test RNA longer than x. Table 4 reports the predicted percentages of binding and not-binding sequences for each model.

5 Discussion and Conclusion

The identification of RNAs associated with a given regulatory RBP is costly and time consuming when realized experimentally. Moreover it highly depends on the cell type or on the organism used in the experiment. So far there is no other approach to detect specific RNA-protein binding *in vivo*. Machine learning techniques have been successfully applied on biological data and SVMs are known to be a powerful method to solve classification problems obtaining high performance values. Hence we proposed to classify RNA sequences, as binding or as not binding to CELF1 protein, using SVMs trained on CLIP-seq data. Two experiments are described to test the prediction ability of our model. The first experiment applies SVM and localSVM on a CLIP-seq dataset. Due to the unbalanced data the application of a balancing algorithm was necessary. Results in Table 3 show a high sensitivity and specificity (for the balanced dataset) for both SVM and localSVM, indicating that binding information from SELEX and CLIP-seq experiments can be efficiently used to train SVMs. To avoid bias, potentially introduced by the data balancing, we designed a second experiment in which we grouped sequences by their length to obtain homogeneous and equalized datasets. The results in Table 4 indicate that sub-sequences of length $x = 400$ allow to train a SVM to discriminate binding from non-binding RNAs more accurately than shorter sequences. Codifying short sequences in tetramers causes numerous features to be zero, therefore they contain probably not enough information for a proper classification. One advantage of the SVM defined on sub-sequences is that it allows to focus more rapidly on the potential localization of the binding site for CELF1. So far it is unclear whether the use of PSSM scores alone performs better or worse than the described SVM. To determine which approach performs better a thorough analysis between SVM and PSSM is in progress. In future we plan also to adapt our model to other species with the goal to predict CELF1-binding RNAs in several cell-lines. An SVM trained on experimental data and applied on other cell types can reduce costly CLIP-seq experiments. For instance the detection of CELF1 target RNAs in pathological cells. Furthermore, because CELF1 and its binding specificity are highly conserved among vertebrates, it will be possible to apply the approach to other genomes and to determine whether any CELF1 centered gene networks are conserved during evolution. The availability of experimentally validated CLIP-seq datasets for several proteins will allow us to extend our model to more RBPs.

In this work we attempted the use of *in vivo* derived information to train a SVM with promising good results. Its application to unseen and unknown data can individuate binding RNAs reducing the number of time-consuming and expensive laboratory experiments.

References

1. AAAI Press: Fitting a mixture model by expectation maximization to discover motifs in biopolymers. AAAI Press (1994)
2. Auweter, S., Oberstrass, F., Allain, F.: Sequence-specific binding of single-stranded rna: is there a code for recognition? Nucleic Acid Research 34(17), 4943–4959 (2006)
3. Chang, C.C., Lin, C.J.: LIBSVM: a library for support vector machines (2001), http://www.csie.ntu.edu.tw/~cjlin/libsvm
4. Chawla, N.V., Bowyer, K.W., Hall, L.O., Kegelmeyer, W.P.: Smote: Synthetic minority over-sampling technique. Journal of Artificial Intelligence Research 16, 341–378 (2002)
5. Cheng, C.W., Chia-Yu, S., Hwang, J., Sung, T., Hsu, W.: Predicting rna-binding sites of proteins using support vector machines and evolutionary information. BMC Bioinformatics 9 (2008)
6. Dreyfuss, G., Kim, V.N., Kataoka, N.: Messenger-rna-binding proteins and the messages they carry. Nature Reviews Molecular Cell Biology 3, 195–205 (2002)
7. Green, E., Brenner, S., Regents, U.: motifbs. a program to generate dna or rna position-specific scoring matrices and to search databases of sequences with these matrices (2003), http://compbio.berkeley.edu/people/ed/motifBS.html
8. Gupta, A., Gribskov, M.: The role of rna sequence and structure in rna–protein interactions. Journal of Molecular Biology 409(4), 574–587 (2011)
9. Hafner, M., Landthaler, M., Burger, L., Khorshid, M., Hausser, J., Berninger, P., Rothballer, A., Ascano, M.J., Jungkamp, A.C., Munschauer, M., Ulrich, A., Wardle, G.S., Dewell, S., Zavolan, M., Tuschl, T.: Transcriptome-wide identification of rna-binding protein and microrna target sites by par-clip. Cell 141(1), 129–141 (2010)
10. Hebsgaard, S.M., Korning, P.G., Tolstrup, N., Engelbrecht, J., Rouze, P., Brunak, S.: Splice site prediction in arabidopsis thaliana pre-mrna by combining local and global sequence information. Nucleic Acid Research 24(17), 3439–3452 (1996)
11. Jeong, E., Chung, I.F., Miyano, S.: A neural network method for identification of rna-interacting residues in protein. Genome Informatics 15(1), 105–116 (2004)
12. Jones, S., Daley, D.T., Luscombe, N.M., Berman, H.M.: Protein-rna interactions: a structural analysis. Nucleic Acid Research 29(4), 943–954 (2001)
13. Klug, S.J., Famulok, M.: All you wanted to know about selex. Molecular Biology Reports 20(2), 97–107 (1994)
14. Liu, Z.P., Wu, L.Y., Wang, Y., Zhang, X.S., Chen, L.: Prediction of protein–rna binding sites by a random forest method with combined features. Bioinformatics 26(13), 1616–1622 (2010)
15. Maetschke, S., Yuan, Z.: Exploiting structural and topological information to improve prediction of rna-protein binding sites. BMC Bioinformatics 10(341) (2009)
16. Marquis, J., Paillard, L., Audic, Y., Cosson, B., Danos, O., Bec, C.L., Osborne, H.B.: Cug-bp1/celf1 requires ugu-rich sequences for high-affinity binding. Biochemical Journal 400(2), 291–301 (2006)
17. Mersch, B., Gepperth, A., Suhai, S., Hotz-Wagenblatt, A.: Automatic detection of exonic splicing enhancers (eses) using svms. BMC Bioinformatics 9(1), 369 (2008)
18. Segata, N.: Falkm-lib v1.0: a library for fast local kernel machines. Tech. rep., DISI, University of Trento, Italy (2009), Software available at http://disi.unitn.it/~segata/FaLKM-lib
19. Terribilini, M., Lee, J., Yan, C., Jernigan, R.L., Honavar, V., Dobbs, D.: Prediction of rna binding sites in proteins from amino acid sequences. RNA (12), 1450–1462 (2006)

20. Le Tonquèze, O., Gschloessl, B., Namanda-Vanderbeken, A., Legagneux, V., Paillard, L., Audic, Y.: Chromosome wide analysis of cugbp1 binding sites identifies the tetraspanin cd9 mrna as a target for cugbp1-mediated down-regulation. Biochemical and Biophysical Research Communications 394(4), 884–889 (2010)
21. Vapnik, V.N.: The Nature of Statistical Learning Theory. Springer, Heidelberg (1995)
22. Wang, L., Brown, J.: Bindn: a web-based tool for efficient prediction of dna and rna binding sites in amino acid sequences. Nucleic Acid Research 34, 243–248 (2006)
23. Zien, A., Raetsch, G., Mika, S., Schoelkopf, B., Lengauer, T., Mueller, K.R.: Engineering support vector machine kernels that recognize translation initiation sites. Bioinformatics (2000)

Parameter Influence in Genetic Algorithm Optimization of Support Vector Machines

Paulo Gaspar, Jaime Carbonell, and José Luís Oliveira

Abstract. Support Vector Machines provide a well established and powerful classification method to analyse data and find the minimal-risk separation between different classes. Finding that separation strongly depends on the available feature set. Feature selection and SVM parameters optimization methods improve classification accuracy. This paper studies their joint optimization and attribution improvement. A comparison was made using genetic algorithms to find the best parameters for SVM classification. Results show that using the RBF kernel returns better results on average, though the best optimization for some data sets is highly dependent on the choice of parameters and kernels. We also show that, overall, an average 26% relative improvement with 8% std was obtained.

Keywords: SVM, Genetic Algorithm, Optimization, Feature Selection, Kernel.

1 Introduction

After their introduction in the world of classifiers, support vector machines (SVM) have been widely studied and used for their reliable performance in non-probabilistic classification and regression [1, 2, 3]. Multiple applications to real problems have been reported to use SVM based models, such as time series forecasting [4], hand writing recognition [5], text categorization [2], bankruptcy prediction [6], face identification and recognition [7], biological and medical aid [1, 3], among many other areas.

Paulo Gaspar · José Luís Oliveira
IEETA, University of Aveiro. Campus Universitário de Santiago, 3810 - 193 Aveiro, Portugal
e-mail: {paulogaspar, jlo}@ua.pt

Jaime Carbonell
Language Technologies Institute, Carnegie Mellon University, 5000 Forbes Avenue, Pittsburgh, PA 15213, USA
e-mail: jgc@cs.cmu.edu

M.P. Rocha et al. (Eds.): 6th International Conference on PACBB, AISC 154, pp. 43–51.
springerlink.com

In order to classify data, SVMs first find the maximal margin which separates two (or more) classes and then outputs the hyperplane separator at the center of the margin. New data is classified by determining on which side of the hyperplane it belongs and hence to which class it should be assigned. However, some input spaces might not be linearly separable, and therefore kernel functions might be used to map the input space into a higher-dimensional feature space where the data can be more easily separated [8]. The selected kernel often depends on the classification problem, and it usually has specific parameters than can be controlled to fine-tune its performance.

Sometimes data are not fully separable even with powerful kernels. Therefore some tolerance is considered when calculating the separating hyperplane through the use of slack variables that measure the degree of misclassification for each instance. A penalty function is used to penalize that misclassification when looking for the best separation [8]. SVM training consists of the following minimization problem[1]: $\min_{w,\xi} \{\frac{1}{2}||w||^2 + C\sum_i \xi_i\}$ where w represents the hyperplane normal vector and C is the weight of the penalty function which is formulated by the sum of all ξ slack variables [8]:

Moreover, the available feature set in the data plays a crucial role when separating data into two classes as the SVM bases its algorithm in using the values of the features to plot each entry of data in a high dimensional space. It is well established that features have a large influence on how well data can be separated into two classes, depending on the correlation of each feature to its class [9, 10]. Also, some features might not contribute, or contribute negatively to the classification process. Thus, many SVM classification optimizations focus on the process of feature selection (FS) [11, 12].

In short, there are generally three approaches to optimize SVM classification accuracy[2]: by choosing the correct kernel (yielding the lowest generalization error); by tuning Kernel and SVM parameters such as the penalty weight and σ sigma value (RBF kernel); and by performing feature selection. These factors can be combined to significantly improve the performance of SVM classifications. Using genetic algorithms, the parameters can be encoded together with features into a single vector, and the SVM model used as a fitness measure of each individual (i.e. each variation of that vector) [11, 6, 13, 14].

Nonetheless, current literature lacks a structured comparison of the influence of each parameter and their optimization potential. Many studies already demonstrated that genetic algorithms and their optimization capabilities can successfully improve SVM predictions. This is often achieved separately: by addressing to the hyperparameters (SVM and kernel parameters), such as the penalty weight, RBF sigma value and polynomial kernel order [6, 13, 14, 15]; or by performing selection of the best sub-set of features that improves classification [11]. However, except for Huang and Wang [16], the SVM optimization approach does not simultaneously address both hyperparameters and feature selection. Huang and Wang show that

[1] Subject to: $c_i(w \cdot x_i - b) \geq 1 - \xi_i, 1 \leq i \leq n$.

[2] Accuracy in a broad sense, since we are actually looking to improve f-measure, which is more sensitive to unbalanced classes.

the simultaneous optimization of RBF kernel parameters and feature selection significantly improves classification accuracy with SVMs. Nonetheless, in [16] they use only RBF and its parameters, and no comparison between sets of parameters is made to assess the influence of each parameter in the optimization process. Also, the classification accuracy is used in their objective function, which can lead to a weak form of characterizing a classification result when using unbalanced data sets [17]. In [15], Sterlin et al also present a relevant study of SVM optimization using genetic algorithms to compare the optimization influence in several data sets. However, no conclusions are made regarding the use of different sets of parameters or different kernels. Also, there is a lack of understanding of how individual parameters and feature selection affect the classification.

To approach those open questions, in this paper, we analyse the influence of different combinations of parameters in SVM binary classification. We achieve that by evolving a genetic algorithm population using several subsets of the available customizable parameters.

2 Methods

In order to optimize the parameters, an evolutionary approach was made by using genetic algorithms. The selected parameters are placed in a vector form and a random population of parameter values is created and evolved.

For that, the optimization and classification toolboxs from MatLab[3] was used to configure and run the genetic algorithm and SVM. Also, a series of configurations were made to the genetic operators: **a)** The creation operator was set to randomly create a population of 20 individuals that represented the SVM parameters, Kernel parameters and Feature Selection. Those parameters are encoded in a vector, bound to maximum and minumum values, except for feature selection which was represented using a bit string. **b)** For the selection operator, a roulette wheel strategy was used, hence selecting individuals proportionately to their fitness value. Therefore, individuals with higher fitness were more prone to being selected. **c)** The crossover operator was applied using uniform crossover, with a 50% chance of selecting either parent. Thus, one offspring is created by randomly combining two individuals. **d)** And for the mutation operator, the following strategy was followed: in the feature selection bit string, the chances of toggling a bit were proportional to the current generation such that it started with a 6% mutation probability and ended with 0%, accordingly to: $pMutation = (1 - \frac{generationNumber}{maxGenerations})/15$

However, for the SVM and Kernel parameters, the mutation was performed by randomizing the original value using a Gaussian distribution with mean on the original value and a standard deviation that varies with generations, according to the following: $stdDev = (1 - \frac{generationNumber}{maxGenerations}) \times (UpperBound - LowerBound)$. Upper and Lower bounds represent the maximum and minimum values that the parameter

[3] MatLab 7.10.0 (R2010a). All the scripts and data that were used can be downloaded at `http://bioinformatics.ua.pt/support/svm/data.zip`

can have. This allows starting the mutation process with large standard deviations (95% of the bounds) and ending with very specific values.

Moreover, the fitness operator used the encoded parameters to run the SVM train and classification and then calculate F-measure. The F-measure was returned and used as an assessment of the encoded parameters. Also, the fitness function always generated different training and testing sets in each call to avoid over-fitting.

Furthermore, the C parameters varied between 0.001 and 10000. the σ between 0.001 and 1000, and the polynomial order between 2 and 10. The genetic algorithm was always run with a limit of 50 generations. The values of population size (20 individuals), number of generations (50) and mutation probabilities were empirically chosen to maximize the final result and minimize the execution time.

The workflow of the genetic algorithm is as follows, for each experiment:

1 Create population by building random (in range) vectors
2 Use the crossover operator to obtain 1 new vector from each 2 parents, then use the mutation and fitness operators to change and evaluate the offspring.
3 Use the selection operator to choose individuals for the next generation
4 Repeat the process starting from step 2, for 50 generations
5 The best (highest f-measure) individual in the population is returned.

3 Experiments

Several heterogeneous data sets were selected from the UCI machine learning repository [18]. These sets come from different areas such as wine quality (DS1), heart disease (DS2), adult income (DS3), abalone data (DS4), SPECT heart images (DS5) and stock data (DS6, not obtained from the UCI). The sets were modified to remove incomplete entries, and transform all information into numerical data. Also, all data sets were prepared for classification by selecting two classes in each data set. The data sets have different sizes (in number of entries) and different number of features, as shown in Table 1.

To build a baseline, the data sets were first classified by a simple SVM using a linear kernel and the default penalty weight (constant C) of 1, with 40 fold validation. The validation was performed by separating the data into two sub-sets (train and

Table 1 Data sets used. Skew is calculated as the ratio of negative classes.

Data Sets information and base result						
	Source	**Set Size**	**Nr Features**	**Sparsity**	**Average F_1**	**Std**
DS1	Wine Quality	1000	11	77.3%	0.4376	0.0398
DS2	Heart Disease	297	13	53.9%	0.8039	0.0074
DS3	Adult Income	1000	12	75.5%	0.4941	0.0348
DS4	Abalone	2835	7	53.9%	0.4298	0.0198
DS5	SPECT heart images	267	22	20.6%	0.8790	0.0166
DS6	Stock data	793	27	59.8%	0.6456	0.0305

test), with each sub-set containing half of the positive and negative class entries of the original data set. By creating different train and test sets on each run, we avoid biasing the average result to the selected sets. The average value of the harmonic mean of precision and recall (f-measure) of the classifications of each data set was considered for the subsequent optimization experiments (Table 1).

4 Optimization Parameters

Important parameters were selected to perform the optimization of the SVM classification process [19]: the penalty weight constant, the set of input features [14] and kernel parameters (such as the RBF σ value and Polynomial kernel order) [20]. Several combinations of these parameters were applied to SVMs using genetic algorithms to fine-tune them and achieve the best classification (using F-measure to compare results). Furthermore, binding the penalty constant with each entry of the training set allows another type of parameter, a penalty array. Hence, a different weight is associated with each slack variable, allowing larger flexibility in the generalization (though possibly adding some over-fitting). The SVM minimization equation then becomes the following: $\min_{w,\xi} \{\frac{1}{2}||w||^2 + \sum_i C_i \xi_i\}$.

When associating the penalties with each slack variable, the penalty array becomes deeply dependent on the training set that is used. Therefore, changing the train and test sets on each call to the SVM prevents the genetic algorithm from correctly evolve the penalty array. Thus, in one experiment (SP5) the training set is also used in the parameter set (Set Selection). Three main kernels, linear, radial basis function and polynomial, were the focus of the experiments. The combinations of selected parameters (SP) are depicted in Table 2.

Table 2 Used combinations of parameters. C is the SVM penalty weight; FS stands for feature selection; C array are weights for each slack variable; the σ and *poly order* are kernel parameters.

SVM Parameters				
kernel		**Selected parameters**		
Linear kernel	**SP1**	C		
	SP2	FS		
	SP3	C	FS	
	SP4	C array	FS	
	SP5	C array	FS	Set Selection
RBF kernel	**SP6**	C	σ	
	SP7	C	σ	FS
Polynomial kernel	**SP8**	C	poly order	
	SP9	C	poly order	FS

5 Results

Each SP was subject to the genetic algorithm, and the best result of the experiments was compared to the base result in order to assess the value of that combination of parameters in the overall classification improvement (Table 3).

Table 3 Results of the experiments on several data sets. For each data set, the F1-measure of each experiment is presented as a relative improvement of the baseline result, e.g. SP1 showed a 0.4706 relative improvement, thus, the final result was 0.4376×1.4706.

	F-Measure Relative Improvement										
	SP1	**SP2**	**SP3**	**SP4**	**SP5**	**SP6**	**SP7**	**SP8**	**SP9**	**DS-Average**	**Std Dev**
DS1	0.4706	0.2204	0.4724	0.3455	0.5444	0.6637	0.6568	0.5356	0.4158	*0.4806*	*0.1423*
DS2	0.0940	0.0932	0.1342	0.0955	0.1224	0.0976	0.0909	0.0663	0.0715	*0.0962*	*0.0215*
DS3	0.3250	0.2223	0.3302	0.2104	0.3493	0.4114	0.4719	0.3446	0.4538	*0.3465*	*0.0909*
DS4	0.2373	0.1462	0.2561	0.2274	0.2673	0.0654	0.1115	0.2356	0.2711	*0.2020*	*0.0749*
DS5	0.0303	0.0799	0.0279	0.0665	0.0849	0.0597	0.0604	0.0571	0.0694	*0.0596*	*0.0196*
DS6	0.3600	0.1121	0.3665	0.2729	0.5072	0.3829	0.4643	0.3590	0.4799	*0.3672*	*0.1205*
SP-Average	*0.2529*	*0.1457*	*0.2646*	*0.2030*	*0.3126*	*0.2801*	*0.3093*	*0.2664*	*0.2936*		
Std Dev	*0.1668*	*0.0627*	*0.1618*	*0.1059*	*0.1913*	*0.2461*	*0.2530*	*0.1854*	*0.1873*		

The average value of relative improvements made by each selected parameters experiment in each data set is shown to better compare the several parameter sets. Also, the standard deviation allows assessing the dispersion of the improvement over the data sets. When clustering Table 3 by selected parameters set one can visualize a bars chart where the effects of each optimization experiment are observed (depicted in Figure 1). Moreover, to assess the statistical significance of the results, the baseline F-scores were compared with the final F-scores using a 2-tailed paired t-test. The p-values are shown in Table 4.

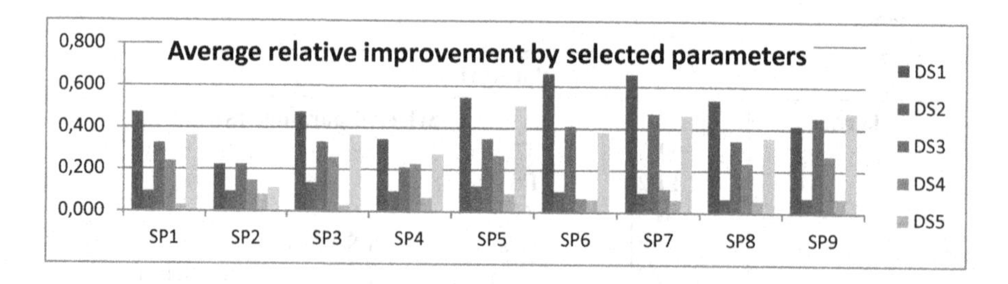

Fig. 1 Each selected parameter set (SP) is presented along with its results for each data set (DS). The vertical axis represents the relative improvement in F-score in relation to the baseline value.

Table 4 The p-values were calculated using a two-tailed t-test. All the values are below the α significance level of 5%, and therefore, they all reject the null hypothesis.

p-values of the 2-tailed paired t-test								
SP1	**SP2**	**SP3**	**SP4**	**SP5**	**SP6**	**SP7**	**SP8**	**SP9**
0,0047	0,0003	0,0032	0,0008	0,0016	0,0181	0,0131	0,0056	0,0039

6 Discussion

Optimization of SVMs using genetic algorithm is a common procedure to obtain better classification results. However, authors refer different strategies, approaching several parameters that can be fine-tuned within the SVM. Overall, in out experiments, we obtained an average 26% relative improvement to the simple SVM classification F-measure, with an 8% standard deviation.

Moreover, SP2 results had the smallest standard deviation, suggesting that feature selection (FS) optimizations tend to offer improvements less dependent on the data set. On SP1, the penalty weight (a common target of improvement) optimization shows much more dependence on the data set than FS, but also better results. On the other hand, the mean standard deviation for all experiments is high (0.17), leading to the conclusion that the best optimization depends on the data set. It can also reflect that there is a trade-off between F-score improvement and generalization, such that the more a model is optimized to improve F-score the higher the over-fitting. This is perceived in the correlation[4] that exists (0.90) between the SP-averages and their corresponding standard deviations.

Also, for a fixed data set, the improvement does not depend much on the selected parameters but is rather limited by the data itself (the mean standard deviation for data sets is small: 0.08). Moreover, a strong negative correlation (-0.97) was found between the overall DS-averages and the corresponding initial baseline values, translating that sets with lower base results tend to be significantly more improved than higher base results. For the RBF kernel, the number of available features have a meaningful inverse impact on the result (-0.71 correlation with number of features). Also, the RBF kernel experiments show the highest standard deviations.

7 Conclusions

Overall, thought optimizing the C penalty weight is a widely used technique, there are better configurations to achieve higher improvements, even considering low standard deviation. Surprisingly, jointly optimizing feature selection and an array of C penalty weights (SP4) revealed itself as a good method to significantly improve the model F-score without sacrificing generalization (average 0.20 improvement with 0.04 std). If generalization in the sense of non-dependence on the data set is the goal, then optimizing kernel parameters should be avoided, since they yielded

[4] All correlations were calculated as person correlations.

the higher standard deviations among data sets. On the other hand, if over-fitting is not an issue, then using a C array along with set selection (SP5) significantly improves the model for any data set (with an acceptable standard deviation that is not very high).

Also important to note is that, experiments where FS was added to the parameter set, the average result always increased (2% on average) and the standard deviation remained unchanged. Thus, feature selection generally improves SVM classification while maintaining generalization, which the goal of prediction models.

Acknowledgements. The research leading to these results has received funding from the European Community's Seventh Framework Programme (FP7/2007-2013) under grant agreement no. 200754 - the GEN2PHEN project.

References

1. Guyon, I., Weston, J., Barnhill, S., Vapnik, V.: Gene selection for cancer classification using support vector machines. Machine learning 46(1), 389–422 (2002)
2. Siolas, G., et al.: Support vector machines based on a semantic kernel for text categorization, p. 5205. IEEE Computer Society (2000)
3. Akay, M.F.: Support vector machines combined with feature selection for breast cancer diagnosis. Expert systems with applications 36(2), 3240–3247 (2009)
4. Cao, L.: Support vector machines experts for time series forecasting. Neurocomputing 51, 321–339 (2003)
5. Bahlmann, C., Haasdonk, B., Burkhardt, H.: On-line handwriting recognition with support vector machines-a kernel approach (2002)
6. Wu, C., Tzeng, G., Goo, Y., Fang, W.: A real-valued genetic algorithm to optimize the parameters of support vector machine for predicting bankruptcy. Expert Systems with Applications 32(2), 397–408 (2007)
7. Guo, G., Li, S., Chan, K.: Face recognition by support vector machines. fg, 196 (2000)
8. Cortes, C., Vapnik, V.: Support-vector networks. Machine learning 20(3), 273–297 (1995)
9. Yu, L., Liu, H.: Feature selection for high-dimensional data: A fast correlation-based filter solution. In: International Workshop then Conference on Machine Learning, vol. 20(2), p. 856 (2003)
10. Hall, M.: Correlation-based feature selection for machine learning. Ph.D. dissertation, Citeseer (1999)
11. Holger, F., Chapelle, O.: Feature selection for support vector machines by means of genetic algorithms. In: Proceeding of the 15th IEEE International Conference on Tools with Artificial Intelligence (ICTAI 2003), pp. 142–148 (2003)
12. Chen, Y., Lin, C.: Combining svms with various feature selection strategies. Feature Extraction, 315–324 (2006)
13. Friedrichs, F., Igel, C.: Evolutionary tuning of multiple SVM parameters. Neurocomputing 64, 107–117 (2005)
14. Liepert, M.: Topological Fields Chunking for German with SVMs: Optimizing SVM-parameters with GAs. In: Proceedings of the International Conference on Recent Advances in Natural Language Processing (RANLP), Bulgaria, Citeseer (2003)
15. Sterlin, P., Jenatton, R., Paris, E.: Optimizing a SVM Classifier with a Genetic Algorithm: Experiments and Comparison. Neural Networks 8, 7 (2007)

16. Huang, C., Wang, C.: A GA-based feature selection and parameters optimization for support vector machines. Expert Systems with applications 31(2), 231–240 (2006)
17. Akbani, R., Kwek, S., Japkowicz, N.: Applying Support Vector Machines to Imbalanced Datasets. In: Boulicaut, J.-F., Esposito, F., Giannotti, F., Pedreschi, D. (eds.) ECML 2004. LNCS (LNAI), vol. 3201, pp. 39–50. Springer, Heidelberg (2004)
18. http://archive.ics.uci.edu/ml/
19. Chapelle, O., Vapnik, V., Bousquet, O., Mukherjee, S.: Choosing multiple parameters for support vector machines. Machine Learning 46(1), 131–159 (2002)
20. Chung, K., Kao, W., Sun, C., Wang, L., Lin, C.: Radius margin bounds for support vector machines with the RBF kernel. Neural Computation 15(11), 2643–2681 (2003)

Biological Knowledge Integration in DNA Microarray Gene Expression Classification Based on Rough Set Theory

D. Calvo-Dmgz, J.F. Galvez, Daniel Glez-Peña, and Florentino Fdez-Riverola

Abstract. DNA microarrays have contributed to the exponential growth of genetic data from years. One of the possible applications of this large amount of gene expression data diagnosis of diseases like cancer using classification methods. In turn, explicit biological knowledge about gene functions has also grown tremendously over the last decade. This work integrates explicit biological knowledge in classification process using Rough Set Theory, making it more effective. In addition, the proposed model is able to indicate which part of biological knowledge has been used building the model and classifing new samples.

Keywords: DNA microarray classification, Biological Knowledge, Principal Component Analysis, Discriminant Fuzzy Pattern, Rough Sets, Basic Category.

1 Introduction and Related Work

During the last decade, Molecular Biology has produced vast amounts of information from DNA. DNA microarrays consist of an arrayed series of thousands of microscopic spots of DNA oligonucleotides. Nowadays, microarrays are essential tools in genomic research labs because high-density DNA microarrays allow researchers to take "snapshots" of the state of cells at level of transcription at a genomic scale [1]. One of the possible applications of DNA microarray technology is its use as a tool for decision in cancer diagnosis. Watching at expression levels in thousands of genes simultaneously, experiments with DNA microarrays provide a better understanding of molecular mechanisms that governing diseases, to achieve more reliable diagnosis [2].

D. Calvo-Dmgz · J.F. Galvez · Daniel Glez-Peña · Florentino Fdez-Riverola
ESEI: Escuela Superior de Enxeñería Informática, University of Vigo,
Ed. Politécnico, Campus Universitario As Lagoas s/n 32004 Ourense, Spain
e-mail: {dcalvo,galvez,dgpena,riverola}@uvigo.es

M.P. Rocha et al. (Eds.): 6th International Conference on PACBB, AISC 154, pp. 53–61.
springerlink.com

Classification techniques with DNA microarray data can be a useful tool to disease diagnosis. Many classification models have been applied to microarray data, some of the most relevant include SVMs [3], k-nearest neighbors [4], artificial neural networks [5], linear discriminant analysis [2], Random Forests [6], Naïve Bayes [7], etc., achieving a good level of accuracy. In addition, *Rough Set Theory* has a strong mathematical background and is useful to produce classification rules that determining the class of a sample [8] as well as to determine which attributes are more relevant for classification.

However, the large amount of data it generates makes necessary to select relevant genes for the diagnosis of a particular disease. There are some interesting proposals on the use of prior biological knowledge with the aim of improving results in classifying samples. Some of them use data structures to represent knowledge in the form of networks and other as sets of genes. Among those using sets of genes, there are recent works such as *supergenes* [9], *NPR* [10], *GRDA* [11] and *mPAM/mPLS* [12].

This paper presents a rough set-based model for DNA microarray classification using *CAI model* [13] to work with uncertainty, based on *Variable Precision Rough Set Theory* [14].

2 Rough Set, Variable Precision Rough Set and CAI Model

Rough Set Model was introduced by Z. Pawlak in 80's to satisfy the need for a formal framework to manage imprecise knowledge expressed in terms of data acquired from experiments [15]. *Imprecise* refers to the fact that the granularity of knowledge causes *indiscernibility*. These imprecise concepts can be defined approximately with available knowledge using two precise concepts called *lower approximation* ($\underline{R}X$) and *upper approximation* ($\overline{R}X$). Any set defined by its lower and upper approximations is called Rough Set.

Pawlak defined an *information system* as a pair $I = (\mathbb{U}, \mathbb{A})$ be an , where $\mathbb{U}$ is a non-empty finite set of objects, $\mathbb{A}$ is a non-empty finite set of attributes such that $A = \{\mathbb{C} \bigcup \mathbb{D}\}$ where $\mathbb{C}$ is the set of condition attributes and $\mathbb{D}$ is the set of decision attributes and there is a mapping $a : \mathbb{U} \to V_a \forall a \in A$, where V_a is the value set of a.

The relation $IND(R)$ is called a *R-indiscernibility relation*. Two objects are indiscernible if they have same values for the same set of attributes.

$$IND(R) = \{(x,y) \in \mathbb{U} \times \mathbb{U} | \forall a \in R, a(x) = a(y)\} \quad (1)$$

Let $X \subseteq \mathbb{U}$ be a target set that we wish to represent using attribute subset R. $\underline{R}X$ is the set of elements of $\mathbb{U}$ that can be certainly classified as elements of X with the knowledge R. $\overline{R}X$ is the set of elements of $\mathbb{U}$ which can be possibly classified as elements of X, using knowledge R.

Based on these concepts, $\mathbb{U}$ can be divided in three regions: $POS_R(X) = \underline{R}X$ as positive region of X, $NEG_R(X) = \mathbb{U} - \overline{R}X$ as negative region of X and $BN_R(X) = \overline{R}X - \underline{R}X$ as boundary region of X.

The boundary region consists of those objects that can neither be ruled in nor ruled out as members of the target set X.

Boundary region can be perceived as an area where classification can't be done without ambiguity. This region can be considered also as an area where classification isn't possible under a certain level of error. With this in mind, Rough Set model can be extended to characterize a set in terms of uncertain information under some levels of certainty. This idea is based in *Variable Precision Rough Set* [14].

The standard definition of the set inclusion relation is too rigoruos to represent any *almost* complete set inclusion. So, the extended notion should be able to allow for some degree of misclassification in the large correct classification. The measure $c(X,Y)$ of the *relative degree of misclassification* (2) of the set X with respect to set Y is defined as:

$$\begin{array}{ll} c(X,Y) = 1 - |X \cap Y| / |X| & \text{if } |X| > 0 \text{ or} \\ c(X,Y) = 0 & \text{if } |X| = 0 \end{array} \quad (2)$$

The *majority inclusion relation* (3) under an admissible umbral of classification error β (which must be within the range $0 \leq \beta \leq 0.5$ is defined as:

$$X \subseteq_{\beta} Y \Leftrightarrow c(X,Y) \leq \beta \quad (3)$$

All concepts defined in RS are redefined using *majority inclusion relation* in VPRS.

The CAI (*Conjuntos Aproximados con Incertidumbre*) [13] model is derived from the VPRS model. As the VPRS model, CAI works also with uncertain information but with the aim of improve the classification power in order to introduce stronger rules. In the CAI model, uncertainy is introduced at two different levels: the constituting blocks of knowledge (elementary categories) and the overall knowledge, through the relationship of majority inclusion. So that, two different knowledge bases P and Q are equivalent or approximately equal, and denoted by $P \approx_{\beta} Q$, if the majority of their constituting blocks are similar.

3 Introducing Biological Knowledge

One of the problems with high-dimensional data sets is that not all the genes (attributes, variables) are useful for classifying a sample in a class of disease (phenomena of interest) [16].

Introducing biological knowledge in microarray data analysis can reduce data dimensionality, targeting only at genes that are involved in a metabolic process, genomic process, ontology, etc.

Given a universe of discourse $\mathbb{U}$ (e.g. composed by genes expressions from a microarray), often a concept of interest in not explicitly expressed. Instead is expressed by joining a series of subsets of the universe of interest, defined independently.

Therefore, it is called *interpretation context* [17] to any family of subsets $F = \{F_1,\ldots,F_i,\ldots,F_n\}$, with $F_i \subseteq \mathbb{U}$ where all interpretation context defines a concept (subset) of interest, formed by the union of all categories of F and denoted by $\bigcup F$. Any interpretation context $F = \{F_1,\ldots,F_i,\ldots,F_n\}$ imposes a structure on the concept of interest given by $\bigcup F$, formed by basic categories characterized by non-overlapping and form a coating of $\bigcup F$. Formally, given an interpretation context $F = \{F_1, \dots, F_i, \dots, F_n\}$ and given $N = \{1, 2, \dots, n\}$, it's called *basic category* to any set constructed from F as follows:

$$m_S = \left(\bigcap_{i \in S} F_i\right) - \left(\bigcup_{i \in N-S} F_i\right), \text{con } \emptyset \neq S \in \wp(N) \tag{4}$$

4 Classification Process

The classification process is divided into five steps:

1. Supergene generation
2. Attribute discretization
3. Feature selection
4. Calculation of reducts and decision rules
5. Rule application in classification process

The idea of supergenes was introduced by X. Chen and L. Wang [9] and it's a construction that summarizes information from a set of genes like *gene categories, pathways, gene sets*, or, in this case, *basic categories*. The information summarized from genes is generated using the *principal component analysis* (PCA) method [18] taking the first component. Once supergenes representing information from each basic cateogory have been generated, they are used as predictors of the sample class instead using genes.

Rough Sets algorithms works with *Nominal* attributes but gene expression levels are *floating point* values, so it's necessary to discretize values to make Rough Sets applicable to data from DNA microarray. Discretization transforms a continuous range of values in a defined number of bins. Each bin will contain all values of a subrange of values and will represent this range with a discrete value. *Fuzzy discretization* is used to discretize supergene values of the output from the previous step. We have used DFP, an *Bioconductor* package for the programming language and statistical environment R [19].

In real data analysis such as microarray data, the data set may contain a number of insignificant features (redundances). Ideally, the selected features should have high relevance with the classes and high significance in the feature set. A feature set with high relevance and high significance enhances the predictive capability [20]. Rough set theory is used to select the most relevant features from supergene Data set. The method of *Max β-relevance* based on CAI model [13] has been defined after failing to apply the method of *Max-relevance, Max-significance* because the apply the significance and relevance of all supergenes was zero.

Define $\widehat{r}_\beta(s_i,\mathbb{D})$ as the β-relevance of the supergene s_i with respect to the class labels $\mathbb{D}$. The β-relevance of s_i with respect to $\mathbb{D}$ can be calculated as:

$$\widehat{r}_\beta(s_i,\mathbb{D}) = \frac{|POS_{s_i,\beta}(\mathbb{D})|}{|\mathbb{U}|}, \text{ where } 0 \leq \widehat{r}_\beta(s_i,\mathbb{D}) \leq 1 \tag{5}$$

The purpose of maximum β-relevance is to select the features with more power for classification and reject those that interfere with the process.

Among the selected supergenes redundancy still exist, therefore it is neccesary to perform a reduct calculation. Intuitively, a *β-reduct* of the set of supergenes $\mathbb{C}$ is its essential part, which is sufficient to define all the basic concepts of data considered, with an classification error less than or equal to β.

The process of determining the reducts of an information system is know to be very expensive in terms of execution time. The method used to simplify decision tables under CAI model [21] consists of the following steps: First, β-reducts of condition attributes (supergenes) are computated, i.e., remove superfluous condition attributes; then, superfluous attribute values are removed (β-reducts of categories). After these steps, a set of decision rules is obtained. Decision rules can be expressed as:

$$(s_i = v_i) \wedge (s_j = v_j) \wedge \ldots \wedge (s_k = v_k) \rightarrow (d = c) \tag{6}$$

A decision rule can be also expressed as:

$$\{(s_i,v_i),\ldots,(s_j,v_j),\ldots(s_k,v_k),(d,c)\} \tag{7}$$

where pairs of the form $(s_i = v_i)$ or (s_i,v_i) denote pairs of <conditional attribute, value> and $(d = c)$ or (d,c) denotes a pair of <decision attribute, value>.

Let $Cover(R_i)$ denote the number of objects of the training set that *support* the decision rule R_i, let $NOC(R_i)$ denote the number of objects with the same decision label as the decision rule R_i, and let $E(R_i)$ denote the classification error (in training set) of the decision rule R_i. Define $s(R_i)$ as the score of the decision rule R_i. The score of R_i can be calculated as:

$$s(R_i) = \frac{Cover(R_i)}{|NOC(R_i)|}(1 - E(R_i)^2) \tag{8}$$

The purpose of decision rule score is to sort rules, placing first rules withm more coverage of objects (samples) and with less error.

5 Experimental Results

The performance of the proposed classification technique is extensively studied and compared with some existing classic classification methods: *Sequential Minimal Optimization* for *Support Vector Machines* [22], *K-nearest neighbors* [23], and *Random Forests* [24]. Classification methods for comparation are implemented in Weka

(*Waikato Environment for Knowledge Analysis*). Experiments run in LINUX environment running on an Intel Core i7 2.80 GHz with 8 MB cache and 4 GB of RAM.

The performance of different algorithms is analyzed doing the experimentation on a *breast cancer* microarray dataset. A set of basic categories, with one set of genes for each basic category is introduced.

Different methods were compared using two breast cancer datasets. The first breast cancer dataset contains expression levels of 12650 genes in 286 breast tumor samples. Data set GSE2990 [25] has been extracted from the public database GEO [26]. Samples are classified according to their estrogen receptor (ER) status: active (ER+) or inactive (ER-), an interesting factor in determining the aggressiveness necessary during treatment. The data set has been normalized and used to evaluate and compare classification methods using a 10-fold cross validation. The second dataset (GSE2034) [27] was also extracted from GEO database, and has been used to determine the optimal value of β .

The basic category dataset was created using two different sources. 33 metabolic pathways related to cancer, identified by SABioscience enterprise and 5 gene sets related to breast cancer from the OMM database[28].

All this gene sets were combined to create 130 basic categories used as knowledge input for the proposed method. In addition, union of all gene sets selected from the above sources were used to restrict the gene expression levels used for classification in the rest of techniques.

As mentioned, the input dataset contains thousands of genes that can be reduced to 130 supergenes using 130 basic categories.

At this point we have a decision table with 130 attributes of condition (supergenes) and 1 decision attribute (ER+ or ER-) for 286 samples. So, the next step was to get the reducts and decision rules from this decision tables. A β factor of error in the calculation of the *positive region* was introduced. This is intended to obtain relevant attributes greater than zero.

The optimal value for β will be the minimum one that will allow to obtain β-relevant attributes. The optimal value in this case is $\beta = 0.1$. With this value, you get about 6 or 7 relevant attributes. These supergenes, which correspond to some of the basic categories, are the most relevant to the class.

Once a few supergenes have been selected (more β-relevant supergenes), reducts and decision rules are computed in the last step. Decision rules serve as a fundamental core of the classifier using the score proposed as a method for sorting rules. Performance of this classifier is evaluated and compared with other conventional methods.

As was stated above, the classification methods are evaluated and compared through *accuracy* and *Cohen's kappa coefficient*. Accuracy indicates the percentage of correctly classified samples over total samples. Kappa coefficient indicates the agreement between expected classes and obtained classes from each classifier [29].

Table 1 shows average values for accuracy and kappa, for each of the classification methods. It also shows the variability of each measure by its standard deviation. *Accuracy* of the proposed method is quite similar to the other classic classification

Table 1 Accuracy and kappa of classification methods

	Accuracy		Kappa	
	Average	Std. Dev.	Average	Std. Dev.
k-NN	86.3742%	±8.1268	0.48083	±0.2593
SMO	83.0409%	±6.6363	0.41626	±0.1989
Random Forests	86.3742%	±7.0865	0.47729	±0.1814
Rough Sets	83.1578%	±8.0386	0.49898	±0.1878

methods but staying slightly below the accuracy of *k-NN* and *Random Forest*. Average *kappa* value of proposed method has overcome the rest of classifiers. This is a very hopeful fact encouraged to continue working on this line because the scenario favored by the data set has been the worst possible.

6 Conclusions and Future Work

The developed method has obtained successful results, beings near of the accuracy of other classic classifiers and overcoming them in kappa. The added value of this method is the possibility that provides the expert to introduce explicit biological knowledge in the process of constructing the classifier in the form of basic categories, making it more biologically explainable and probably more robust when combining samples from different laboratories. In addition, the proposed model not only got a similar performance than other classifiers, it's also able to identify which features (in the form of basic categories or supergenes) are more relevant for classification.

As future work, we will test the performance of this method using another data set, hoping that it will be better than the other classification techniques as we can see in Maji's work [20]. We also will try to use heuristic algorithms for computing reducts. This can have two purposes: increase the number of relevant genes, because a heuristic algorithm can find reducts faster with a larger number of attributes; or it makes feasible to search reducts directly on the entire set of attributes. Finally, another future direction is to continue working on the method to sort decision rules.

References

1. McLachlan, G.J., Do, K.A., Ambroise, C.: Analyzing Microarray Gene Expression Data. John Wiley & Sons, Inc., Chichester (2004)
2. Dudoit, S., Fridlyand, J., Speed, T.P.: Comparison of discrimination methods for the classification of tumors using gene expression data. Journal of the American Statistical Association 97(457), 77–87 (2002)
3. Furey, Cristianini, Duffy, Bednarski, Schummer, Haussler: Support vector machine classification and validation of cancer tissue samples using microarray expression data. Bioinformatics 16, 906–914 (2000)

4. Ramaswamy, S., Tamayo, P., Rifkin, R., Mukherjee, S., Angelo, M., Ladd, C., Reich, M., Mesirov, P., Poggio, T., Gerald, W., Loda, M., Lander, E.S., Golub, T.R.: Multi-class cancer diagnosis using tumor gene expression signatures. Proceedings of the National Academy of Sciences of the United States of America 98, 15149–15154 (2001)
5. Meltzer, P.S., Khan, J., Wei, J.S., Ringnér, M., Saal, L.H., Ladanyi, M., Westermann, F., Berthold, F., Schwab, M., Antonescu, C.R., Peterson, C.: Classification and diagnostic prediction of cancers using gene expression profiling and artificial neural networks. Nature Medicine 7, 673–679 (2001)
6. Díaz-Uriarte, R., de Andrés, S.A.: Gene selection and classification of microarray data using random forest. BMC Bioinformatics 7, 3 (2006)
7. Demichelis, F., Magni, P., Piergiorgi, P., Rubin, M.A., Bellazzi, R.: A hierarchical naïve bayes model for handling sample heterogeneity in classification problems: an application to tissue microarrays. BMC Bioinformatics 7, 514 (2006)
8. Pawlak, Z.: Rough Sets, Theoretical aspects of reasoning about data. Kluwer Academic Publishers (1991)
9. Chen, X., Wang, L.: Integrating biological knowledge with gene expression profiles for survival prediction of cancer. Computational Biology 16(2), 265–278 (2009)
10. Wei, Z., Li, H.: Nonparametric pathway-based regression models for analysis of genomic data. Biostatistics 8, 265–284 (2007)
11. Tai, F., Pan, W.: Incorporating prior knowledge of gene functional groups into regularized discriminant analysis of microarray data. Bioinformatics 23(23), 3170–3177 (2007)
12. Tai, F., Pan, W.: Incorporating prior knowledge of predictors into penalized classifiers with multiple penalty terms. Bioinformatics 23(14), 1775–1782 (2007)
13. Galvez, J.F., Diaz, F., Carrion, P., Garcia, A.: An Application for Knowledge Discovery Based on a Revision of VPRS Model. In: Ziarko, W.P., Yao, Y. (eds.) RSCTC 2000. LNCS (LNAI), vol. 2005, pp. 296–303. Springer, Heidelberg (2001)
14. Ziarko, W.: Variable precision rough set model. Computer and System Sciences 46, 39–59 (1993)
15. Pawlak, Z.: Rough sets. International Journal of Computer and Information Sciences 11(5), 341–356 (1982)
16. Fodor, I.: A survey of dimension reduction techniques. tech. rep., Lawrence Livermore National Laboratory (May 2002)
17. Glez-Pena, D.: Modelo para la integratión de conocimiento biológico explícito en técnicas de clasificación aplicadas a datos procedentes de microarrays de ADN. PhD thesis, University of Vigo (2009)
18. Pearson, K.: On lines and planes of closest fit to systems of points in space. Philosophical Magazine 2, 559–572 (1901)
19. Glez-Pena, D., Alvarez, R., Diaz, F., Fdez-Riverola, F.: Dfp: a bioconductor package for fuzzy profile identification and gene reduction of microarray data. BMC Bioinformatics 10(1), 37 (2009)
20. Maji, P., Paul, S.: Rough set based maximum relevance-maximum significance criterion and gene selection from microarray data. Int. J. Approx. Reasoning 52(3), 408–426 (2011)
21. Galvez, J.F., Olivieri, D., Carrion, P.: An improved algorithm for determining reducts in rough set models (2003)
22. Platt, J.: Fast training of support vector machines using sequential minimal optimization. In: Schölkopf, B., Burges, C.J.C., Smola, A.J. (eds.) Advances in Kernel Methods — Support Vector Learning, pp. 185–208. MIT Press, Cambridge (1999)
23. Fix, E., Hodges, J.L.: Discriminatory analysis – nonparametric discrimination: Consistency properties. Tech. Rep. Project 21-49-004, Report No. 4, 261-279, USAF School of Aviation Medicine, Randolph Field, Texas (1951)

24. Breiman, L.: Random forests. Machine Learning 45(1), 5–32 (2001)
25. Sotiriou, C., Wirapati, P., Loi, S., Harris, A., Fox, S., Smeds, J., Nordgren, H., Farmer, P., Praz, V., Haibe-Kains, B., Desmedt, C., Larsimont, D., Cardoso, F., Peterse, H., Nuyten, D., Buyse, M., Van de Vijver, M.J., Bergh, J., Piccart, M., Delorenzi, M.: Gene expression profiling in breast cancer: understanding the molecular basis of histologic grade to improve prognosis. Journal of the National Cancer Institute 98(4), 262–272 (2006)
26. Edgar, R., Domrachev, M., Lash, A.E.: Gene expression omnibus: Ncbi gene expression and hybridization array data repository. Nucleic Acids Research 30(1), 207–210 (2002)
27. Wang, Y., Klijn, J., Zhang, Y., Sieuwerts, A., Look, M., Yang, F., Talantov, D., Timmermans, M., Meijervangelder, M., Yu, J.: Gene-expression profiles to predict distant metastasis of lymph-node-negative primary breast cancer. The Lancet 365, 671–679 (2005)
28. Amberger, J.S., Bocchini, C.A., Scott, A.F., Hamosh, A.: Mckusick's online mendelian inheritance in man (OMIM®). Nucleic Acids Research 37(Database-Issue), 793–796 (2009)
29. Ben-David, A.: Comparison of classification accuracy using cohen's weighted kappa. Expert Syst. Appl. 34(2), 825–832 (2008)

Quantitative Assessment of Estimation Approaches for Mining over Incomplete Data in Complex Biomedical Spaces: A Case Study on Cerebral Aneurysms

Jesus Bisbal, Gerhard Engelbrecht, and Alejandro F. Frangi

Abstract. Biomedical data sources are typically compromised by fragmented data records. This incompleteness of data reduces the confidence gained from the application of mining algorithms. In this paper an approach to approximate missing data items is presented, which enables data mining processes to be applied on a larger data set. The proposed framework is based on a *case-based reasoning* infrastructure which is used to identify those data entries that are more appropriate to support the approximation of missing values. Moreover, the framework is evaluated in the context of a complex biomedical domain: *intracranial cerebral aneurysms*. The dataset used includes a wide diversity of advanced features obtained from clinical data, morphological analysis, and hemodynamic simulations. The best feature estimations achieved errors of only 7%. There are, however, large differences between the estimation accuracy achieved with different features.

1 Introduction

Data mining is often being applied to uncover hidden patterns in vast amounts of data, and for decision support. However, the major potential benefits of such an approach are hampered by the fact that data sources are very commonly incomplete [8, 7, 5]. Missing values compromise the amount of data which is available in practice for further anlysis. This, in turn, reduces the confidence of the outcomes produced by the mining algorithms.

Jesus Bisbal
DTIC, Universitat Pompeu Fabra, Spain
e-mail: jesus.bisbal@upf.edu

Gerhard Engelbrecht · Alejandro F. Frangi
CISTIB, Universitat Pompeu Fabra, and CIBER-BBN, Barcelona, Spain
e-mail: {gerhard.engelbrecht,alejandro.frangi}@upf.edu

M.P. Rocha et al. (Eds.): 6th International Conference on PACBB, AISC 154, pp. 63–71.
springerlink.com

The research reported in this paper applied classification algorithms to the specific context of *intracranial cerebral aneurysms*. An aneurysm is a localized, blood-filled dilation of a blood vessel caused by disease or weakening of the vessel wall. If left untreated, it can burst leading to severe haemorrhage and sudden death [1]. The complexity of the potentially influencing factors leading to rupture of an aneurysm is, to a large extend, still unknown. Classification algorithms can help predict whether a given clinical case is likely to rupture or not [2, 12]. The presence of missing values, however, significantly limits the outcomes of such an approach.

This paper reports the design of a framework that allows the application of alternative approaches to estimate appropriate values for missing entries. The framework is also evaluated using a complex dataset produced during the EU funded @neurIST project [1], which focused on the management of cerebral aneurysms.

The remainder of this paper is organized as follows. The next Section reviews existing work related to this research. Section 3 defines the set of metrics used in selecting similar cases to the one which contains the value being estimated. Section 4 describes the dataset used to evaluate this research. The common approaches used to estimate missing values are outlined in Section 5. The experimental results of the framework evaluation are detailed in Section 6. The final section summarises the paper and describes a number of future directions for this research.

2 Background and Related Work

Research on data mining over incomplete datasets can be broadly classified in two types of approaches. On the one hand, those that apply a pre-processing step to the incomplete dataset, where missing values are estimated [3, 5, 7], and then state-of-the-art mining algorithms are applied to the completed data. The challenge here relies on devising sound estimation approaches that are robust enough to be applied as a generic mechanism. A model of the missing values is defined, and inferences are based on the likelihood under that model.

On the other hand, a different type of approaches develop mining algorithms that can cope with missing entries in the dataset [11, 9, 4], without any explicit attempt to approximate the value for these entries. Many efforts have concentrated on building [11, 8, 6] and applying [4] decision trees in the presence of uncertain data. The Bayesian classifier model was also used in this context [9].

The present paper addresses the problem of missing values, not of uncertain values, thus is inspired by [5, 7]. As an initial attempt, the estimated values are computed by averaging a set of values from a set of 'similar cases'. In this context, it relies on a *case-based reasoning framework* through which the most similar 'cases' to the one being estimated can be robustly identified. A similar problem was addressed in [11, 9] to compare their methods with standard classifiers build by approximating missing values with averages. These averages are computed from the entire value set, as opposed to a carefully selected set of values. In [7] case similarity was computed using the euclidean distance only. Here, in contrast, thanks to our generic CBR infrastructure, the most appropriate metric to be used has also been thoroughly evaluated (see Section 6.1).

3 Metrics for Content-Based Retrieval

Similarity is considered higher when the computed distance between cases is smaller. Different metrics can be appropriate depending on the values and the semantics of feature vectors. Table 1 summarises a set of well-known metrics [10], those that have been evaluated and compared in our experiments. In these definitions, consider two cases as two points in a n-dimensional space, $x = (x_1, x_2, \ldots, x_n)$ and $y = (y_1, y_2, \ldots, y_n)$. Only continuous features are considered here.

Table 1 Set of Metrics Used to Evaluate Case Similarity

Name	Definition	Observations
Minkowski (of order p)	$\left(\sum_{i=1}^{n} \lvert x_i - y_i\rvert^p\right)^{\frac{1}{p}}$	When $p = 1$, *Manhattan distance* (*taxicab* or L_1 norm). For $p = 2$, it becomes the *Euclidean* distance. When $p \to \infty$, *Chevyshev distance* (or L_∞ norm).
Cosine	$1 - \frac{\sum_{i=1}^{n} x_i \cdot y_i}{\sqrt{\sum_{i=1}^{n} x_i^2} \cdot \sqrt{\sum_{i=1}^{n} y_i^2}}$	Common in the information retrieval literature. Often outperforms Euclidean distance. Defined by the *cosine of the angle* formed by the vectors.
Mahalanobis	$\sqrt{(x-y)^T S^{-1} (x-y)}$	Being S^{-1} the inverse of the covariance matrix. Takes into account the correlations of the data set. It computes the 'center of mass' of all points. Used to compute the dissimilarity between points.

4 Data Collection: Cerebral Aneurysms

The dataset used for experimental evaluation of this work originated in the @neurIST project [1], focused on advancing the management of cerebral aneurysms using information technology. In total, data of 1420 subjects suspected aneurysms were collected. Due to incomplete clinical data collection and image acquisition along with limited effort in processing and simulating individual cases a total of 157 cases with complete information are available. These cases represent the ground truth used for evaluation in Section 6.

The potential set of features for each case is extremely large, which includes:

Clinical Information: Data commonly used for clinical care, such as demographics (e.g. age, gender), lifestyle (e.g. smoker status), and aneurysm anatomical location and type.

Morphological Analysis: Image processing to obtain morphological descriptors about an aneurysm, such as aspect ratio, non-sphericity index, volume, or moment invariants.

Hemodynamic Analysis: Blood flow simulations used to compute characteristics of the blood flow and pressure, as well as their impact on the vessel wall.

The set of features selected in the experiments reported in this paper are those that have been found to be the most relevant ones in order to predict the rupture of a cerebral aneurysm. Please refer to [2] for a detailed description of the complete feature set considered in this study.

5 Value Estimation Approaches

The problem of dealing with missing values in a dataset in the context of data mining can be addressed by inferring possible values for these missing values, and then applying mining algorithms to the resulting 'completed' data set. Several approaches have been proposed to infer those values [5]:

Mean imputation: Where the means from sets of recorded values are substituted.
Regression imputation: Where the missing values are substituted by predicted values from the regression on known (recorded) feature values.
Model-based imputation: Where a model of the missing values is defined, and inferences are based on the likelihood under that model, with model parameters estimated by procedures such as *maximum likelihood.*

To the best of the authors' knowledge, these approaches have not been evaluated in the context of complex biomedical datasets, considering feature types of widely different nature, as detailed in Section 4. *Mean Imputation* has been used in this paper as the first and conceptually simpler approach. Other approaches will be evaluated and compared in future work.

The work reported here follows an approach inspired by the field of *case-based reasoning* (CBR). It performs a search in the n-dimensional space defined by the feature set under study to identify those cases which are *more similar* to that case for which a missing value needs to be inferred. Similarity is defined according to a given metric (e.g. see Section 3).

5.1 Accuracy of Value Estimation

The error introduced by the estimation process can be computed from two different perspectives, namely *Direct* and *Indirect* measures.

Direct Error Measure: It takes into account the error introduced when estimating each individual missing value. Let $X_k = \{x_1, x_2, \dots, x_n\}$ denote the original values in a given feature k for some cases in a dataset, and $\widehat{X_k} = \{\widehat{x_1}, \widehat{x_2}, \dots, \widehat{x_n}\}$ the estimated values for the same cases. The *relative error* is computed as follows:

$$Error_k\left(X_k, \widehat{X_k}\right) = \frac{\sum_{i=1}^{n} |x_i - \widehat{x_i}|}{\sum_{i=1}^{n} |x_i|} \tag{1}$$

This produces normalised error calculations which are comparable across different features. This is important in our experiments, as the feature values range over many orders of magnitude.

Indirect Error Measure: This research is performed in the context of applying data mining algorithms to complex biomedical domains. Therefeore, it must compare the (classification) accuracy of mining algorithms applied over a dataset with estimated values, with the same algorithms applied over a complete dataset (no estimated values),

6 Experimental Evaluation

The value estimation framework developed in this research has been evaluated on the dataset described in Section 4, from two different perspectives: selecting the best metric to compute case-similarity, and quantifying the error introduced by the estimation process.

6.1 *Selection of Metric for CBR Infrastructure*

The evaluation of the metrics defined in Section 3[1] is shown in Figure 1. This figure illustrates the error produced when using each of the different metrics in identifying the set of most similar cases. As described in Section 5, these cases are used in order to compute the average value, used as the estimated value. Clearly, if the set of cases used for the estimation are actually not very similar to the estimating case, the errors are expected to be large. It has, therefore, been limited which cases to include. A *similarity threshold* is defined, and only those cases the similarity of which fall above that threshold are included. This percentage is calculated by considering that the largest of all distances for one case represents the 0% similarity, and the rest are calculated proportionally. Thus, the x-axis in this figure indicates the given threshold applied for this calculation, which varies from 0% to 90%. The y-axis indicates the average error produced by the estimation process, following Equation 1. This is computed for each missing value, averaged over all missing values for that given feature, and then averaged over the estimation of all features. Clearly, as the similarity threshold approaches 0%, it means that all cases will be included for each estimated value, even those very different from the estimated case, therefore the estimation error becomes larger.

This figure shows rather large average errors, because it includes also very badly performing features. Section 6.2 details the accuracy for each feature independently. In any event, Figure 1 justifies that the *Cosine* distance outperforms, on average, all other metrics, and thus will be the metric used for the experiments reported next.

[1] For illustration purposes only, a Minkowsky distance for $p = 2.72$ is shown. More extreme cases of this distance are already represented by the Euclidean and Chebyshev distances.

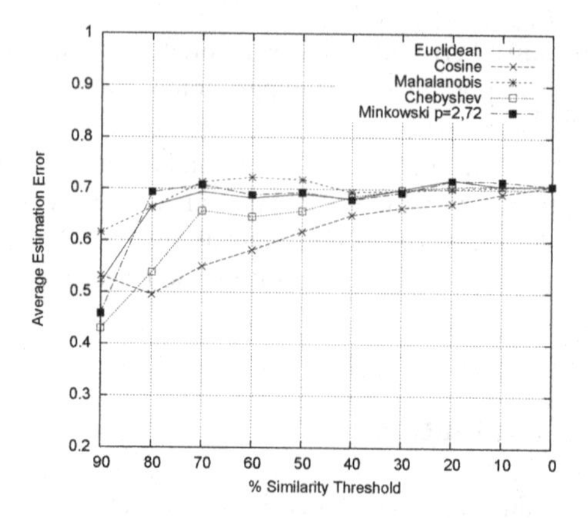

Fig. 1 Average Estimation Error over All Features for Each Metric

6.2 Estimation Error in the Aneurysms Dataset

The results of the experiments to evaluate the estimation errors are shown in Figure 2, where the x-axis also indicate the 'similarity threshold', and the y-axis the errors introduced by the estimation process.

There are very large differences in the estimation errors, depending on which feature is being estimated. The results are, thus, divided into three curves. Figure 2a displays the set of features which can be most accurately estimated, with an error as low as 7%. Figure 2b shows features with estimation errors still too high for use in biomedical domains. However, it is expected that further research (e.g. using other estimation approches) could improve very significantly the estimation accuracy for at least some of these features. For those features shown Figure 2c[2], in contrast, there is probably an inherent limitation within the data itself, and may not be meaningfully estimated. It must be noted that these features take their values in a range of several orders of magnitude. Considering that the whole dataset used in these experiments includes only 157 instances, it may not be possible to improve significantly these obtained results.

Finally, Table 2 shows the *indirect error measure* referred to in Section 5.1. It applied a mining algorithm to classify cases as either aneurysm ruptured or not ruptured. Following [2], a supporting vectors machine classifier was also been used in these experiments.

We generated several datasets, by removing a given percentage (5% to 25%) of values from the original dataset. We then estimated these missing values, and applied the classification algorithm to the resulting estimated dataset. Column '$AC_{Original}$' refers to the accuracy of the classifier over the original dataset, and column '$AC_{Estimated}$' over the estimated dataset. This column is expressed as the percentage of accuracy with respect to the value in column '$AC_{Original}$', where 100%

[2] Please note that the scale is slightly different than in Figure 2b.

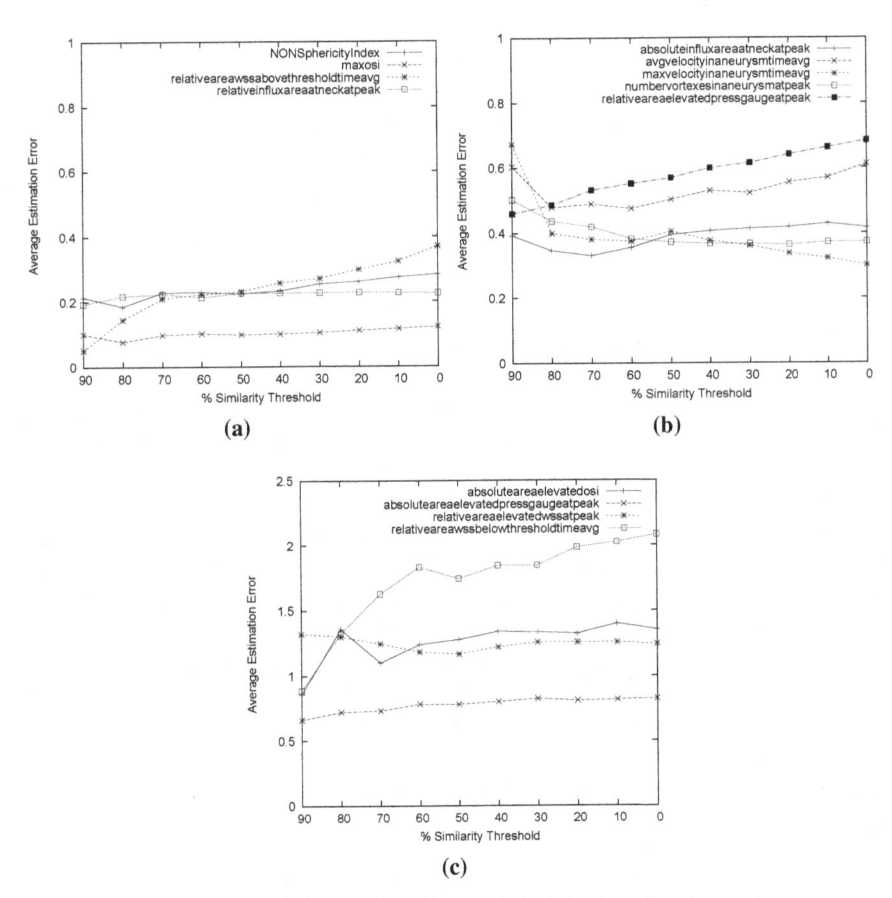

Fig. 2 Estimation Errors: (a) Best, (b) Medium, and (c) Worst Performing Features

would indicate the values in both columns would be identical. As expected, mining over estimated data produces lower (e.g. below 100%) accuracy rates than for the original (complete) dataset. Accuracy generally decreases as the percentage of missing values increases. These results must be compared with column 'AC_{MI}' (for Mean Imputation). It shows the classification accuracy over the estimated dataset (also as the percentage of '$AC_{Original}$'), but using the state-of-the-art *global replacement approach*. With this approach[3] all missing values are replaced by the average value of their feature, considering all available values for the feature, as opposed to a few carefully selected values as done in our research. Our approach clearly outperforms global replacement, even for large percentage of missing values. Also, it shows that good classification accuracy can still be achieved after estimation.

[3] Computed in this case by Weka's package built-in functions, `www.cs.waikato.ac.nz/ml/weka`

Table 2 Classification Accuracy on Estimated Datasets

% Removal	$AC_{Original}$	$AC_{Estimated}$	AC_{MI}
5%	68.5	93.4%	84.3%
10%	68.5	91.9%	84.2%
15%	68.5	92.3%	81.2%
20%	68.5	93.5%	82.7%
25%	68.5	94.2%	81.2%

7 Conclusions and Future Work

This paper has reported on an initial evaluation of a framework to estimate missing values in the context of a complex biomedical domain, that of intracranial cerebral aneurysms. The dataset used in this evaluation included 157 different instances or cases. It also includes a large feature set, including clinical information, morphological analysis, and hemodynamic simulations. The experiments showed that for some features, values can be estimated to a satisfactorily degree of accuracy (7% of error). Good classification accuracy can still be obtained using the estimated dataset.

Future work will focus on the imputation model. The *Mean Imputation* used in this paper estimates a missing value as the average of a set of selected values. While promising results have already been achieved, this approach seems clearly inappropriate for some features. More advanced approaches, such as *regression imputation* or *model-based imputation* [3, 5], should lead to more accurate estimations for some features. These approaches will need to exploit potential dependencies between features in order to improve estimations.

Acknowledgements. This work has been partially funded by the integrated project VPH-Share, grant number 269978 under the 7th Framework Programme of the European Union.

References

[1] Benkner, S., Arbona, A., et al.: @neurIST: Infrastructure for advanced disease management through integration of heterogeneous data, computing, and complex processing services. IEEE Trans. Inf. Tech. Biomed. 14, 1365–13270 (2010)

[2] Bisbal, J., Engelbrecht, G., Villa-Uriol, M.C., Frangi, A.F.: Prediction of Cerebral Aneurysm Rupture Using Hemodynamic, Morphologic and Clinical Features: A Data Mining Approach. In: Hameurlain, A., Liddle, S.W., Schewe, K.-D., Zhou, X. (eds.) DEXA 2011, Part II. LNCS, vol. 6861, pp. 59–73. Springer, Heidelberg (2011)

[3] de Waal, T., Pannekoek, J., Scholtus, S.: Handbook of Statistical Data Editing and Imputation. Wiley (2011)

[4] Hawarah, L., Simonet, A., Simonet, M.: Dealing with missing values in a probabilistic decision tree during classification. In: Zighed, D.A., Tsumoto, S., Ras, Z.W., Hacid, H. (eds.) Mining Complex Data. SCI, vol. 165, pp. 55–74. Springer, Heidelberg (2009)

[5] Little, R.J.A., Rubin, D.B.: Statistical Analysis with Missing Data. Wiley (2002)

[6] Magnani, M., Montesi, D.: Uncertainty in Decision Tree Classifiers. In: Deshpande, A., Hunter, A. (eds.) SUM 2010. LNCS, vol. 6379, pp. 250–263. Springer, Heidelberg (2010)
[7] Parthasarathy, S., Aggarwal, C.: On the use of conceptual reconstruction for mining massively incomplete data sets. IEEE Transactions on Knowledge and Data Engineering 15(6), 1512–1521 (2003)
[8] Qin, B., Xia, Y., Li, F.: DTU: A Decision Tree for Uncertain Data. In: Theeramunkong, T., Kijsirikul, B., Cercone, N., Ho, T.-B. (eds.) PAKDD 2009. LNCS, vol. 5476, pp. 4–15. Springer, Heidelberg (2009)
[9] Qin, B., Xia, Y., Li, F.: A bayesian classifier for uncertain data. In: Proc. 2010 ACM Symp. Applied Comp., pp. 1010–1014 (2010)
[10] Rajaraman, A., Ullman, J.: Mining of Massive Datasets. Cambridge University Press (2011)
[11] Tsang, S., Kao, B., Yip, K., Ho, W.S., Lee, S.D.: Decision trees for uncertain data. IEEE Transactions on Knowledge and Data Engineering 23, 64–78 (2011)
[12] Valencia, C., Villa-Uriol, M.C., Pozo, J.M., Frangi, A.F.: Morphological descriptors as rupture indicators in middle cerebral artery aneurysms. In: Conf. Proc. IEEE Eng. Med. Biol. Soc., pp. 6046–6049 (2010)

ASAP: An Automated System for Scientific Literature Search in PubMed Using Web Agents

Carlos Carvalhal, Sérgio Deusdado, and Leonel Deusdado

Abstract. In this paper we present ASAP - Automated Search with Agents in PubMed, a web-based service aiming to manage and automate scientific literature search in the PubMed database. The system allows the creation and management of web agents, specifically parameterized thematically and functionally, that automatically and periodically crawl the PubMed database, oriented to search and retrieve relevant results according the requirements provided by the user. The results, containing the publications list retrieved, are emailed to the agent owner weekly during the activity period programmed for the web agent. The ASAP service is devoted to help researchers, especially from the field of biomedicine and bioinformatics, in order to increase their productivity, and can be accessed at: http://esa.ipb.pt/~agentes.

1 Introduction

In the last two decades the areas of biomedicine and bioinformatics registered an unparalleled growth on research investment, generating an unprecedented amount of new knowledge and its consequent expression in terms of scientific bibliography. PubMed[1] is the largest public database for biomedical literature, currently

Carlos Carvalhal · Leonel Deusdado
Technology and Management School, Polythecnic Institute of Bragança, 5301-857 Bragança, Portugal

Sérgio Deusdado
CIMO - Mountain Research Center, Polythecnic Institute of Bragança, 5301-855 Bragança, Portugal
e-mail: sergiod@ipb.pt

[1] http://www.ncbi.nlm.nih.gov/pubmed/

M.P. Rocha et al. (Eds.): 6th International Conference on PACBB, AISC 154, pp. 73–78.
springerlink.com

comprises more than 21 million citations from Medline, life science journals, and online books. This huge volume of literature represents a great progress in knowledge and information accessibility, but also originates difficulties for filtering relevant results, and moreover, increases the time needed to search efficiently a permanently updated database.

Aiming to increase researchers' productivity, it is desirable that scientific literature search incorporate computational help, from crawlers and web agents mainly [1], to automate routine processes like periodical updates on new publications on specific subjects, generating alerts to the user. Having this automatism, researchers may receive automatically the results from a web agent that performs periodically, based on terms and requirements specifically defined for each agent, an accurate personalized search [2] of potentially relevant bibliography, only requiring from the scientists the minimum necessary time to analyze the results.

This reasons motivated us to create and develop a web-based service, named ASAP - Automated Search with Agents in PubMed, which is publically available and allows the creation and management of web agents to automate bibliography searching in the PubMed database.

2 Related Work

PubCrawler [3] was developed in 1999, at the Trinity College Dublin, and is a free "alerting" service that scans daily updates to the NCBI Medline (PubMed) and GenBank databases. PubCrawler helps keeping scientists informed of the current contents of Medline and GenBank, by listing new database entries that match their research interests. This service is available at: http://pubcrawler.gen.tcd.ie/

PubCrawler results are presented as an HTML web page, similar to the results of a NCBI PubMed or Entrez query. This web page can be located on the PubCrawler WWW-service, on the user's computer (the stand-alone program), or can be received via e-mail. The web page sorts the results into groups of PubMed/GenBank entries that are zero-days-old, 1-day-old, 2-days-old, etc., up to a user-specified age limit.

@Note [4] was developed at the University of Minho and is a platform that aims at the effective translation of the advances between three distinct classes of users: biologists, text miners and software developers. Among other features, can work as an information retrieval module enabling PubMed search and journal crawling. Using the EUtils service provided by Entrez-NCBI, the information retrieval module can retrieve results based on keyword-based queries.

3 Developed Work

The ASAP system was developed to assist researchers on the time consuming task of periodical or casual scientific literature search, providing them the automation of this task. ASAP performs focused crawling on the PubMed database to automate publications search and emails the results to the user on a weekly basis.

ASAP is mainly composed by three modules, that are interconnected to perform the fundamental workflow of the system, namely the web-based services to manage the data that supports the system, the database that organizes and made available the data, and the information retrieval module - this last module also integrates the results communication function. The functional architecture of the ASAP system is presented in Fig.1.

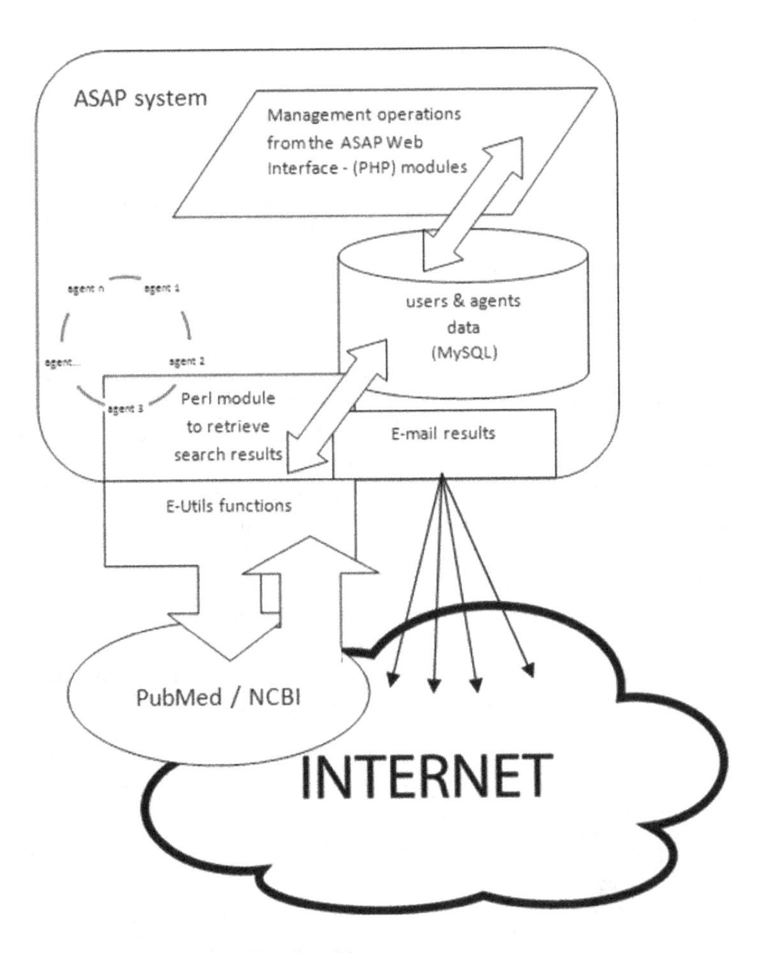

Fig. 1 The ASAP system's functional architecture.

In order to facilitate the adoption of the ASAP service in our community we decided to keep the processes of registering and web agents creation as simple as possible. ASAP is a web-based service and is already available online and translated in several languages. The ASAP homepage, accessible at http://esa.ipb.pt/~agentes, has an easy interface as depicted in Fig.2.

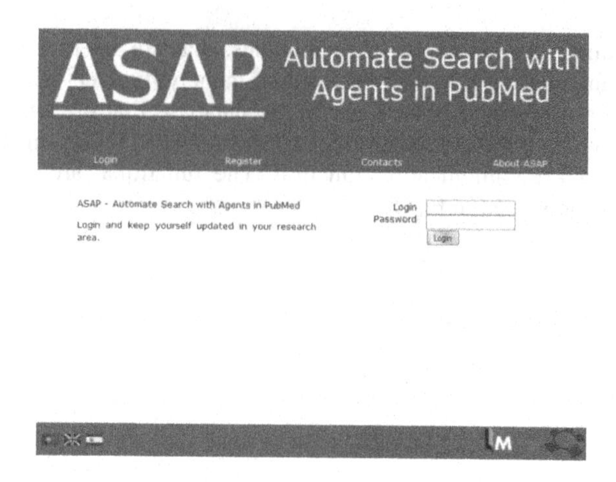

Fig. 2 The ASAP service homepage.

ASAP registered users can create agents to perform, in an automated fashion, scientific literature search based on keywords on the PubMed database, replicating the same search and retrieving the same results as the available ones at the original PubMed's site, since it uses the EUtils functions [5] provided by the Entrez framework [6] from NCBI. The EUtils (Entrez Programming Utilities) are used to invoke the crawling processes in the NCBI databases. EUtils comprise a set of seven different methods (eInfo, eGQuery, eSearch, eSummary, ePost, eFetch and eLink) providing an excellent way of access to the NCBI scientific content from external applications.

Each agent can be programmed thematically and functionally. The theme of interest is defined by specifying up to three keywords for each agent. The functional part is modeled by defining the date interval which matches the user interest. A snapshot of the interface used to create agents is presented in Fig. 3.

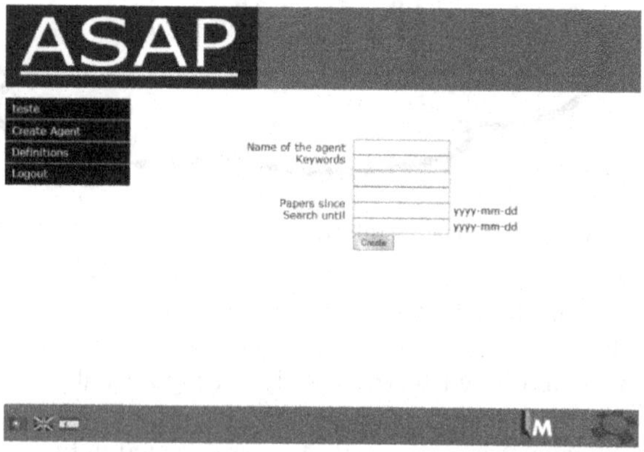

Fig. 3 Agent creation interface.

ASAP modules to manage the users and agents database were written in PHP. The database was created using MySQL.

The information retrieval module, including the notification part, was written in Perl. This module performs a different search for each valid agent and emails the results to the owner, this process is weekly initiated by a daemon.

4 Results and Discussion

ASAP results are delivered weekly to the agents' owners, verified the premise of the agent's validity/caducity. The crawler employs the keywords from each agent in the database, building the queries, sent through the EUtils functions, to search the PubMed repository, and returns the retrieved results. From these results it compiles e-mails for each agent, containing the details of the publications list, including the respective abstracts. The newest results appear at the top of the list to facilitate its verification. A link to the original results is also provided for extended view or deeper prospection of the results.

Without effort, the researchers receive in their mailboxes a list of results for each agent and only employ supplementary time on literature searching if novelties are perceived. In this way, a productivity gain is obtained as part of the searching task is automated.

ASAP is actually under tests within the IPB (Polytechnic Institute of Bragança) community and the subjective feedback, as far as it is possible to evaluate by now, is encouraging.

5 Conclusions and Future Work

We presented ASAP - Automated Search with Agents in PubMed, a web-based service aiming to manage and automate scientific literature search in the PubMed database. The system allows the creation and management of web agents that automatically and periodically crawl the PubMed database to search and retrieve relevant results according the filtering parameters provided by the user, returning him automatic result updating. ASAP is publicly and freely available, is provided “as is” without warranty of any kind, but we employ our best efforts to protect the confidentiality and security of the users’ data.

In future version the parameterization possibilities can be extended to meet the expectations of advanced users, namely to enable enhanced filtering capabilities, but on the other hand, the simplicity and lightness of the service could be compromised. Therefore, we expect to upgrade ASAP but keeping its simplicity as a mandatory requirement, since we believe most users are not interested in web agents if they are difficult to create and maintain.

References

[1] Kobayashi, M., Takeda, K.: Information retrieval on the web. ACM Computing Surveys 32, 144–173 (2000)

[2] Micarelli, A., Gasparetti, F., Sciarrone, F., Gauch, S.: Personalized Search on the World Wide Web. In: Brusilovsky, P., Kobsa, A., Nejdl, W. (eds.) Adaptive Web 2007. LNCS, vol. 4321, pp. 195–230. Springer, Heidelberg (2007)

[3] Hokamp, K., Wolfe, K.H.: PubCrawler: keeping up comfortably with PubMed and GenBank. Nucleic Acids Research 32(Web Server), W16–W19 (2004)

[4] Lourenço, A., et al.: @Note: a workbench for biomedical text mining. Journal of Biomedical Informatics 42(4), 710–720 (2009)

[5] Bethesda: Entrez Programming Utilities Help, National Center for Biotechnology Information, US (2011)

[6] Sayers, E.W., et al.: Database resources of the National Center for Biotechnology Information. Nucleic Acids Research 39, D38–D51 (2010)

Case-Based Reasoning to Classify Endodontic Retreatments

Livia Campo, Vicente Vera, Enrique Garcia, Juan F. De Paz,
and Juan M. Corchado

Abstract. Within the field of odontology, an analysis of the probability of success of endodontic retreatment facilitates the diagnostic and decision-making process of medical personnel. This study presents a case-based reasoning system that predicts the probability of success and failure of retreatments to avoid extraction. Different classifiers were applied during the reuse phase of the case-based reasoning process. The system was tested on a set of patients who received retreatments, and a set of variables considered to be of particular interest, were selected.

Keywords: case-based reasoning, classification, odontology.

1 Introduction

Predictive systems in medicine are relevant for determining the probability of success or failure of specific treatments. Their use currently extends to include various fields such as the study of cancer, or other fields such as odontology[1][2][3]. The decisions made by odontologists have been traditionally based on past experience of previous treatment cases, whereby experience in itself has been a necessary factor in the decision making process. There are normally too many variables to consider, which has in fact resulted in the high failure rate of retreatments.

Livia Campo · Vicente Vera · Enrique Garcia
Department of Estomatology II, Complutense University of Madrid
Plaza Ramón y Cajal, s/n 28040, Madrid, Spain
e-mail: lcampo@estumail.ucm.es, viventevera@odon.ucm.es,
aegarcia@odon.ucm.es

Juan F. De Paz · Juan M. Corchado
Department of Computer Science and Automation, University of Salamanca
Plaza de la Merced, s/n, 37008, Salamanca, Spain
e-mail: {fcofds,corchado}@usal.es

M.P. Rocha et al. (Eds.): 6th International Conference on PACBB, AISC 154, pp. 79–86.
springerlink.com

Consequently, it has become necessary to create a system that facilitates the decision making process of odontologists, and results in the decisions that minimize the failure of endodontic treatments and retreatments.

Endodontics makes up 20% of all treatments performed in dental clinics and has a 90% rate of success. The remaining 10% includes endodontic treatments that were unsuccessful to a greater or lesser degree, of which 40% are the result of root crown fractures, which in turn represent 5% of all dental fractures. The bacterial recolonization of the root canal and the subsequent appearance of radiological symptoms represent 15% of endodontic failure [8] [9] [10]. Many different alternative methods for analyzing data in odontology have already been investigated. The techniques applied in these fields are usually limited to the study of variables. A set of variables of interest is determined, followed by statistical tests and graphical representations of data to extract the relevant variables. Statistical analysis is limited to the application of specific tests such as chi square [12], Mann-Whitney [18] or Kruskal-Wallis [11]. These tests identify which variables present different characteristics in different groups; the value of the variables can subsequently be taken into consideration for the final classification. Nevertheless, it is necessary to create a process that can combine all the information gathered in order to perform a final classification. Previous Works in the field of bioinformatics CBR (Case-Based Reasoning) systems have been successfully applied to predict leukemia. This study proposes a reasoning system to predict the success of retreatments. A set of variables are recovered for a group of patients. This data set is used to generate a CBR system that incorporates different classification techniques during the reuse phase, in order to generate a classification for a new element. Traditional statistical techniques are applied during the revision phase to facilitate the interpretation of the results by selecting the variables that present different characteristics from those in the groups of individuals.

This article is divided as follows: section two describes the multi-agent systems and planning mechanisms used for assigning dynamic tasks; section three presents the proposed model; section four describe the results obtained and the conclusions respectively.

2 Prediction System

The use of predictive techniques in medicine and especially in the field of odontology has been studied since the late 80s, having primarily used the statistical analysis of clinical data.

In 2001 Chungal N.M. published data related to a study of teeth extracted after unsuccessful endodontic treatments at the University of Connecticut School of Dental Medicine. The patients included in this study were treated between 1988 and 1992 in the graduate program and had experienced unsuccessful endodontic treatment within the previous four years. Variables were taken from both the clinical trial and x-rays taken at the time of the endodontic treatment. The data obtained in this case were studied with contingency tables and the chi-squared test.

The risk factors were compared using t-tests for independent groups, or with non-parametric tests (Mann-Whitney or Kruskal-Wallis) [1].

Using the same characteristics, Givol, N. published the results in 2001 of his study performed in patients from private clinics in Israel. In this case, all the possible clinical variables prior and subsequent to the endodontic treatment were fathered from 5217 patients treated between 1992 and 2008. The data were also studied using statistical tests: chi-squared [2].

In July of 2011, Song, M. presented the data relative to a study performed on patients from the Department of Conservative Dentistry at the Dental College of Yonsei University, Seoul, Korea between August 2004 and December 2008. Included in this study were patients who had undergone unsuccessful endodontic treatment and were in need of periapical surgery. Song considered clinical and x-ray data from prior to the treatment, demographic data, and data subsequent to the failed treatment. To analyze the factors that could predict the endodontic failure, he applied a chi-squared statistical study [3].

Of the previously cited studies, none used artificial intelligence or case base reasoning; nor did any use predictive tools other than the application of statistical studies to analyze risk factors. The use of this type of system offers, therefore, a wide area of study within the field of odontology and in particular with the prediction of unsuccessful endodontic treatments.

3 Proposed Reasoning System

The purpose of CBR is to solve new problems by adapting solutions that have been used to solve similar problems in the past [4]. The primary concept when working with CBRs is the concept of case. A case can be defined as a past experience, and is composed of three elements: a problem description which describes the initial problem, a solution which provides the sequence of actions carried out in order to solve the problem, and the final state which describes the state achieved once the solution was applied. A CBR manages cases (past experiences) to solve new problems. The way cases are managed is known as the CBR cycle, and consists of four sequential steps which are recalled every time a problem needs to be solved: retrieve, reuse, revise and retain. Each of the steps of the CBR life cycle requires a model or method in order to perform its mission. The algorithms selected for the retrieval of cases should be able to search the case base and select the problem and corresponding solution most similar to the new situation. Once the most important variables have been retrieved, the reuse phase begins, in which the solutions for the retrieved cases are adapted and a new solution is generated. The revise phase consists of an expert revision for the proposed solution. Finally, the retain phase allows the system to learn from the experiences obtained in the three previous phases, consequently updating the cases memory.

During the recovery phase, existing cases in which a retreatment was performed are selected from the case memory. This eliminates all cases that involve only an initial treatment.

During the reuse phase, previously retrieved cases are selected and an associated classifier is built. In this case, the technique selected to carry out the classification phase corresponds to a Bayesian network. The new case is then introduced and classified according to the classifier built in this phase.

The Bayesian networks are constructed by following the Friedman-Goldsmidtz [5] algorithm. Having two different classes, two Bayesian networks will be generated, one for each of the classes.

The TAN classifier is constructed based on the plans recovered that are most similar to the current plan, distinguishing between efficient and inefficient plans to generate the model (the tree). Thus, by applying the Friedman-Goldsmidtz [5] algorithm, the two classes that are considered are efficient and inefficient. The Friedman-Goldsmidtz algorithm makes it possible to calculate a Bayesian network based on the dependent relationships established through a metric. The metric considers the dependent relationships between the variables according to the classifying variable. In this case, the classified variable is efficient and the remaining variables indicate whether a service is or is not available. The metric proposed by Friedman can be defined as:

$$I(X,Y \mid Z) = \sum_{x \in X} \sum_{y \in Y} \sum_{z \in Z} P(x,y,z) \cdot \log\left[\frac{P(x,y \mid z)}{P(x \mid z) \cdot P(y \mid z)}\right] \tag{1}$$

Based on the previous metric, the probabilities are estimated according to the frequencies of the data. The Friedman-Goldsmidtz [23] algorithm is broken down into the following steps:

- Calculate the value of $I(X,Y \mid C)$ for the different variables/attributes X, Y that may be interconnected in the original graph, class C varies between the similar efficient and inefficient cases.
- Construct a complete nondirected graph
  - Establish the different attributes/variables as nodes.
  - Within the connections, establish the values obtained in the first step as weights. For the arcs that do not have connections, set the value as 0.
- Create a maximum tree based on the Kruskal [6] algorithm.
- Convert the nondirected tree into a directed tree. The initial connection and the selection of the next node to connect will indicate the direction of the connections.
- Finally, construct the TAN model by adding a node that represents class C and an arc that connects to C for each of the attributes.

The revise phase includes statistical techniques to extract relevant variables during the classification process. There are a lot of variables therefore it is necessary and automatic method for extracting the relevant information for helping an expert during the reviewing process. The chi-square [12], the Yates correction tests [15], the chi-square with the Monte Carlo simulation [17], and Fisher's exact test [16] are applied to select the variables of interest that characterize the various pathologies. It is important to note that in order for the expected frequency to be less than 5, the result may be incorrect; consequently Yates correction would be applied in an attempt to mitigate this issue. The statistical results from chi-squared are also provided, applying the Monte Carlo simulation to verify the results. Finally, an exact Fisher test is applied, which is the recommended method when the sample size is small and it is not possible to ensure that 80% of the data from a contingency table have a value greater than 5. Medical studies such as [14] use a process similar to the one presented for selecting variables that affect malformations; other biomedical studies include [15] [16] [17]. There are many alternatives for correcting data, such as that in [13]. The Figure 1 shows the CBR cycle and the techniques for each step.

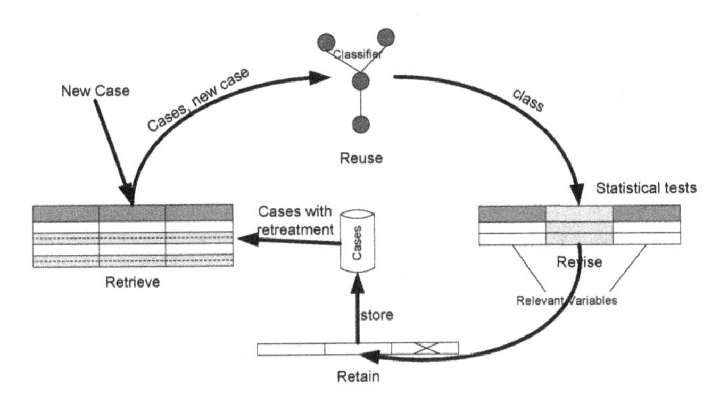

Fig 1 ROC curves with the classification accuracy.

4 Results and Conclusions

The selected cases were chosen from the patient files at the Faculty of Odontology, Masters of Endodontics, at the Complutense University of Madrid. All patients received root canal treatments between September 2000 and May 2011. Among all the patients treated during this time, we selected 35 cases that satisfied the inclusion criteria and were interested in a follow up appointment. None of patients from the selected cases who came for a follow up treatment refused to participate in the study.

A total of 18 women and 17 men were selected whose ages ranged from 18 to 85 years. The average age of the patients was 54.6 years and they all satisfied the inclusion critera as previously established. The selected cases contained all the information needed to complete the 72 variables being considered. These variables take into account all information relevant to the patient: medical and dental history, habits. Data relative to the state of the tooth prior to treatment were also included: the evolution, the clinical technique used, and the post treatment results. Certain initial variables included a high number of categories, which resulted in their recodification to ensure that the final number of categories per variable had around 3 or 4 different values. The stage that was most thoroughly analyzed during the study was the reuse stage. BayesNet, NaiveBayes, AdaBoostM1, Bagging, DecisionStump, J48, IBK, JRip, LMT, Logistic, LogitBoost, OneR, SMO and Stacking were analyzed. The following table shows the number of correct classifications obtained for each of the methods applying the leave one out technique to the CBR system, since the number of cases was not very high and cross validation could not therefore be applied. The rate of correct classifications for the system was 89%.

Table 1 Correct classifications

Classifier	Correct	Classifier	Correct
BayesNet	88.57	JRip	60.00
NaiveBayes	82.86	LMT	65.71
AdaBoostM1	68.57	Logistic	80.00
Bagging	68.57	LogitBoost	68.57
DecisionStump	42.86	OneR	65.71
J48	77.14	SMO	77.14
IBK	82.86	Stacking	74.29

The precision of the Bayes Net increased to 0.89 and recall to 1. Precision and recall are defined as follows.

$$precision = t_p / (t_p + f_p) \quad recall = t_p / (t_p + f_n)$$

t_p true positive, f_p false positive, f_n false negative.

A graphical representation with ROC curves was made with the previous results. The ROC curves facilitate the analysis of different classifiers according to the area represented beneath the curve. The bigger the area, the better the classifier. The main advantage is the ability to distinguish the relevance of false negatives compared to false positives. In this case, a positive is understood as a successful retreatment, given that the point is to avoid determining that an extraction is required if it were not actually necessary in the end. Figure 2 shows the ROC curve for each of the methods and the final result obtain. As shown, the result for the Bayesian network was satisfactory since the area beneath the curve is high and there are no false negatives (no extractions were predicted for successful cases).

To facilitate the revise phase, a revision was made to determine the difference between the values of the variables for the categories of successful retreatments and extractions. To perform this analysis, the Chi square, Yates correction, chi square with Monte Carlo simulation, and the Fisher's exact tests were applied. Table 2 displays the set of variables that were considered relevant by any of the three methods. We can see how the selection of variables coincides to a great degree for the different methods.

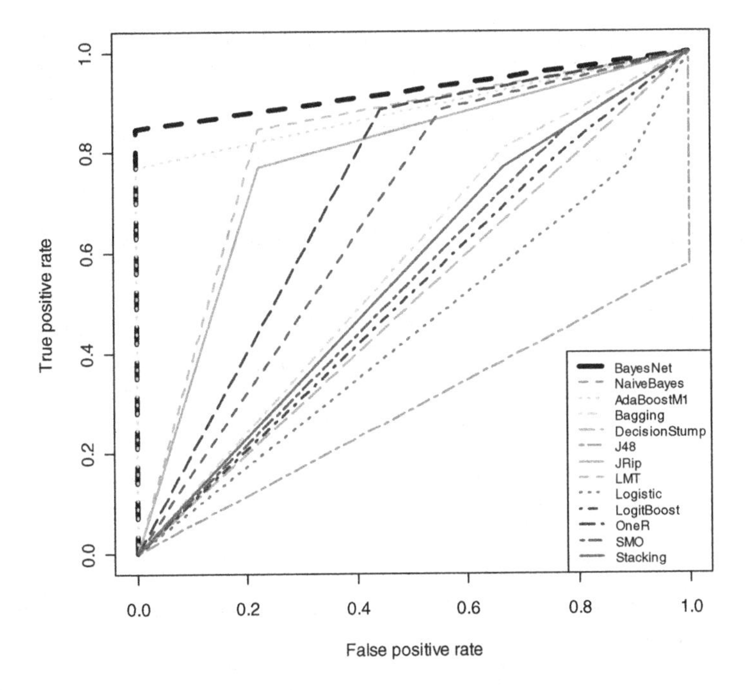

Fig 2 ROC curves with the classification accuracy

Table 2 Relevant variables

Variable	P value		
	Chi-Squared		Exact Fisher Test
	Yates	Monte Carlo	
Reason for treatment	0.007548049	0.004997501	0.008995502
Conde. Lateral or vertical	0.073673116	0.037481259	0.044117647
Clamps	0.052565842	0.052473763	0.033983008
Type of pain	0.038628647	0.033983008	0.034482759
Cause of fracture	0.00781095	0.008995502	0.002998501
Type of fracture	0.005016858	0.00049975	0.001470355
Location	0.022018021	0.013493253	0.018990505
Signs of fissure/fracture	0.008699709	0.005997001	0.004272424
Probing	0.005016858	0.001999	0.001470355
Visible fissure	0.022879307	0.009995002	0.013070078
Level	0.076170975	0.037481259	0.036300838
Other	0.027888372	0.044977511	0.015492254
Retreatment	0.000876579	0.00049975	0.000411132

With the CBR analysis, the data obtained were relevant because by ordering the established variables, particularly those with the highest risk factor, we could predict the final solution for treatment and retreatment in 89% of the cases without obtaining any false negatives. Furthermore, the system makes it possible to extract the relevant variables that can distinguish the different types of individuals. Nevertheless, more cases are required to contrast the results with greater accuracy.

Acknowledgments. This work has been supported by the MICINN TIN 2009-13839-C03-03.

References

1. Chugal, N.M., Clive, J.M., Spangberg, L.S.: A prognostic model for assessment of the outcome of endodontic treatment: Effect of biologic and diagnostic variables. Oral Surg. Oral Med. Oral Pathol. Oral Radiol. Endod. 91(3), 342–352 (2001)
2. Givol, N., et al.: Risk management in endodontics. J. Endod. 36(6), 982–984
3. Song, M., et al.: Prognostic factors for clinical outcomes in endodontic microsurgery: a retrospective study. J. Endod. 37(7), 927–933
4. Kolodner, J.: Case-Based Reasoning. Morgan Kaufmann (1993)
5. Friedman, N., Geiger, D., Goldszmidt, M.: Bayesian Network Classifiers. Machine Learning 29, 131–163 (1997)
6. Castro, J.L., Navarro, M., Sánchez, J.M., Zurita, J.M.: Loss and gain functions for CBR retrieval. Information Science 179(11), 1738–1750 (2009)
7. Joyanes, L., et al.: Knowledge Management. University of Paisley, Salamanca (2001)
8. Jurisica, I., Glasgow, J.: Applications of case-based reasoning in molecular biology. Artificial Intelligence Magazine 25(1), 85–95 (2004)
9. Canalda, C., Brau, E.: Endodoncia: técnicas clínicas y bases científicas, vol. 2. Masson, Barcelona (2006)
10. Casanellas, J.M.: Restauración del Diente Endodonciado, 1st edn. Pues, Madrid (2006)
11. Kruskal, W., Wallis, W.: Use of ranks in one-criterion variance analysis. Journal of American Statistics Association (1952)
12. Kenney, J.F., Keeping, E.S.: Mathematics of Statistics, Pt. 2, 2nd edn. Van Nostrand, Princeton (1951)
13. Martín Andrés, A., Silva Mato, A.: Optimal correction for continuity and conditions for validity in the unconditional chi-squared test. Computational Statistics & Data Analysis 26(1), 609–626 (1996)
14. Himmetoglu, O., Tiras, M.B., Gursoy, R., Karabacak, O., Sahin, I., Onan, A.: The incidence of congenital malformations in a Turkish population. International Journal of Gynecology & Obstetrics 55(2), 117–121 (1996)
15. Shaul, D.B., Scheer, B., Rokhsar, S., Jones, V.A., Chan, L.S., Boody, B.A., Malogolowkin, M.H., Mason, W.H.: Risk Factors for Early Infection of Central Venous Catheters in Pediatric Patients. Journal of the American College of Surgeons 186(6), 654–658 (1998)
16. Yang, X., Huang, Y., Crowson, M., Li, J., Maitland, M.L., Lussier, Y.A.: Kinase inhibition-related adverse events predicted from in vitro kinome and clinical trial data. 43(3), 376–384 (2010)
17. Nilsson, B.: A compression algorithm for pre-simulated MonteCarlop-value functions: Application to the ontological analysis of microarray studies. Pattern Recognition Letters 29(6), 768–772 (2008)
18. John, M., Priebe, C.E.: A data-adaptive methodology for finding an optimal weighted generalized Mann–Whitney–Wilcoxon statistic. Computational Statistics & Data Analysis 51(9), 4337–4353 (2007)

A Comparative Analysis of Balancing Techniques and Attribute Reduction Algorithms

R. Romero, E.L. Iglesias, and L. Borrajo

Abstract. In this study we analyze several data balancing techniques and attribute reduction algorithms and their impact over the information retrieval process. Specifically, we study its performance when used in biomedical text classification using Support Vector Machines (SVMs) based on Linear, Radial, Polynomial and Sigmoid kernels. From experiments on the TREC Genomics 2005 biomedical text public corpus we conclude that these techniques are necessary to improve the classification process. Kernels get some improvements about their results when attribute reduction algorithms were used. Moreover, if balancing techniques and attribute reduction algorithms are applied, results obtained with oversampling are better than subsampling.

Keywords: Biomedical text mining, classification techniques, balanced data, attribute reduction.

1 Introduction

Nowadays there are a lot of information related to biomedicine. This information is stored in public resources commonly used by the healthcare community and other more specific but with less use.

In the first group we could include textual information like literature, medical reports and medical records of patients. Each of these sources of information has different characteristics such as the presence or absence of an external structure the document, mix between free text and structured data, various length of documents, languages, etc. These differences about gender, dominance, structure and scale difficult to develop systems that provide a robust information.

R. Romero · E.L. Iglesias · L. Borrajo
Univ. of Vigo, Computer Science Dept
e-mail: {rrgonzalez,eva,lborrajo}@uvigo.es

M.P. Rocha et al. (Eds.): 6th International Conference on PACBB, AISC 154, pp. 87–94.
springerlink.com

It should also be considered the huge volume of data available. For example, the Medline bibliographic database, the most important and consulted in the biomedical domain, stores references to magazine articles from 1950 to present, containing over 18 million citations. This volume of items makes impossible for any expert managing it, even although all that amount of information was of interest.

Trying to solve these problems (range of structures in the documents and huge number of items) over the last years text mining techniques are being applied in the field of biomedicine. Particularly there is a great interest by the researchers in the study of new document automatic categorization and classification techniques. However, to the best of our knowledge, results are not enough good mainly due to the imbalanced nature of the biomedical papers.

The data imbalance problem exists in a broad range of experimental data, but only recently it has attracted close attention for researchers [1, 13]. Data imbalance occurs when the majority class is represented by a large portion of all the examples, while the minority class has only a small percentage [11]. When a text classifier encounters an imbalanced document corpus, the performance of machine learning algorithms often decreases [2, 7, 8, 9].

Another problem affecting the performance of classifiers is the handling of large data volumes. In our case, the biomedical documents are represented as vectors of many attributes (relevant words), so it is necessary to reduce their size to be manageable by such algorithms.

The purpose of this research is to analyze the effects of using data balancing techniques and attribute reduction methods in biomedical text classification, both separately and together. The work begins with a description of the document classification process proposed (Section 2) to proceed with the selection, implementation and analysis of experimental tests (Section 3) and show the most relevant conclusions (Section 4).

2 Information Retrieval Model

The proposed classification process is divided into four tasks:

1. **Annotation:** This task processes the documents extracting the most important information, and creates the test and train sparse matrices. Each of the rows of the resulting matrices are the meaningful representation of a document.

   In this research the GATE tool [4] is used with an annotation plugin named Abner-Tagger [10]. The Abner-Tagger plugin allows to use different dictionaries in order to train its entity recognition model.

   During the matrix generation a normalization technique named tf-idf is used. This process weights the attributes taking in account its relevance in each document and the corpus.
2. **Operation set:** To carry out the classification the train and test matrices must include the same number of attributes in a particular order. When dictionaries used in the annotation process are large -as in our case-, the relevant attributes to the train matrix may not be the most relevant to the test matrix.

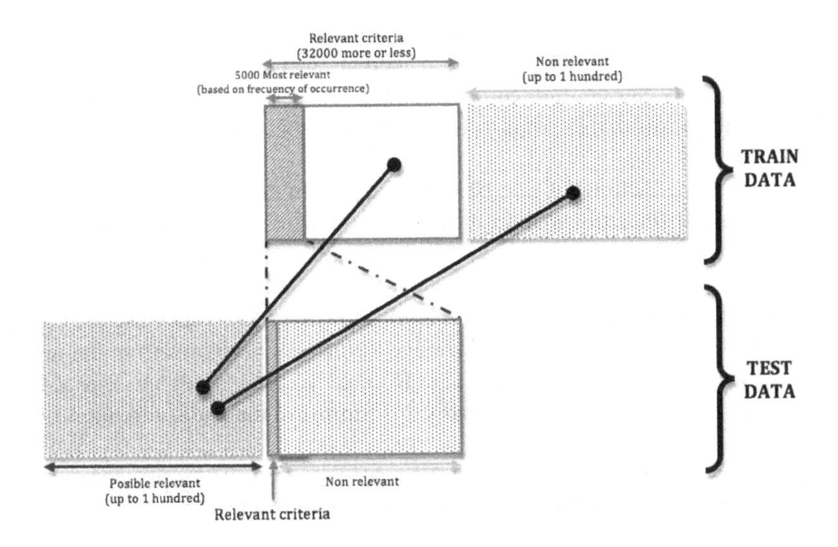

Fig. 1 Operation set task

In Fig. 1 an example is shown. As observed, the relevant attributes to the first array represent a small proportion of the relevants to the test matrix. Consequently, the second matrix is virtually empty, making incorrect classification. It is necessary to reannotate the second matrix (test data) with the same entries obtained by the first matrix (train data).

The objective of this task is to reduce the number of attributes of the train and test matrices using the set operators union, intersect and minus. As a result, matrices more manageable and with more correlated attributes are obtained.

3. **Instance filtering:** The objective of this task is to balance the number of instances that belong to each class.

   The instance filtering represents a powerful tool in cases of over-training regard to a determinate class. There are two techniques that perform these actions, named *Oversampling* and *Subsampling*. These concepts are commonly used in signal processing but it is possible to apply the same principles to text processing. In case of signal processing they are usually used in order to make high or low filters. In text mining it can be used in order to balance the documents that belong to a single class.

   The Subsampling algorithm decreases artificially the number of samples that belong to majority class. It corresponds to decrease the number of non-relevant documents taking into account the relevant class. To remove instances a random selection algorithm is used in order to accomplish the distribution factor between each class. Although we test different distribution values, $\{10, 5, 4, 3, 2, 1\}$, we got best results setting these parameters to an uniform distribution (factor equal to 1). So this distribution is used for the experiments.

Similarly, the Oversampling technique redistributes the number of instances that belongs to minority class taking into account the majority one.

Both concepts can be applied simultaneously, increasing or decreasing the number of instances that belongs to each class. In this case we are talking about *Resampling*.

The balancing techniques can be optimized trying to find the most relevant attributes, taking in account the relevance per document, class or both. Some classifiers, as KNN, do not work well if the number of attributes is too big and consume a high amount of memory.

4. **Attribute reduction:** This task applies algorithms in order to decrease the number of attributes. The objective is to achieve a subset of attributes from the original set that provides the next goals: (i) to improve the classification process, (ii) to build more efficient models, and (iii) to improve the understanding of the models generated.

   In order to accomplish these tasks Weka [5] is used. Weka provides some classes which implement Subsampling, Oversampling and Resampling, and different attribute selection algorithms as: *Cfs* (that evaluates the worth of a subset of attributes by considering the individual predictive ability of each feature along with the degree of redundancy between them), *ChiSquared* (that evaluates the worth of an attribute by computing the value of the chi-squared statistic with respect to the class), *GainRatio* (that evaluates the worth of an attribute by measuring the gain ratio with respect to the class), and *InfoGain* (that evaluates the worth of an attribute by measuring the information gain with respect to the class).
5. **Classification:** This task applies different reasoning models to classify texts in relevant or not relevant. For this work, we implement four different Support Vector Machines based on Linear, Radial, Polynomial and Sigmoid kernels, respectively. Support Vector Machines are based on the idea that by increasing the dimensionality of the data, it gets easier to separate. In fact, the SVM uses an N-dimensional space, where N is the number of samples in the training set. This approach allows the SVM to classify problems with arbitrary complexity. The problem is that outliers can easily sabotage the classifier and its generalization capabilities can be questionable. In our experiments, hard margins for the classifier are used (i.e the result is relevant or non relevant with no mention of degree of membership).

   In order to support the software we employ a library called LibSVM [3] that implements a kernel method based on costs (C-SVM). Furthermore, we use some parameters like *probability estimates* to generate probabilities instead of $[-1,+1]$ values for SVM output classification, or *normalize* to scale attribute values between $[-1,+1]$.

   The kernels are represented by the following equations:

$$Linear : U \cdot V \quad (1)$$

$$Polynomial : (Gamma \cdot u \cdot v + Coef)^{Degree} \quad (2)$$

$$Exponential : exp(-Gamma \cdot |u-v|^2) \quad (3)$$

$$Sigmoid: tanh(Gamma \cdot u' \cdot v + Coef) \tag{4}$$

In the polynomial kernel (2) we test the *Degree* and *Gamma* values between 1 to 7. With exponential (3) and sigmoid kernels (4) we made tests using values for the *Gamma* parameter between 1 to 15. Other parameters that appear in kernel equations like *Coef*, which means a single coefficient, was set to the default value 0.0. Finally, the cost parameter associated to this SVM, namely C-SVM, was set to 1.0.

3 Experimental Results

Taking into account the previous information retrieval model some experiments were performed. They are grouped in three sets, attribute selection and balancing techniques separately, subsampling technique and attribute reduction algorithms used at the same time, and attribute reduction combined with oversampling.

To perform tests the TREC Genomics 2005 biomedical text public corpus [12] has been used. In 2005, TREC provided a set of evaluation tasks to know the state of the art of applying information extraction techniques to problems in biology. Specifically, the goal of the TREC Genomics Track was to create test collections for evaluation of information retrieval and related tasks in the genomics domain [6]. The corpus size is 4591008 MEDLINE records.

In Information Retrieval, it is usual to use statistical measures in order to make a comparative between tests. The most common measures are *recall*, *precision* and *F-measure*. Recall allows to know the percentage of relevant documents which are correctly identified or classified. Precision represents the fraction of retrieved documents that are relevant. F-measure represents the harmonic mean between recall and precision.

Regarding to the task of attribute reduction, note that the number of attributes representative of the corpus (originally 7711 attributes) was reduced to 163 by applying the Cfs algorithm. ChiSquared, GainRatio and InfoGain got 278 attributes but in a different order.

Results of the first group are showed in Fig. 2. In all figures we use acronyms to explain some SVM parameters: *N* as Normalize meaning, *P* as Probabilistic meaning, *NP* is equal to both Normalize and Probabilistic, and $G[X]$ is the Gamma parameter which X means different values.

The best results are achieved by SVM Linear Probabilistic based on the subsampling technique, achieving as maximum value 0.179. In addition we can observe a clear division between kernels algorithms and techniques. Linear and Polynomial kernels got better results using balancing techniques; by contrast, Radial and Sigmoidal kernels work much better in tests based on attribute reduction. At last, to remark that attribute reduction algorithms based on to maximize the relation between each feature with its class (ChiSquared, InfoGain, GainRatio) are well correlated with Sigmoid and Radial kernel functions.

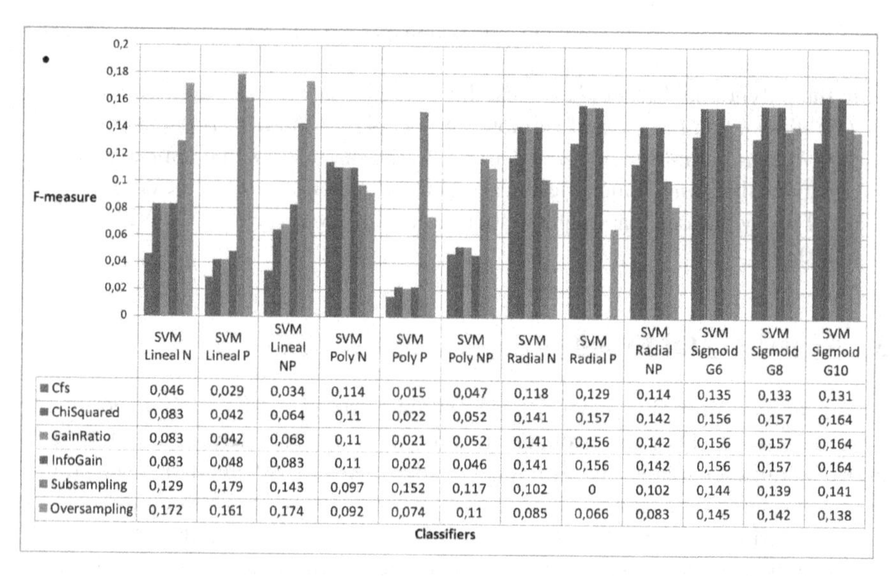

	SVM Lineal N	SVM Lineal P	SVM Lineal NP	SVM Poly N	SVM Poly P	SVM Poly NP	SVM Radial N	SVM Radial P	SVM Radial NP	SVM Sigmoid G6	SVM Sigmoid G8	SVM Sigmoid G10
Cfs	0,046	0,029	0,034	0,114	0,015	0,047	0,118	0,129	0,114	0,135	0,133	0,131
ChiSquared	0,083	0,042	0,064	0,11	0,022	0,052	0,141	0,157	0,142	0,156	0,157	0,164
GainRatio	0,083	0,042	0,068	0,11	0,021	0,052	0,141	0,156	0,142	0,156	0,157	0,164
InfoGain	0,083	0,048	0,083	0,11	0,022	0,046	0,141	0,156	0,142	0,156	0,157	0,164
Subsampling	0,129	0,179	0,143	0,097	0,152	0,117	0,102	0	0,102	0,144	0,139	0,141
Oversampling	0,172	0,161	0,174	0,092	0,074	0,11	0,085	0,066	0,083	0,145	0,142	0,138

Fig. 2 Attribute reduction and balancing techniques using in separate mode

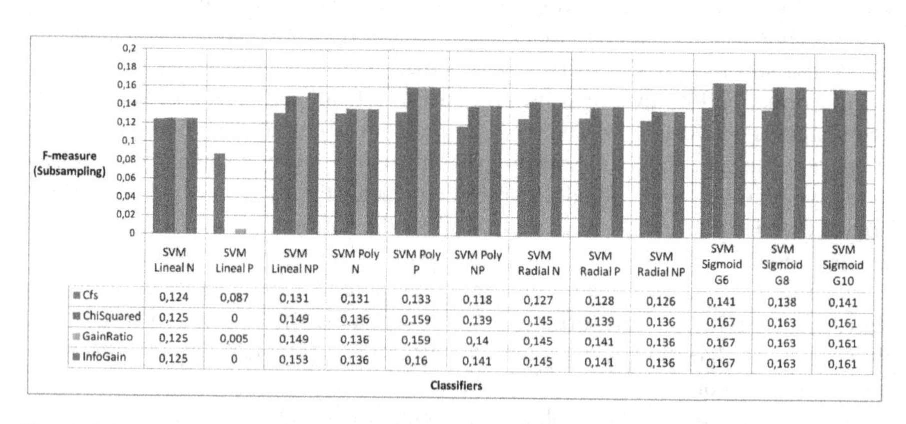

	SVM Lineal N	SVM Lineal P	SVM Lineal NP	SVM Poly N	SVM Poly P	SVM Poly NP	SVM Radial N	SVM Radial P	SVM Radial NP	SVM Sigmoid G6	SVM Sigmoid G8	SVM Sigmoid G10
Cfs	0,124	0,087	0,131	0,131	0,133	0,118	0,127	0,128	0,126	0,141	0,138	0,141
ChiSquared	0,125	0	0,149	0,136	0,159	0,139	0,145	0,139	0,136	0,167	0,163	0,161
GainRatio	0,125	0,005	0,149	0,136	0,159	0,14	0,145	0,141	0,136	0,167	0,163	0,161
InfoGain	0,125	0	0,153	0,136	0,16	0,141	0,145	0,141	0,136	0,167	0,163	0,161

Fig. 3 Subsampling technique and attribute reduction algorithms at the same time

As commented, second group of tests are based on a combination between subsampling technique and attribute selection. We apply an attribute selection over the corpus and the balancing technique later. We get best results through Sigmoid kernel with Gamma equal to 6. Regarding attribute algorithms, those which are based on analysing each attibute value respect to its class (ChiSquared, GainRatio and InfoGain) obtain best results. Applying these algorithms our results are normalized in almost all cases, except for SVM Linear Probabilistic which got very poor results.

If we compare tests in Fig. 2 and Fig. 3 we can conclude that the combination between reduction algorithms and balancing techniques stables the results. There are hardly differences between the kernels.

The last group, oversampling and attribute reduction combinations are plotted in Fig. 4. Best results, 0.169, are achieved by SVM Sigmoid and Gamma equal to 6, combined with ChiSquared, GainRatio or InfoGain. SVM Linear Probabilistic get better results again. As a result, we conclude that combinations with oversampling are more effective than subsampling regardless of kernel.

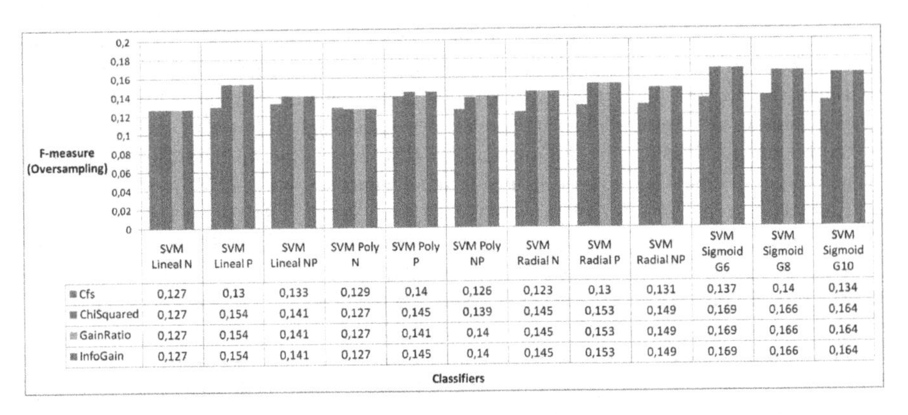

	SVM Lineal N	SVM Lineal P	SVM Lineal NP	SVM Poly N	SVM Poly P	SVM Poly NP	SVM Radial N	SVM Radial P	SVM Radial NP	SVM Sigmoid G6	SVM Sigmoid G8	SVM Sigmoid G10
Cfs	0,127	0,13	0,133	0,129	0,14	0,126	0,123	0,13	0,131	0,137	0,14	0,134
ChiSquared	0,127	0,154	0,141	0,127	0,145	0,139	0,145	0,153	0,149	0,169	0,166	0,164
GainRatio	0,127	0,154	0,141	0,127	0,141	0,14	0,145	0,153	0,149	0,169	0,166	0,164
InfoGain	0,127	0,154	0,141	0,127	0,145	0,14	0,145	0,153	0,149	0,169	0,166	0,164

Fig. 4 Oversampling technique and attribute reduction algorithms at the same time

4 Conclusions

In this study we analyze several balancing techniques, attribute reduction algorithms and their impact over an information retrieval model. From the experiments we conclude that these techniques are necessary to improve the classification process. However results differ in some cases. Best results are achieved by SVM Linear Probabilistic under subsampling technique.

In addition we show that kernels got some improvements about their results when attribute reduction algorithms are used. On the other hand, if both techniques and algorithms are applied, we can see that obtained results with oversampling are better than subsampling, even in case of Linear Kernel.

As a final conclusion, and following results about second and third groups commented in experimental results, the classification process is not directly dependant by the kernels used. Therefore, the classifier becomes more permissive and more effective.

Acknowledgements. This work has been partially funded by the Spanish Ministry of Science and Innovation, the Plan E from the Spanish Government and the European Union from the ERDF (TIN2009-14057-C03-02).

References

1. Barandela, R., Sánchez, J.S., García, V., Rangel, E.: Strategies for learning in class imbalance problems. Pattern Recognition 36(3), 849–851 (2003)
2. Borrajo, L., Romero, R., Iglesias, E.L., Redondo Marey, C.M.: Improving imbalanced scientific text classification using sampling strategies and dictionaries. Journal of Integrative Bioinformatics 8(3) (2011)
3. Chang, C.-C., Lin, C.-J.: LIBSVM: a library for support vector machines (2001)
4. Cunningham, H., Wilks, Y., Gaizauskas, R.J.: Gate - a general architecture for text engineering. In: COLING, pp. 1057–1060 (1996)
5. Garner, S.R.: Weka: The waikato environment for knowledge analysis. In: Proc. of the New Zealand Computer Science Research Students Conference, pp. 57–64 (1995)
6. Hersh, W., Cohen, A., Yang, J., Bhupatiraju, R.T., Roberts, P., Hearst, M.: Trec 2005 genomics track overview. In: Voorhees, E.M., Buckland, L.P. (eds.) Proceedings of the Fourteenth Text REtrieval Conference, TREC 2005, Special Publication 500-266, pp. 14–25. National Institute of Standards and Technology, NIST (2005)
7. Kang, P., Cho, S.: EUS SVMs: Ensemble of Under-Sampled SVMs for Data Imbalance Problems. In: King, I., Wang, J., Chan, L.-W., Wang, D. (eds.) ICONIP 2006. LNCS, vol. 4232, pp. 837–846. Springer, Heidelberg (2006)
8. Romero, R., Iglesias, E.L., Borrajo, L.: Building biomedical text classifiers under sample selection bias. In: Abraham, A., Corchado, J.M., Rodríguez-González, S., Santana, J.F.D.P. (eds.) DCAI. AISC, vol. 91, pp. 11–18. Springer, Heidelberg (2011)
9. Romero, R., Iglesias, E.L., Borrajo, L., Marey, C.M.R.: Using dictionaries for biomedical text classification. Advances in Intelligent and Soft Computing 93, 365–372 (2011)
10. Settles, B.: Abner: An open source tool for automatically tagging genes, proteins, and other entity names in text. Bioinformatics 21(14), 3191–3192 (2005)
11. Tan, S.: Neighbor-weighted k-nearest neighbor for unbalanced text corpus. Expert Systems with Applications 28(4), 667–671 (2005)
12. Voorhees, E.M., Buckland, L.P. (eds.): Proceedings of the Fourteenth Text REtrieval Conference,TREC 2005, vol. Special Publication 500-266. National Institute of Standards and Technology, NIST (2005)
13. Weiss, G.M.: Mining with rarity: a unifying framework. SIGKDD Explor. Newsl. 6, 7–19 (2004)

Sliced Model Checking for Phylogenetic Analysis

José Ignacio Requeno, Roberto Blanco, Gregorio de Miguel Casado,
and José Manuel Colom

Abstract. Model checking provides a powerful and flexible formal framework to state and verify biological properties on phylogenies. However, current model checking techniques fail to scale up the big amount of biological information relevant to each state of the system. This fact motivates the development of novel cooperative algorithms and slicing techniques that distribute not the graph structure of a phylogenetic system but the state information contained in its nodes.

Keywords: phylogenetic analysis, model checking, state slicing, concurrent verification.

1 Introduction

Phylogenetic trees are abstractions for characterizing evolutionary relationships which biologists can infer using a variety of software packages. In this context, the exploitation of the temporal nature of phylogenetic information via temporal logics has allowed for the introduction of model checking techniques and then for a wide pool of software tools which can be used for automating property verification [2]. Model checking is an automated verification technique that, given a finite state model of a system and a formal property, systematically checks whether this property holds for (a given state in) that model.

Nevertheless, a memory bottleneck appears when representing phylogenetic taxa as states in a Kripke structure, that is, when coding DNA or protein sequences as strings [14]. In addition, the characterization of properties by temporal logics produces large sets of atomic propositions to be verified sparsely. In this regard, conventional monolithic techniques conceived for improving performance in other model

José Ignacio Requeno · Roberto Blanco · Gregorio de Miguel Casado · José Manuel Colom
Department of Computer Science and Systems Engineering (DIIS)/Aragon Institute of Engineering Research (I3A), Universidad de Zaragoza, C/ María de Luna 1, 50018 Zaragoza, Spain
e-mail: {nrequeno,robertob,gmiguel,jm}@unizar.es

M.P. Rocha et al. (Eds.): 6th International Conference on PACBB, AISC 154, pp. 95–103.
springerlink.com

checking applications outline the need for new criteria to increase scalability in the verification process according to the features of the biological data.

This paper proposes a novel distributed verification approach which improves the conventional model checking algorithm by introducing the concept of "state slicing", in opposition to other distributed approaches based on "graph partition". After this introduction, Section 2 summarizes the essentials of phylogenetic model checking and presents the classic model checking algorithms. Next, Section 3 formalizes our approach and develops the the sliced model checking algorithms. Finally, Section 4 gathers the conclusions drawn from this research.

2 Phylogenetic Model Checking

Phylogenetics and model checking can be bridged after reflecting about the processes of modeling and specification. Formally, a phylogenetic tree with root r is a rooted labeled tree $T = (V,E)$ where each vertex represents a population of compatible individuals described by their genome (D), and the vertices are arranged along the paths of the tree according to a evolutive process [2].

At the same time, a Kripke structure stands for the behavior of a system through a graph and can be used to query properties [7]. It represents a finite transition system through a tuple $M = (S,S_0,R,L)$, where S is a finite set of states, $S_0 \subseteq S$ is the set of initial states, R is a total transition relation between states and L is the labeling function that associates each state with the subset of *atomic propositions* (*AP*) that are true in it [2]. The Kripke structure models a system with infinite behaviors or *paths*: infinite sequences of successive states $\pi = s_0 s_1 s_2 \dots$ such that $s_0 \in S_0$ and $(s_i, s_{i+1}) \in R, i \in \mathbb{N}$. The set of possible executions (paths) in a Kripke structure can be unfolded into its *computation tree*.

A certain phylogenetic tree can be interpreted as a computation of the process of evolution. The set of atomic propositions represents states that stand for distinct populations of individuals sharing a common sequence.

Definition 1 (Branching-time Phylogeny, [2]). A rooted labeled tree $P = (T,r,D)$ is univocally defined by the Kripke structure $M = (V,\{r\},R,L)$, where:

- R is the transition relation composed of the set of tree edges (directed from r) plus self-loops on leaves: $R = E \cup \{(v,v) : \nexists (v,w) \in E \wedge v,w \in V\}$, and
- $AP = \{seq[i] = \sigma \mid \sigma \in \Sigma, \forall i \in \{1,\dots,l\}\}$ (where seq is an array of variables of type Σ), is the set of phylogenetic atomic propositions where we have a pair (variable, value) such as $seq[i]$ is the variable of the string seq at the i-th position and σ is its character value, and
- L is the standard labeling function defined by AP, under which a state v mapped to $D(v) = seq$ with $seq = \sigma_1 \sigma_2 \dots \sigma_l$ satisfies the family of properties $seq[i] = \sigma_i, 1 \leq i \leq l$, plus a unique state identifier in case of several states share the same atomic propositions.

Algorithm 1. Algorithm $Sat(M, \phi)$

Require: $M = (S, S_0, R, L)$ is a Kripke structure and ϕ is a PTL formula
Ensure: A subset of states of S that satisfies ϕ
 if $\phi \equiv \top$ **then return** S {Set of states from the Kripke structure M}
 else if $\phi \equiv p \in AP$ **return** $\{s : p \in L(s)\}$
 else if $\phi \equiv \neg\psi$ **return** $S \setminus Sat(M, \psi)$
 else if $\phi \equiv \psi_1 \vee \psi_2$ **return** $Sat(M, \psi_1) \cup Sat(M, \psi_2)$
 else if $\phi \equiv \mathbf{EX}(\psi)$ **return** $\{s : (s, s') \in R, s' \in Sat(M, \psi)\}$
 else if $\phi \equiv \mathbf{EG}(\psi)$ **return** $fixpoint(M, Sat(M, \psi), \emptyset, S)$
 else if $\phi \equiv \mathbf{E}(\psi_1 \mathbf{U} \psi_2)$ **return** $fixpoint(M, Sat(M, \psi_2), Sat(M, \psi_1), \emptyset)$
 end if

Algorithm 2. Algorithm $fixpoint(M, Sat(M, \phi), Sat(M, \psi), Init)$

Require: $M = (S, S_0, R, L)$ is a Kripke structure
Require: $Sat(M, \psi)$ and $Init$ are the sets of initial states (returned by calls to Sat algorithm) and $Sat(M, \phi)$ is the set of final states
Ensure: A set of states that represents paths going from $Init$ (or $Sat(M, \psi)$) to $Sat(M, \phi)$
 $New \leftarrow Init$
 repeat
  $Old \leftarrow New$
  $New \leftarrow Sat(M, \psi) \cup (Sat(M, \phi) \cap \{s : (s, s') \in R, s' \in Old\})$
 until $New = Old$
 return New

Definition 2 (Phylogenetic Tree Logic, [2]). An arbitrary temporal logic formula ϕ is defined by the following minimal grammar, where $p \in AP$:

$$\phi ::= p \mid \neg\phi \mid \phi \vee \phi \mid \mathbf{EX}(\phi) \mid \mathbf{EG}(\phi) \mid \mathbf{E}[\phi \mathbf{U} \phi] \tag{1}$$

Additionally, a Phylogenetic Tree Logic (PTL) is proposed as a *temporal logic* for the specification of evolutionary properties. Formulas are checked against a model M considering all paths π that originate from a certain state s_0. For example, one of the most frequent basic phylogenetic property asks whether the sequences of a set of organisms exhibit a *back mutation* (for more examples, see [2]). In terms of PTL, it looks for a node (taxon) somewhere in the tree where a mutation appeared $(\mathbf{EF}(seq[col] \neq \sigma))$ but is reverted in the future $(\mathbf{EF}(seq[col] = \sigma))$:

$$BM(col, \sigma) \equiv (seq[col] = \sigma) \wedge \mathbf{EF}[(seq[col] \neq \sigma) \wedge \mathbf{EF}(seq[col] = \sigma)] \tag{2}$$

Current model checking tools load the model and the temporal logic formulas, evaluate them and compute a counterexample if so required [11]. Moreover, evaluation results may reveal new meaningful information that can be reused in subsequent refinements of the (phylogenetic) model.

Verification of temporal formulas is formalized under the framework of set theory; we follow this convention throughout the paper. In fact, the traditional model checking algorithm is usually presented as a recursive function which computes the set of states in a given Kripke structure which satisfy a PTL formula (Algorithm 1).

In order to evaluate the temporal operators $\mathbf{EG}(\psi)$ and $\mathbf{E}(\psi_1 \mathbf{U} \psi_2)$, the greatest and least fixpoints are computed, respectively. Both fixpoint sets can be obtained as the result of a breadth-first search (Algorithm 2). In particular, the call $\mathit{fixpoint}(M, Sat(M, \psi), \emptyset, S)$ produces the greatest fixpoint, and the least fixpoint is produced by $\mathit{fixpoint}(M, Sat(M, \psi_2), Sat(M, \psi_1), \emptyset)$.

3 Sliced Model Checking

Memory usage is a major limiting factor in the verification of complex systems. For a long time it has been possible to perform model checking on systems with states in excess of 10^{120} [5, 1]. Symbolic model checking is perhaps one of the most common methods to achieve this [5], together with general-purpose memory saving techniques such as abstractions, partial reductions or symmetries [7, 3].

One of the first attempts to infer global properties of the system through an incremental bottom-up strategy was compositional reasoning [6]. It attempts to structure the system in components and progressively verify local properties for each part (e.g., in a genomic sequence, genes could be considered its components). For example, in the assume-guarantee paradigm [12] the method establishes assumptions about the environment of each component and uses these to aid in the verification of each property. Alternative approaches focused on the point of view of decomposition of temporal logic formulas are exemplified by [8].

Recently, some methods [9] and [13] operate by partitioning the system according to the Kripke structure and distributing the chunks among available computing units (both storage of the partial Kripke structure and computation of satisfiability of logic formulas [4]). These methods attack the size of the structure (number of states) and not the complexity of each state, which is the limiting factor in phylogenetic model checking. Therefore, they could be used, yet by themselves they would be ineffective. Currently, close to 2×10^6 species have been catalogued [10], but genomes exhibit an enormous range of sizes and complexity, with *Homo sapiens* ranking at 3×10^9 base pairs.

The state slicing presented in this section can be efficiently applied to the verification of the back mutation property (Eq. 2). Each slice verifies a subformula referred to a part of the DNA sequence that is the only information stored in each state of the slice. These verifications can be done independently an so a computational speed up proportional to the number of slices can be obtained. That is, given an aligment of length l, each sequence is distributed in l independent slices where $\bigvee_{col \in \{1 \ldots l\}} M, s_0 \models BM(col, \sigma)$ are verified asynchronously.

Furthermore, it is common for the AP set of a system to encode a heterogeneous collection of properties, whose correlations can be arbitrary in both strength and complexity. The AP set is originally unstructured in the sense that different types of labels may be considered. Thus, the "normalization" of properties and variables along the graph turns out to be a critical issue.

Definition 3 (Homogeneous Kripke Structure). Let Σ be an alphabet. An homogeneous Kripke structure $M = (S, S_0, R, L)$ over AP is a Kripke structure,

- $AP = \bigcup_{i=1}^{l} AP_i$ is the set of atomic propositions,
- $AP_i = \{seq[i] = \sigma \mid \sigma \in \Sigma\}$ with seq an array of variables with type Σ,
- $L : S \rightarrow AP_1 \times \ldots \times AP_l$, where $L(s) = (seq[1] = \sigma_1, \ldots, seq[l] = \sigma_l)$.

In Def. 1, the labeling function L tags each state with a unique tuple (composed of several atomic propositions) in a classic Kripke structure. Additionally, an extension to multiple labels per state ($L : S \rightarrow 2^{AP}$) can be considered. Note that the Cartesian product in the labeling function involves an implicit "and" operator; that is, the label states that a temporal formula is valid on the current state iff it asserts $[(seq[1] = \sigma_1) \wedge \ldots \wedge (seq[l] = \sigma_l)]$. Multilabels would be or-like though.

In opposition to traditional distributed model checking algorithms, which divide the system into disjoint Kripke substructures, we present a new slicing criterion. The notion of *AP normalization* in a homogeneous Kripke structure motivates the definition of a projection (reading) function that takes (cuts) a subset of the variables from the AP structure of state labels.

Definition 4 (Projection Function). A projection function with subindex $i \in \{1, \ldots, l\}$ over AP selects a subset of elements such that:

- $proj_i : AP \rightarrow AP_i$ with $i \geq 1$ such that $proj_i(seq) = (seq[i] = \sigma)$.
- $proj_I : AP \rightarrow AP_{i_l} \times \ldots \times AP_{i_m}$ such that $proj_I(seq) = (seq[i_1] = \sigma_{i_1}, \ldots, seq[i_m] = \sigma_{i_m})$ is a projection function with a set of indices $I = \{i_1, i_2, \ldots, i_m\}$, $m \leq l$

The projection operator allows to slice the nodes of the Kripke structure instead of splitting the graph into regions. We obtain several lighter copies of the original one, where each slice of the AP represents some characteristics of the initial system.

Definition 5 (Sliced Kripke Structure). Let $\{I_i \mid 1 \leq i \leq n, I_i \subseteq I\}$ be a partition of the set of indices $I = \{1, \ldots, l\}$. We say that $M = (S, S_0, R, L)$ can be obtained from the composition of n Kripke structures, named slices of M, each one defined as $M_i = (S, S_0, R, L_i)$, $\forall i \in \{1, \ldots, n\}$, and $L_i : S \rightarrow proj_{I_i}(AP)$, where $L(s) = L_1(s) \cdot L_2(s) \cdot \ldots \cdot L_n(s)$, $\forall s \in S$, with "$\cdot$" the operator function that concatenates tuples.

The sliced model checking algorithm thus obtained is a parallel version of the one already introduced in Algorithm 1. It recursively parses the temporal logic formula passed as input and returns the satisfiability set of states that validate it. The global result is then composed from local results. Nevertheless, temporal logic formulas cannot be always evaluated independently over the respective slices of the Kripke structure. We may find two types of formulas when we apply slicing.

First-order logic formulas can be partitioned and executed independently (asynchronously) across every slice because they don't need to remember visited states (as opposed to PTL path operators). Thus, the solution will be composed by set operations in a blocking barrier synchronization at the end of the different execution threads. This case represents an upper bound to the speed up of our proposal.

Algorithm 3. Algorithm $Sat_i\,(M_i, \phi)$

Require: $M_i = (S, S_0, R, L_i)$ is a sliced Kripke structure and ϕ is a PTL formula
Ensure: A subset of states of S that satisfies ϕ

```
if φ ≡ ⊤ then return S
else if φ ≡ p ∈ AP then return {s : p_i ∈ L_i(s)}
else if φ ≡ ¬ψ then return S \ Sat_i(M_i, ψ)
else if φ ≡ ψ1 ∨ ψ2 then return Sat_i(M_i, ψ1) ∪ Sat_i(M_i, ψ2)
{Read the satisfation sets computed in the rest of threads}
else if φ ≡ EX(ψ) then
   [ broadcast(Sat_i(M_i, ψ)); read(Sat_j(M_j, ψ)) ∀j ∈ {1..i−1, i+1, ..n} ]
   Q ← ∩_{j=1}^{n} Sat_j(M_j, ψ)
   return {s : (s, s') ∈ R, s' ∈ Q}
else if φ ≡ EG(ψ) then
   [ broadcast(Sat_i(M_i, ψ)); read(Sat_j(M_j, ψ)) ∀j ∈ {1..i−1, i+1, ..n} ]
   Q ← ∩_{j=1}^{n} Sat_j(M_j, ψ)
   return fixpoint_i(M_i, Q, ∅, S)
else if φ ≡ E(ψ1 U ψ2) then
   [ broadcast(Sat_i(M_i, ψk)); read(Sat_j(M_j, ψk)), k = {1,2} ∀j ∈ {1..i−1, i+1, ..n} ]
   Q_k ← ∩_{j=1}^{n} Sat_j(M_j, ψk), k = {1,2}
   return fixpoint_i(M_i, Q2, Q1, ∅)
end else if
```

Algorithm 4. Algorithm $fixpoint_i\,(M_i, Sat_i\,(M_i, \phi), Sat_i\,(M_i, \psi), Init)$

Require: $M_i = (S, S_0, R, L_i)$ is a Kripke structure
Require: $Sat_i\,(M_i, \psi)$ and $Init$ are the sets of initial states (returned by calls to Sat_i algorithm) and $Sat_i\,(M_i, \phi)$ is the set of final states
Ensure: A set of states that represents paths going from $Init$ (or $Sat_i\,(M_i, \psi)$) to $Sat_i\,(M_i, \phi)$

```
New_i ← Init
reset(Diff) {atomic, sets Diff ← ∅ on first call alone}
repeat
   Old_i ← New_i ∪ Diff
   New_i ← Sat_i(M_i, ψ) ∪ (Sat_i(M_i, φ) ∩ {s : (s, s') ∈ R, s' ∈ Old_i})
   Diff_i ← New_i \ Old_i
   {Read the local border set of concurrent threads in a mutual exclusion}
   [ broadcast(Diff_i); read(Diff_j) ∀j ∈ {1..i−1, i+1, ..n} ]
   Diff ← ∩_{j=1}^{n} Diff_j
until Diff = ∅
return New_i
```

Conversely, the partition of temporal logic formulas including PTL path operators (**EX, EG, EU**) needs the refinement of current algorithms so that the different slices work together in a synchronized way for the computation of the satisfiability set: some paths may be valid in one slice but not in another.

At the beginning of the model checking process, n threads are instantiated by an *Init* function. The i subindex indicates that the i-th model checker thread works with the associated i-th sliced Kripke structure. As the sliced Kripke structures are lighter than the original one, we can load them in the local memory of the model checker and the algorithms work for a shared memory system. A token ring network enables on-the-fly computation of the intersection sets.

Three synchronization points are added in Algorithm 3 (highlighted with boxes). They allow the intersection of the satisfiability sets returned by the projection of the algorithm's recursive call for the evaluation of formulas with nested PTL path operators. The obtained reduced satisfiability set is equivalent to that obtained by the classic model checking. Furthermore, it is used as seed for the following operations: computation of the predecessors' set in **EX** and fixpoint call in **EG** and **EU**.

Next, the fixpoint Algorithm 4 is adapted to a shared memory system. The algorithm uses a global shared variable *Diff* to store the common border set, that is, the set of newer states computed in the last iteration loop. In every loop step, the border set adds new states to the satisfiability set. The global border is then obtained from the intersection of the local variables $Diff_i$ of each verification thread. This intersection requires a blocking barrier synchronization: a thread cannot advance until the remaining threads compute their partial sets. Hence, their intersection returns the border states that are simultaneously valid in every sliced Kripke structure.

Finally, in this context we can apply two extra optimizations. The first one (Proposition 1) involves the **EX** operator. This optimization is applicable universally and allows an additional level of asynchronicity (parallelism) as it separates PTL formula into two disjoint ones. Furthermore, in the particular case of phylogenetic Kripke structures, Proposition 2 reduces the set of states to be checked, improving the speed of the verification process.

Proposition 1. *The verification of* $M, s_0 \models \mathbf{EX}(\psi \vee \gamma)$ *is asynchronous* ($M, s_0 \models \mathbf{EX}(\psi) \vee M, s_0 \models \mathbf{EX}(\gamma)$).

Proposition 2. *Let* M *be a phylogenetic Kripke structure and* $L(M)$ *the set of its leaf states. The verification of* $M, s_0 \models \mathbf{EG}(\phi)$ *over the root of a phylogenetic tree Kripke structure is equivalent to the verification of* $\exists l \in L(M)\,;\; M, l \models \mathbf{E}(\phi)$ *over the leafs.*

4 Conclusions

This paper has presented an alternative, fully distributed approach to model checking based on "state slicing" as opposed to traditional "graph partition" techniques. Assuming that biological data (as atomic propositions) can be fragmented into minimal meaningful blocks that fit into the local memory of a model checker, each thread executes the verification process over a slice (projection) produced from the original data while sharing a common Kripke structure skeleton. Subsequently, the final result comes from the composition of local results.

The sliced algorithms presented here require a synchronization process across all the slices, which limits the speed up when using a conservative, fully synchronized computation along paths. Nevertheless, more aggressive strategies such as speculative computation can be considered in order to improve CPU usage during the idle time of the synchronization process. In addition, typical formulas are large collections of instances of a simpler formula for different assignments of values to its free variables. In these, many recurrent subformulas are found, a fact which can be exploited to avoid recomputation. Slicing has been proposed as a suitable technique that allows complex PTL formulas to be accepted by standard tools.

Finally, the implementation of the modified algorithms can be undertaken by introducing standard interprocess communication languages (e.g., MPI or OpenMP) in open-source model checkers like SPIN or NuSMV. The core of the conventional algorithms remains unaltered, and only specific points in the algorithm have to be modified to manage the verification of slices introduced on the original *AP* set.

Acknowledgements. This work was supported by the Spanish Ministry of Science and Innovation (MICINN) [TIN2008-06582-C03-02] and [TIN2011-27479-C04-01], the Spanish Ministry of Education [AP2008-03447] and the Government of Aragon [B117/10].

References

1. Baier, C., Katoen, J.-P.: Principles of model checking. The MIT Press, Cambridge (2008)
2. Blanco, R., de Miguel Casado, G., Requeno, J.I., Colom, J.M.: Temporal logics for phylogenetic analysis via model checking. In: 2010 IEEE Int. Conf. on Bioinformatics and Biomedicine Workshops, pp. 152–157. IEEE (2010)
3. Bošnački, D., Edelkamp, S.: Model checking software: on some new waves and some evergreens. Int. J. Software Tool Tech. Tran. 12, 89–95 (2010)
4. Boukala, M.C., Petrucci, L.: Distributed CTL Model-Checking and counterexample search. In: 3rd Int. Workshop on Verification and Evaluation of Computer and Communication Systems (2009)
5. Burch, J.R., Clarke, E.M., Long, D.E., McMillan, K.L., Dill, D.L.: Symbolic model checking for sequential circuit verification. IEEE Trans. Comput-Aided Design Integr. Circuits Syst. 13, 401–424 (1994)
6. Clarke, E.M., Long, D.E., McMillan, K.L.: Compositional model checking. In: Procs. of the 4th Annual Symposium on Logic in Computer Science, pp. 353–362. IEEE (1989)
7. Clarke, E.M., Grumberg Jr, O., Peled, D.A.: Model Checking. The MIT Press, Cambridge (2000)
8. Gabbay, D.M., Pnueli, A.: A Sound and Complete Deductive System for CTL* Verification. Log. J. IGPL. 16, 499–536 (2008)
9. Inggs, C., Barringer, H.: CTL* model checking on a shared-memory architecture. Form. Method. Syst. Des. 29, 135–155 (2006)
10. Lecointre, G., Le Guyader, H.: The Tree of Life: A Phylogenetic Classification. Harvard University Press Reference Library. Belknap Press of Harvard University Press (2007)
11. Păsăreanu, C.S. (ed.): Model Checking Software. LNCS, vol. 5578. Springer, Heidelberg (2009)

12. Pnueli, A.: In transition from global to modular temporal reasoning about programs. In: Logics and Models of Concurrent Systems, pp. 123–144. Springer, New York (1985)
13. Stern, U., Dill, D.L.: Parallelizing the Murφ Verifier. Form. Method. Syst. Des. 18, 117–129 (2001)
14. Requeno, J.I., Blanco, R., de Miguel Casado, G., Colom, J.M.: Phylogenetic Analysis Using an SMV Tool. In: Rocha, M.P., Rodríguez, J.M.C., Fdez-Riverola, F., Valencia, A., et al. (eds.) PACBB 2011. AISC, vol. 93, pp. 167–174. Springer, Heidelberg (2011)

PHYSER: An Algorithm to Detect Sequencing Errors from Phylogenetic Information

Jorge Álvarez-Jarreta, Elvira Mayordomo, and Eduardo Ruiz-Pesini

Abstract. Sequencing errors can be difficult to detect due to the high rate of production of new data, which makes manual curation unfeasible. To address these shortcomings we have developed a phylogenetic inspired algorithm to assess the quality of new sequences given a related phylogeny. Its performance and efficiency have been evaluated with human mitochondrial DNA data.

Keywords: sequencing errors, phylogeny, human mitochondrial DNA.

1 Introduction

Continuous advances in DNA sequencing technologies since the 1970's have provided the scientific community with unparalleled amounts of biological information at ever decreasing costs (for a technical review, see [7]). Furthermore, the size of public sequence databases has continued the exponential growth of the last 30 years; e.g., the number of records in GenBank is currently doubling approximately every 35 months [4]. In part due to this fast growth of public databases, most of their contents cannot undergo independent curation: the metadata are neither standardized nor homogeneous. Hence, the individual quality of a sequence, measured as its accuracy with respect to the original copy, is a priori unknown.

Jorge Álvarez-Jarreta · Elvira Mayordomo
Dept. de Informática e Ingeniería de Sistemas (DIIS) & Instituto de Investigación en Ingeniería de Aragón (I3A), Universidad de Zaragoza, María de Luna 1, 50018 Zaragoza, Spain
e-mail: {jorgeal,elvira}@unizar.es

Eduardo Ruiz-Pesini
Centro de Invest. Biomédica en Red de Enfermedades Raras & Agencia Aragonesa para la Investigación y el Desarrollo & Dept. de Bioquímica y Biología Molecular y Celular, Facultad de Veterinaria, Universidad de Zaragoza, Miguel Servet 177, 50013 Zaragoza, Spain
e-mail: {eduruiz}@unizar.es

M.P. Rocha et al. (Eds.): 6th International Conference on PACBB, AISC 154, pp. 105–112.
springerlink.com

We consider as sequencing errors the sites where the value differs from its counterpart in the original sequence. Sequencing errors may occur due to contamination —as for the reference sequence of human mitochondrial DNA, the *rCRS* [2]— but also because of modern high-throughput sequencing techniques. These techniques replicate very small segments of DNA and sort them together by local alignments, in which shorter segments are more susceptible to yield false positives. The error rates of current technologies, known in some cases [8], unknown in others, are far away from negligible. In addition, contamination is not a measurable factor in isolation. In this paper we consider the use of evolutionary information for the detection of sequencing errors.

A phylogeny allows grouping each new sequence with its close relatives and measuring similarity between these and their ancestors. Representative mutations of each group are respected almost universally, and exceptions to this conservation are almost certainly due to errors in the sequencing process. Although it is possible to exceptionally discover new subgroups and unusual variations, the probability of these facts will depend on the current state of the phylogeny.

In this paper, we motivate and present PHYSER: a phylogenetic inspired algorithm to assess the quality of new sequences by their location in the reference tree. The parameters of the algorithm are the new sequence (fasta format), the phylogenetic tree (newick format), the pairwise alignment of each sequence in the phylogeny with the reference sequence (fasta format) and the reference sequence (fasta format), which must also be included in the phylogeny. As output, the algorithm provides the classification of the input sequence (qualitative value), the total number of differences which are found between the closest node of the phylogeny and the input sequence (*distance*) and the list of possibly erroneous sites. This design follows the work methodology of the authors of MITOMAP [12]. As a byproduct we can use our algorithm to update the phylogenetic tree with the accepted new sequences, thus offering a good compromise between accuracy and an up-to-date state of the phylogeny between its global updates. Although the algorithm can work properly with any kind of data, we have chosen real human mitochondrial DNA (hmtDNA) data to complete the study, mainly due to the fact that a large validated phylogeny is available. Performance and efficiency of the algorithm have also been tested with hmtDNA.

2 Background

There is, to our knowledge, no previous work on automatic sequence evaluation using evolutionary information. There do exist some tools for the placement of sequences into a phylogeny, which in our case is just a byproduct of the main objective of the algorithm.

One of this placement tools is *pplacer* [9], a software application designed for phylogenetic placement of sequences. It uses some techniques like maximum-likelihood (ML) and Bayesian Information Criterion (BIC) to select the closest node of the reference tree to the sequence. Unfortunately, it has been developed to work with metagenomics, and the input data is difficult to create or handle due to its requirements.

Another placement tool we have found is part of the software toolkit of the *Ribosomal Database Project* (also known as RDP) [10]. Its main drawback is that it only works with ribosomal RNA sequences, so all the processes applied are of specific purpose. Additionally, it is only available online.

3 Detecting Sequencing Errors

The algorithm is based on the fact that we are not able to determine whether a mutation is real or not just looking at the sequence to which it belongs. As a solution, the best option is to use a phylogenetic tree. A phylogenetic tree contains a lot of information about how a specific sequence type has evolved over time, so it is straightforward to verify the new sequence seeing where it should be added in the phylogeny. Obviously, it is necessary that the tree selected is based on a well-known and accepted model and with all its sequences checked, so no errors have been introduced in the construction step.

As mentioned above, the main process of the algorithm is to locate the place in the phylogeny where the input sequence fits better. Before explaining its behavior in a more detailed way, we introduce two main operations needed during the process.

The first is the *Hamming filter*. This operation has as input two sequences of the same length and provides as output their Hamming distance, that is, the total number of sites where the two sequences do not share the same value (excluding gap and 'unknown' states).

The second one is the *Reference filter*. As input, we have to provide two sequences: the algorithm's input sequence and a sequence from the tree. First, the operation gets a list of the sites where the reference sequence (parameter of the algorithm) differs from the tree sequence. Due to the fact that most of the gaps of the reference sequence are introduced in order to align it to the rest of the sequences of the tree, these gap sites are not taken into account. Afterwards, it compares the values of the tree sequence and the input sequence only at the sites included in the previous list. The output of the operation will be the total number of differences obtained in this last comparison.

The algorithm will find the closest node to the input sequence. To do this, it takes one node, which we will call *parent*, and all the nodes that are one level below, its *children*. It applies the *Reference filter* setting as input all the pairs formed by the input sequence and each one of the sequences selected. Normally, one of the *children* nodes will be the closest one of all the pairs handled, so this node will be selected as the new *parent*, and the process will be repeated until the algorithm obtains a leaf as the closest node. The first node selected as *parent* will be the root of the phylogenetic tree.

There are some other situations that may happen instead of the common one presented. We can get two or more nodes as the closest ones. If they are all *children* nodes, the algorithm will explore each new path independently, applying the main process individually. The tests shown in the next section demonstrate that this multipath situation will not last longer than two or three iterations. If one of the nodes is the *parent* node, it will be discarded inasmuch as we prefer to get closer to the leaves. The last situation is featured when the *parent* is the only closest node. The tests have revealed some situations that we denominate *local minima*, where the *parent* results as the closest node, but it is just a local situation: there are other nodes, closer to the leaves of the tree, that are closer to the input sequence than the *parent*. To pass through these *local minima*, the algorithm applies the *Hamming filter* to the same pairs handled in the previous filter. The new results are processed as before, except if we obtain again just the *parent* as the closest node. In this case, we have reached a *global minimum*, so this is the closest node to the input sequence of the whole tree.

The algorithm finally applies the *Reference filter* to the selected node in order to obtain the total number of differences with the input sequence. Two thresholds will determine if the sequence is *Right*, if it has some possible errors (*Alarm*), or if it is most probably *Wrong*. It is important to know that these thresholds will not work properly if the input sequence corresponds to any unexplored species within the phylogeny, or similar cases, which are depicted also as 'holes'.

Intuitively, due to the multipath situations, the algorithm may show more than one node as solution. Looking at the tree we have seen that all these solutions are usually close relatives, that is, nodes with the same parent node or nephew nodes.

4 Tests and Results

We have presented an algorithm that can work with any kind of data. As mentioned previously, we have chosen hmtDNA information for the present tests performed. This data have real good properties: easy to sequence, non-recombinant (very useful for phylogenetic reconstruction) and highly informative (see, e.g., [3]). Moreover, most areas of the human mitochondrial phylogeny are well represented and the characteristic mutations, by which large groups of individuals are related, are organized in extensive hierarchies of mitochondrial haplogroups [13]. In fact, our algorithm draws inspiration from the expert procedures applied to the incremental construction of the MITOMAP phylogeny [12].

For our experiments we have used the last phylogenetic tree created by ZARAMIT project [5], which is composed by 7390 hmtDNA sequences obtained from GenBank. The construction of this phylogeny requires more than one year of sequential CPU time. Therefore, due to the mentioned increase of public databases, the number of additions between feasible reconstructions can be extremely significant. As the reference sequence, we have used the revised Cambridge Reference Sequence (rCRS). The *Reference filter* thresholds have been set to 0 for the *Right* - *Alarm* discrimination, and 3 for the *Alarm* - *Wrong* distinction.

4.1 Behavior Study

In order to study the behavior of the algorithm, we have divided the experiments into three groups, each focusing on obtaining specific results within all the possible cases.

1. **Correct location of the leaves:** The first experiment aims to determine the accuracy of the algorithm. We have selected the 62 sequences in [1] (AY738940 to AY739001) and the 23 sequences in [11] (DQ246811 to DQ246833) for this test. All of them are part of the set of leaves of the phylogenetic tree.

   As result, the algorithm has classified all as *Right*, which implies a success rate of 100%. However, only 53 have been located correctly, i.e. a 60.95% of accuracy. In most of the cases where the algorithm has not been able to locate the sequence correctly, the closest nodes are close relatives of the corresponding leaf. There are just two sequences, DQ246830 and DQ246833, where the *distance* field has revealed an anomaly. In the rest of the sequences this field has reached a maximum of 23, and 9 on average, while in these two sequences, the algorithm has obtained a *distance* of 273. If we look at them at GenBank, we will find that their length is 16320, that is, 249 nucleotides shorter than the reference sequence. Therefore, obtaining those distances, as well as the inability of the algorithm to locate the sequences, are normal consequences.

2. **Non-human mtDNA:** In these experiments we have used sequences from different animals, in order to see how the algorithm handles information that does not fit in a hmtDNA phylogenetic tree. The specific animals and sequences accessions are shown in Table 1.

   The alignment of each of these sequences have been made using MUSCLE [6], a tool for alignment and multialignment processes. First, we have aligned the

Table 1 Sequences of different animals and their classification by the algorithm.

Accession	Animal	Classification	Distance
NC_001643	Chimpanzee	RIGHT	1966
NC_001941	Sheep	RIGHT	4719
NC_005313	Bullet tuna	RIGHT	5775
NC_009684	Mallard duck	RIGHT	5818
NC_007402	Sunbeam snake	RIGHT	6322
NC_002805	Dark-spotted frog	RIGHT	6191
NC_008159	Mushroom coral	RIGHT	8242
NC_009885	Nematode	RIGHT	9074
NC_006281	Blue crab	RIGHT	9251
NC_006160	Whitefly	RIGHT	9302

Table 2 Synthetic sequences created from AY738958 and their classification by the algorithm.

Accession	Mutation	Classification	Closest nodes	Distance
AY738958	*base sequence*	RIGHT	Anc3564, Anc4104	7
SEQ00001	-3106A	RIGHT	Anc4104	6
SEQ00002	-3106A, G8859A	ALARM(1)	Anc4104	7
SEQ00003	-3106A, G8859A, G15325A	ALARM(2)	Anc4104	8
SEQ00004	T6775C	RIGHT	Anc4076, Anc4104	7
SEQ00005	T6775C, G1437A	RIGHT	EU130575	13

input sequence with the reference sequence. Afterwards, we have deleted in both sequences all those sites that correspond to new gaps in the reference sequence. The values of those sites in the input sequence are considered as phylogenetic non-relevant information, thus this step does not generate any degradation.

Notice that the main purpose of this algorithm is to detect sequencing errors. Therefore, the fact that human phylogeny helps in classification of non-human mtDNA as *Right* is not a drawback of our algorithm but more a consequence of the closeness of those species to hmtDNA. Therefore, the most important result drawn from these experiments is the *Distance* field. This field is always provided with the classification due to its relevance in a good interpretation of the output of the algorithm. Amongst all the animals, the closeness of chimpanzee to hmtDNA does not seem surprising. Besides, it is possible to see the relationship with the different classes and clades as far as we go deeper in the evolution process. In most cases, the distance implies that more than 30% of the nucleotides are wrong, which is another sign that the sequence does not fit in the tree. If this happens, it should be checked whether the sequence belongs to a *Homo sapiens*. Remark that the distance among 2 hmtDNA is around 40 nucleotides (0.24%).

3. **Synthetic mutations:** In this case we want to prove that the algorithm can really detect every single relevant mutation. We have taken the sequence AY738958 as base sequence in which we are going to "mutate" some sites and show how the results of the algorithm change. These mutations are shown in the first three columns of Table 2.

   As a usual format in biology, the mutations are represented as follows: first the previous value, after that the site and finally the new value assigned to that site. Table 2 contains also the results for each synthetic sequence created, showing again the results for the sequence AY738958 so we can see how the mutations change them.

   The first three new sequences demonstrate how a single mutation, in the right site, can change a *Right* classification to an *Alarm* result. The last two show how, obviously, one mutation can also change the closest nodes. Usually, as in this case, the result will change from one node to one of its close relatives, so it will not be very relevant. But if three or four mutations or mistakes occur along the sequence in the right sites, we can obtain a closest node really far from the real location of the sequence in the phylogenetic tree.

4.2 *Performance Study*

All the experiments have been executed in a computer with a Core 2 Duo E6750 processor and 8 GB of RAM. The load of the phylogenetic tree and the information needed by the application takes at most 30 seconds. All these data just have to be loaded the first time, when the application starts. The program takes 12 seconds on average to locate the input sequence. Hence, the program has an excellent performance and the user can obtain the results in "real time", providing an accurate feedback showing the sites that have been checked as bad (if any) with the closest nodes. The worst case of the performed experiments has been taking as input the leaf DQ246827, where the program took 16.7 seconds to locate it.

5 Conclusions

We have presented PHYSER: a new algorithm to assess the errors made at sequencing processes, providing as output the level of correctness of the sequence given a phylogeny. Nowadays, this checking of the sequences is made by hand, which implies a large investment of time, regardless the possible human mistakes. Our solution provides a detector of possible errors made at sequencing processes, with an accurate performance, that also gives as result the closest nodes of the phylogeny to the new sequence. If the sequence is good enough, the algorithm automatically includes it into the phylogenetic tree, so the information is always updated.

For future improvements we will aim to develop a new checking of the mutations detected as possible errors, adding a new level of biological viability. This checking will consist of taking into account the reversions, so if one mutation has appeared before in the path we have explored in the phylogeny that implies it is not a bad mutation. Moreover, the conservation rate among the different species will provide an extra criterion to the biological viability of the mutation.

Finally, the current implementation of PHYSER is available by request to the first author.

Acknowledgements. This work was supported by the Spanish Ministry of Science and Innovation (Projects TIN2008-06582-C03-02, TIN2011-27479-C04-01) and the Government of Aragón Dept. de Ciencia, Tecnología y Universidad and the European Social Fund (grupo GISED T27).

We want to thank Roberto Blanco for his assistance with the interaction with the phylogenetic tree from ZARAMIT project.

References

1. Achilli, A., Rengo, C., Magri, C., Battaglia, V., Olivieri, A., Scozzari, R., Cruciani, F., Zeviani, M., Briem, E., Carelli, V., Moral, P., Dugoujon, J.M., Roostalu, U., Loogvöli, E.L., Kivisild, T., Bandelt, H.J., Richards, M., Villems, R., Santachiara-Benerecetti, A.S., Semino, O., Torroni, A.: The molecular dissection of mtDNA haplogroup H confirms that the Franco-Cantabrian glacial refuge was a major source for the European gene pool. American Journal of Human Genetics 75, 910–918 (2004)
2. Andrews, R.M., Kubacka, I., Chinnery, P.F., Lightowlers, R.N., Turnbull, D.M., Howell, N.: Reanalysis and revision of the Cambridge reference sequence for human mitochondrial DNA. Nature Genetics 23, 147 (1999)
3. Bandelt, H.J., Macaulay, V., Richards, M. (eds.): Human mitochondrial DNA and the evolution of Homo sapiens. Springer, Berlin (2006)
4. Benson, D.A., Karsch-Mizrachi, I., Lipman, D.J., Ostell, J., Sayers, E.W.: GenBank. Nucleic Acids Research 38, D46–D51 (2010)
5. Blanco, R., Mayordomo, E.: ZARAMIT: A System for the Evolutionary Study of Human Mitochondrial DNA. In: Omatu, S., Rocha, M.P., Bravo, J., Fernández, F., Corchado, E., Bustillo, A., Corchado, J.M. (eds.) IWANN 2009, Part II. LNCS, vol. 5518, pp. 1139–1142. Springer, Heidelberg (2009)
6. Edgar, R.C.: MUSCLE: multiple sequence alignment with high accuracy and high throughput. Nucleic Acids Research 32, 1792–1797 (2004)
7. Kim, S., Tang, H., Mardis, E.R. (eds.): Genome sequencing technology and algorithms. Artech House, Norwood (2007)
8. Margulies, M., Egholm, M., Altman, W.E., Attiya, S., Bader, J.S., Bemben, L.A., Berka, J., Braverman, M.S., Chen, Y.J., Chen, Z., Dewell, S.B., Du, L., Fierro, J.M., Gomes, X.V., Goodwin, B.C., He, W., Helgesen, S., He Ho, C., Irzyk, G.P., Jando, S.C., Alenquer, M.L.I., Jarvie, T.P., Jirage, K.B., Kim, J.B., Knight, J.R., Lanza, J.R., Leamon, J.H., Lefkowitz, S.M., Lei, M., Li, J., Lohman, K.L., Lu, H., Makhijani, V.B., McDade, K.E., McKenna, M.P., Myers, E.W., Nickerson, E., Nobile, J.R., Plant, R., Puc, B.P., Ronan, M.T., Roth, G.T., Sarkis, G.J., Simons, J.F., Simpson, J.W., Srinivasan, M., Tartaro, K.R., Tomasz, A., Vogt, K.A., Volkmer, G.A., Wang, S.H., Wang, Y., Weiner, M.P., Yu, P., Begley, R.F., Rothberg, J.M.: Genome sequencing in open microfabricated high density picoliter reactors. Nature 437, 376–380 (2005)
9. Matsen, F.A., Kodner, R.B., Armbrust, E.V.: pplacer: linear time maximum-likelihood and bayesian phylogenetic placement of sequences onto a fixed reference tree. BMC Bioinformatics 11, 538 (2010)
10. Olsen, G.J., Overbeek, R., Larsen, N., Marsh, T.L., McCaughey, M.J., Maciukenas, M.A., Kuan, W.M., Macke, T.J., Xing, Y., Woese, C.R.: The ribosomal database project. Nucleic Acids Research 20(supplement), 2199–2200 (1992)
11. Rajkumar, R., Banerjee, J., Gunturi, H.B., Trivedi, R., Kashyap, V.K.: Phylogeny and antiquity of M macrohaplogroup inferred from complete mt DNA sequence of Indian specific lineages. BMC Evolutionary Biology 5, 26 (2005)
12. Ruiz-Pesini, E., Lott, M.T., Procaccio, V., Poole, J., Brandon, M.C., Mishmar, D., Yi, C., Kreuziger, J., Baldi, P., Wallace, D.C.: An enhanced mitomap with a global mtdna mutational phylogeny. Nucleic Acids Research 35, D823–D828 (2007)
13. van Oven, M., Kayser, M.: Updated comprehensive phylogenetic tree of global human mitochondrial DNA variation. Human Mutation 29, E386–E394 (2008)

A Systematic Approach to the Interrogation and Sharing of Standardised Biofilm Signatures

Anália Lourenço*, Andreia Ferreira, Maria Olivia Pereira, and Nuno F. Azevedo

Abstract. The study of microorganism consortia, also known as biofilms, is associated to a number of applications in biotechnology, ecotechnology and clinical domains. A public repository on existing biofilm studies would aid in the design of new studies as well as promote collaborative and incremental work. However, bioinformatics approaches are hampered by the limited access to existing data. Scientific publications summarise the studies whilst results are kept in researchers' private *ad hoc* files.

Since the collection and ability to compare existing data is imperative to move forward in biofilm analysis, the present work has addressed the development of a systematic computer-amenable approach to biofilm data organisation and standardisation. A set of in-house studies involving pathogens and employing different state-of-the-art devices and methods of analysis was used to validate the approach. The approach is now supporting the activities of BiofOmics, a public repository on biofilm signatures (http://biofomics.org).

Keywords: infection-causing microorganisms, biofilms, legacy data, data standardisation.

Anália Lourenço · Andreia Ferreira · Maria Olivia Pereira
IBB - Institute for Biotechnology and Bioengineering, Centre of Biological Engineering, University of Minho, Campus de Gualtar, 4710-057 Braga – Portugal
e-mail: {analia,mopereira}@deb.uminho.pt, af18048@gmail.com

Nuno F. Azevedo
LEPAE, Department of Chemical Engineering, Faculty of Engineering, University of Porto, Porto – Portugal
e-mail: nazevedo@fe.up.pt

* Corresponding author.

M.P. Rocha et al. (Eds.): 6th International Conference on PACBB, AISC 154, pp. 113–120.
springerlink.com

1 Introduction

Microorganisms have evolved various strategies to survive and adapt to the ever changing environmental conditions. The formation of biofilms is an example of such adaptation strategies. Biofilms are structured communities of microorganisms that are able to survive virtually everywhere in Nature because of their ability to adhere to a surface and embed in a protecting, self-produced matrix of extracellular polymeric substances [1-2].

Due to their persistence and resistance to antimicrobial agents, biofilms are at the basis of a range of problems in areas of great importance to human development, such as hygiene and food safety in the food industry [3], nosocomial infections [4-7], acute and chronic infections [8-10], and clogging and contaminations in drinking water systems [11-12]. The interest of the scientific community for these problems is obvious and much research has been devoted to the understanding of initial cell adhesion and biofilm formation phenomena in the last decades.

Similarly to other domains, biofilm research has benefited from the technological evolution occurred in the last decades [13-14]. The development of high-throughput biofilm-forming devices (e.g. the 96-well plate, the microtiter plate with coupons and the Calgary device) has enabled the simultaneous testing of large sets of conditions. The implementation of automated spectrophotometry and microscopy systems (e.g. scanning electron, atomic force, and confocal laser scanning microscopy) has empowered the large scale analysis of biofilm features, such as biofilm biomass, biofilm activity and microbial composition. The "omics" platforms are supporting the study of the transcriptome [15-16], proteome [17-19] and metabolome [20-21] of biofilms.

Therefore, biofilm research is becoming data-intensive and thus, the need for suitable bioinformatics tools, namely databases on existing studies, is compelling [22]. However, scientific publications only summarise the obtained results and data files are not submitted to any public location, remaining on researchers' private archives. Besides limited access, no protocol exists on how to document biofilm studies, i.e. the minimum information required to guarantee self-contained and explanatory documentation. Data files vary widely from laboratory to laboratory, from researcher to researcher and even from a researcher's experiment to the next (Fig.1). Files lack comprehensive documentation on the experimental conditions evaluated (often they are only mentioned by abbreviations) and, even more important, on data quantification terms, such as the units of measure used.

The contribution of the present work is a novel computer-amenable approach to the systematic collection and storage of experimental data related to biofilm studies. Aiming to enable the large-scale classification, retrieval, and dissemination of biofilm signatures, the goal of this approach is to help incorporate existing studies.

The remainder of the paper is organised as follows. Section 2 describes the minimum set of information required to describe a biofilm study unambiguously, and the algorithm of the proposed approach. The proof of concept and the implementation are discussed in Section 3. Finally, some conclusions and future work are summarised.

2 Characterisation of Biofilm Signatures

Usually, biomedical databases derive their contents from literature and previously existing databases, or electronic records of some sort, and manual curation plays a decisive role in quality assessment. Biofilms differ from this scenario in that the data is not deposited in any location, but in the private *ad hoc* files of their owners. Therefore, biofilm data collection is dependent on the willingness of researchers to deposit their data and help in their correct interpretation and standardised structuring.

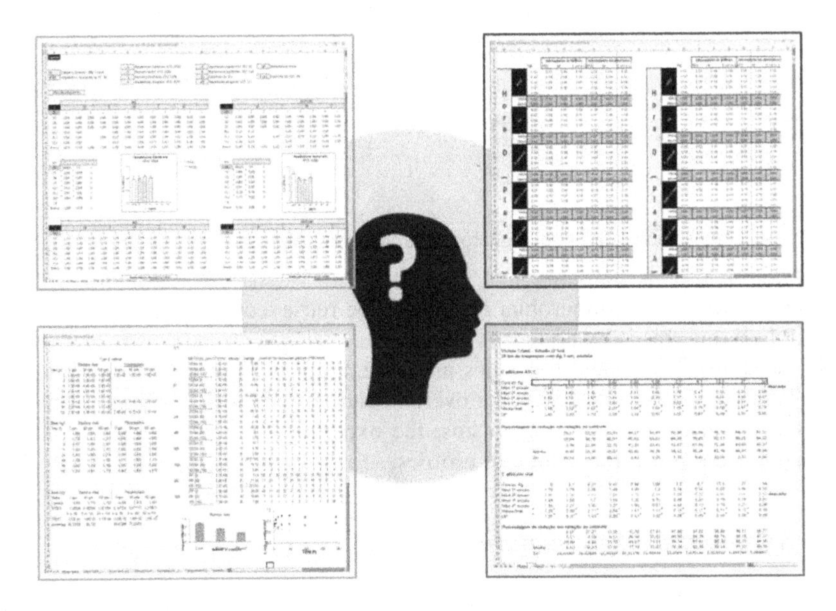

Fig. 1 Examples of biofilm data files from published biofilm studies.

2.1 Modelling Requirements

Biofilm signatures should be self-contained, i.e. there should be enough information about the purpose and execution of the experiment as well as the analysis of results. The signature should include information on the operating procedures (i.e. devices and associated settings that emulate the environment under study), the statistical validation (i.e. the number of replicates and reproductions performed), the methods of analysis and associated units of measure, and, of course, the actual result set (Fig.2). It is important to emphasise that results are fully comparable only for similar methods under identical conditions.

Biofilm signatures should also be self-explanatory, i.e. any researcher should be able to interpret their details. However, controlled vocabulary is scarce. Microorganism characterisation can be obtained from the NCBI Taxonomy Browser [21]

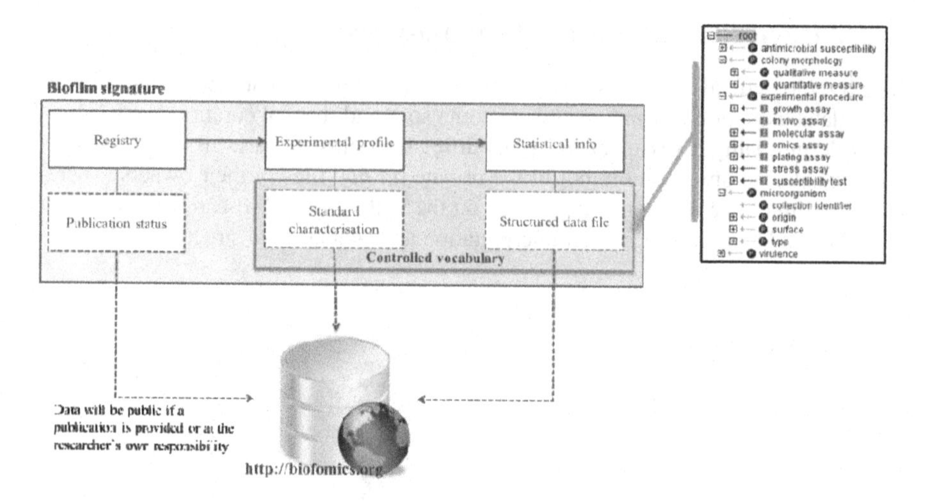

Fig. 2 Conceptual view of biofilm signature characterisation.

and terminology on antimicrobial agents can be retrieved from databases such as CAMP [22], but this is by no means sufficient.

In the scope of our bioinformatics initiative, a novel ontology on biofilm studies is being prepared to describe the overall concept of biofilm signature, i.e. a set of information on the microorganisms, the experimental procedures, the antimicrobial susceptibility and virulence profiles.

2.2 *Systematic Organisation*

Since most researchers already have their data in this format, data submission is for now simplified to data re-organisation and nomenclature standardisation on the familiar format of Microsoft Excel worksheets. Our approach aims to be flexible in terms of the type and number of conditions tested as well as the data series that are to be associated, accounting for the variability inherent to biofilm studies.

The modelling of biofilm signatures is challenged by the increasing complexity and variability of the studies. Therefore, the Excel file is customised according to researcher's specifications. As illustrated in the pseudo-code in Fig.3, experimental conditions are distributed into worksheets such that single value conditions (i.e. the conditions that are constant to the whole experiment) are at the first worksheet and the multi-value conditions (i.e. the conditions that are tested in the experiment) are organised hierarchically in descendent order. This hierarchical structure grows vertically – as many sub-levels as condition tests and experiment reproductions there are – and no restriction is imposed to the length of the data series of each-level (i.e. the number of test replicates).

```
Input: Econd: list of experimental conditions; Ma: list of
methods of analysis
Output: Fexcel customised file

E_cond_single ← singleValueConditions(E_cond)
E_cond_multi ← E_cond − E_cond_single
w_s ← initialiseWorksheet('experiment constants')
for c ∈ sortByType(E_cond_single)
        w_s ← fillIn(c.name, c.value)
end
F_excel ← addWorksheet(w_s)

for m ∈ M_a
    w_m ← initialiseWorksheet(m.name)
    for c ∈ sortDescendNumValues(E_cond_multi)
      w_m ← fillInHierarchy(c.name, c.values)
    end
    F_excel ← addWorksheet(w_m)
end

return F_excel
```

Fig. 3 Pseudo-code of the systematic construction of biofilm structured data files.

3 Proof of Concept and Implementation

A dozen of in-house published studies validated the proposed approach. These studies are quite diverse in nature, addressing problems such as microbial susceptibility to clinically relevant antibiotics, biofilm adhesion to abiotic surfaces and antimicrobial synergisms. They present different levels of complexity and include data quantifications based on various state-of-the-art analytical methods (e. g. crystal violet (CV), 4',6-diamidino-2-phenylindole (DAPI) and fluorescence *in situ* hybridization (FISH)).

The approach was applied successfully and, as a whole, it enabled the standard characterisation of over 10000 data points. As an example, Fig. 4 illustrates a small excerpt of the one of such signatures related to the study on the adhesion of water stressed *Helicobacter pylori* to abiotic surfaces (namely, polypropylene, glass, copper and stainless steel) by Azevedo et al. [25].

Currently, the approach is supporting the activity of BiofOmics database, a general scope biofilms database which is freely accessible to the community (http://biofomics.org). Its implementation is based on open-source, platform-independent software, namely MySQL server (version 14.14, distribution 5.1.52) and PHP 5.1.6.

One of the long term aims of the approach is to be able to perform data mining on the data sets collected. To achieve this goal, it is essential that future experiments follow very strict guidelines, presenting comprehensive and unambiguous

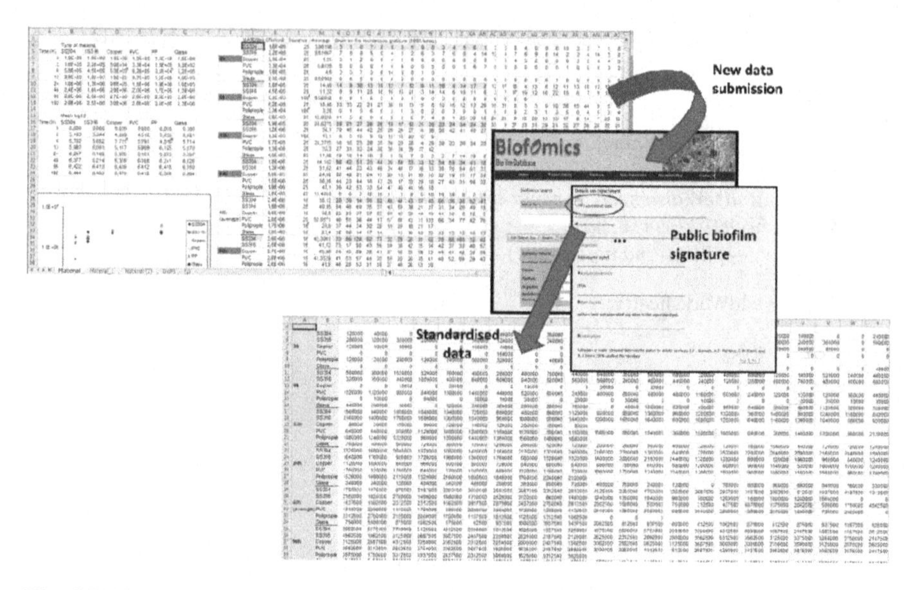

Fig. 4 Different stages of the data presentation of a study on the adhesion of water-stressed Helicobacter pylori to abiotic surfaces by Azevedo et al. [25].

information on how the experiment was performed and its outcomes. Namely, efforts should address the standardization of nomenclature in the biofilm area and the establishment of minimum information guidelines for reporting biofilm experiments. Once this has been achieved one should be able to query for patterns in biofilm signatures and thus empowering analysis abilities.

4 Conclusions

Biofilm research relies increasingly on large collections of data sets. This "big data" dimension calls for the development of novel computational tools, specialised in biofilm data management, interchange and analysis. However, we are unaware of any efforts to standardise and disseminate biofilm data at large scale. Currently, sharing of biofilm data among researchers is poor at best, in detriment of research and community at large.

Data standardisation is an urge to make knowledge more explicit, help detect errors, ensure data reliability, and promote data interchange. In here, we propose a systematic approach to the standardised organisation of biofilm data. The approach has been validated with a number of highly variable, already published experiments and it is already in practice, supporting the operation of the BiofOmics database, a database that facilitates data search and comparison as well as data interchange between laboratories (publicly accessible at http://biofomics.org).

Acknowledgments. The authors thank, among others, Rosário Oliveira, Maria João Vieira, Idalina Machado, Nuno Cerca, Mariana Henriques, Pilar Teixeira, Douglas Monteiro, Melissa Negri, Susana Lopes, Carina Almeida and Hélder Lopes, for submitting their data. The financial support from IBB-CEB, Fundação para a Ciência e Tecnologia (FCT) and European Community fund FEDER (Program COMPETE), project PTDC/SAU-ESA/646091/2006/FCOMP-01-0124-FEDER-007480, are also gratefully acknowledged.

References

1. McBain, A.J.: Chapter 4: In vitro biofilm models: an overview. Adv. Appl. Microbiol. 69, 99–132 (2009)
2. Jain, A., Gupta, Y., Agrawal, R., Khare, P., Jain, S.K.: Biofilms-a microbial life perspective: a critical review. Crit. Rev. Ther. Drug Carrier Syst. 24(5), 393–443 (2007)
3. Van Houdt, R., Michiels, C.W.: Biofilm formation and the food industry, a focus on the bacterial outer surface. J. Appl. Microbiol. 109(4), 1117–1131 (2010)
4. Machado, I., Lopes, S.P., Sousa, A.M., Pereira, M.O.: Adaptive response of single and binary Pseudomonas aeruginosa and Escherichia coli biofilms to benzalkonium chloride. Journal of Basic Microbiology 51, 1–10 (2011)
5. Frei, E., Hodgkiss-Harlow, K., Rossi, P.J., Edmiston Jr., C.E., Bandyk, D.F.: Microbial Pathogenesis of Bacterial Biofilms: A Causative Factor of Vascular Surgical Site Infection. Vasc. Endovascular Surg (2011)
6. Rodrigues, L.R.: Inhibition of bacterial adhesion on medical devices. Adv. Exp. Med. Biol. 715, 351–367 (2011)
7. Høiby, N., Bjarnsholt, T., Givskov, M., Molin, S., Ciofu, O.: Antibiotic resistance of bacterial biofilms. Int. J. Antimicrob. Agents 35(4), 322–332 (2010)
8. Bowler, P.G., Duerden, B.I., Armstrong, D.G.: Wound microbiology and associated approaches to wound management. Clin. Microbiol. Rev. 14(2), 244–269 (2001)
9. Simões, M.: Antimicrobial strategies effective against infectious bacterial biofilms. Curr. Med. Chem. 18(14), 2129–2145 (2011)
10. Smith, A., Buchinsky, F.J., Post, J.C.: Eradicating chronic ear, nose, and throat infections: a systematically conducted literature review of advances in biofilm treatment. Otolaryngol. Head Neck Surg. 144(3), 338–347 (2011)
11. Gião, M.S., Azevedo, N.F., Wilks, S.A., Vieira, M.J., Keevil, C.W.: Persistence of Helicobacter pylori in Heterotrophic Drinking-Water Biofilms. Applied and Environmental Microbiology 74(19), 5898–5904 (2008)
12. Jang, H.J., Choi, Y.J., Ka, J.O.: Effects of diverse water pipe materials on bacterial communities and water quality in the annular reactor. J. Microbiol. Biotechnol. 21(2), 115–123 (2011)
13. Ramage, G., Culshaw, S., Jones, B., Williams, C.: Are we any closer to beating the biofilm: novel methods of biofilm control. Curr. Opin. Infect. Dis. 23(6), 560–566 (2010)
14. Cos, P., Toté, K., Horemans, T., Maes, L.: Biofilms: an extra hurdle for effective antimicrobial therapy. Curr. Pharm. Des. 16(20), 2279–2295 (2010)
15. Wood, T.K.: Insights on Escherichia coli biofilm formation and inhibition from whole-transcriptome profiling. Environ. Microbiol. 11(1), 1–15 (2009)
16. Coenye, T.: Response of sessile cells to stress: from changes in gene expression to phenotypic adaptation. FEMS Immunol. Med. Microbiol. 59(3), 239–252 (2010)

17. Silva, M.S., De Souza, A.A., Takita, M.A., Labate, C.A., Machado, M.A.: Analysis of the biofilm proteome of Xylella fastidiosa. Proteome Sci. 9, 58 (2011)
18. Jiao, Y., D'haeseleer, P., Dill, B.D., Shah, M., Verberkmoes, N.C., Hettich, R.L., Banfield, J.F., Thelen, M.P.: Identification of biofilm matrix-associated proteins from an acid mine drainage microbial community. Appl. Environ. Microbiol. 77(15), 5230–5237 (2011)
19. Cabral, M.P., Soares, N.C., Aranda, J., Parreira, J.R., Rumbo, C., Poza, M., Valle, J., Calamia, V., Lasa, I., Bou, G.: Proteomic and functional analyses reveal a unique lifestyle for Acinetobacter baumannii biofilms and a key role for histidine metabolism. J. Proteome Res. 10(8), 3399–3417 (2011)
20. Xu, H., Lee, H.Y., Ahn, J.: Growth and virulence properties of biofilm-forming Salmonella enterica serovar typhimurium under different acidic conditions. Appl. Environ. Microbiol. 76(24), 7910–7917 (2010)
21. Workentine, M.L., Harrison, J.J., Weljie, A.M., Tran, V.A., Stenroos, P.U., Tremaroli, V., Vogel, H.J., Ceri, H., Turner, R.J.: Phenotypic and metabolic profiling of colony morphology variants evolved from Pseudomonas fluorescens biofilms. Environ. Microbiol. 12(6), 1565–1577 (2010)
22. Azevedo, N.F., Lopes, S.P., Keevil, C.W., Pereira, M.O., Vieira, M.J.: Time to "go large" on biofilm research: advantages of an omics approach. Biotechnol. Lett. 31(4), 477–485 (2009)
23. Sayers, E.W., Barrett, T., Benson, D.A., Bolton, E., Bryant, S.H., Canese, K., Chetvernin, V., Church, D.M., DiCuccio, M., Federhen, S., Feolo, M., Fingerman, I.M., Geer, L.Y., Helmberg, W., Kapustin, Y., Landsman, D., Lipman, D.J., Lu, Z., Madden, T.L., Madej, T., Maglott, D.R., Marchler-Bauer, A., Miller, V., Mizrachi, I., Ostell, J., Panchenko, A., Phan, L., Pruitt, K.D., Schuler, G.D., Sequeira, E., Sherry, S.T., Shumway, M., Sirotkin, K., Slotta, D., Souvorov, A., Starchenko, G., Tatusova, T.A., Wagner, L., Wang, Y., Wilbur, W.J., Yaschenko, E., Ye, J.: Database resources of the Na-tional Center for Biotechnology Information. Nucleic Acids Res. 39, D38–D51 (2011)
24. Thomas, S., Karnik, S., Barai, R.S., Jayaraman, V.K., Idicula-Thomas, S.: CAMP: a useful resource for research on antimicrobial peptides. Nucleic Acids Res. 780, D774–D780 (2010)
25. Azevedo, N.F., Pacheco, A.P., Keevil, C.W., Vieira, M.J.: Adhesion of water stressed Helicobacter pylori to abiotic surfaces. J. Appl. Microbiol. 111, 718–724 (2006)

Visual Analysis Tool in Comparative Genomics

Juan F. De Paz, Carolina Zato, María Abáigar, Ana Rodríguez-Vicente, Rocío Benito, and Jesús M. Hernández

Abstract. Detecting regions with mutations associated with different pathologies is an important step in selecting relevant genes, proteins or diseases. The corresponding information of the mutations and genes is distributed in different public sources and databases, so it is necessary to use systems that can contrast different sources and select conspicuous information. This work presents a visual analysis tool that automatically selects relevant segments and the associated genes or proteins that could determine different pathologies.

1 Introduction

Different techniques presently exist for the analysis and identification of pathologies at a genetic level. Along with massive sequencing, which allows the exhaustive study of mutations, the use of microarrays is highly extended. With regards to microarrays, there are two primary types of chips according to the direction of the analysis that will be carried out: expression arrays and array CGH (Comparative Genomic Hybridization) [18]. CGH arrays (aCGH) are a type of microarray that can analyze information on the gains, losses and amplifications [16] in regions of the chromosomes to detect mutations that can determine some pathologies [13] [11].

Juan F. De Paz · Carolina Zato
Department of Computer Science and Automation, University of Salamanca
Plaza de la Merced, s/n, 37008, Salamanca, Spain
e-mail: {fcofds,carol_zato}@usal.es

María Abáigar · Ana Rodríguez-Vicente · Rocío Benito · Jesús M. Hernández
IBMCC, Cancer Research Center, University of Salamanca-CSIC, Spain
e-mail: {mymary,anita82,beniroc,jhmr}@usal.es

Jesús M. Hernández
Servicio de Hematología, Hospital Universitario de Salamanca, Spain

M.P. Rocha et al. (Eds.): 6th International Conference on PACBB, AISC 154, pp. 121–127.
springerlink.com

Microarray-based CGH and other large-scale genomic technologies are now routinely used to generate a vast amount of genomic profiles. An exploratory analysis of this data is critical in helping to understand the data and form biological hypotheses. This step requires visualization of the data in a meaningful way to visualize the results and to perform first level analyses [15]. At present, tools and software already exist to analyze the data of arrays CGH, such as CGH-Explorer [10], ArrayCyGHt [9], CGHPRO [3], WebArray [17] or ArrayCGHbase [12], VAMP [15]. The problem with these tools is the lack of usability and of an interactive model. For this reason, it is necessary to create a visual tool to analyse the data in a simpler way.

The process of arrays CGH analysis is broken down into a group of structured stages, although most of the analysis process is done manually from the initial segmentation of the data. Once the segmentation is finished, the next step is to perform a visual analysis of the data using different tools, which is a quite slow process. For this reason, the system tries to facilitate the analysis and the automatic interpretation of the data by selecting the relevant genes, proteins and information from the previous classification of pathologies. The system provides several representations in order to facilitate the visual analysis of the data. The information for the identified genes, CNV, pathologies etc. is obtained from public databases.

This article is divided as follows: section 2 describes our system, and section 3 presents the results and conclusions.

2 aCGH Analysis Tool

aCGH is a technique that can detect copy number variations in patients who have undergone different mutations in chromosomic regions. Typically, the variations have been previously catalogued, allowing the existing information to be used to catalogue and evaluate the mutation. In this case study, the cases are defined according to the segments into which the chromosomic regions have been fragmented.

The developed system receives data from the analysis of chips and is responsible for representing the data for extracting relevant segments on evidence and existing data. Working from the relevant cases, the first step consists of selecting the information about the genes and transcripts stored in the databases. This information will be associated to each of the segments, making it possible to quickly consult the data and reveal the detected alterations at a glance. The data analysis can be carried out automatically or manually.

2.1 *Automatic Analysis*

Knowledge extraction algorithms can be categorized as decision trees, decision rules, probabilistic models, fuzzy models, based on functions, statistics, or gain functions. Some of these algorithms include: decision rules RIPPER [4], One-R [7], M5 [8], decision trees J48 [14], CART [2] (Classification and Regression Trees), probabilistic models naive Bayes [6], fuzzy models K-NN (K-Nearest Neighbors) [1] and finally statistical techniques, such as non parametrics Kruskal-Wallis [21] and Mann-Whitney U-test [19] for two groups, and parametrics Chi Squared [22], ANOVA [5]. The gain functions are a particular case of the techniques used in decision trees and decision rules for selecting the attributes, which is why they are not considered separately.

For this particular system, the use of decision trees was chosen to select the main genes of the most important pathologies, specifically J48 [14] in its implementation for Weka [20]. However, if the system needs a generic selection, the gain functions are chosen (specifically, Chi Squared [22], which is also implemented in the Weka library). Chi Squared was chosen because it is the technique that makes it possible to work with different qualitative nominal variables to study factor and its response. The contrast of Chi Squared makes it possible to obtain as output the values that can sort the attributes by their importance, providing an easier way to select the elements. As an alternative, gain functions could be applied in decision trees, providing similar results.

2.2 Visual Analysis

A visual analysis is performed of the data provided by the system and the information recovered from the databases. New visualizations are performed in order to more easily locate the mutations, thus facilitating the identification of mutations that affect the codification of genes among the large amount of genes. Visualization facilitates the validation of the results due to the interactivity and ease of use of previous information. Existing packages such as CGHcall [23] in R do not display the results in an intuitive way because it is not possible to associate segments with regions and they do not allow interactivity.

The system provides a visualization to select the regions with more variants and relevant regions in different pathologies. The visualizations make is possible to extract information from databases using a local database.

2.3 Reviewing Process

Once the relevant segments have been selected, the researchers can introduce information for each of the variants. The information is stored in a local database. These data are considered in future analyses although they have to be reviewed in detail and contrasted by the scientific community. The information is shown in future analyses with the information for the gains and losses. However, because only the information from public databases is considered reliable, this information is not included in the reports.

3 Results and Conclusions

The system was applied to two different kinds of CGH arrays: BAC aCGH, and Oligo aCGH. The information obtained from the BAC aCGH after segmenting and normalizing is represented in table 1. As shown in the figure, there is one patient for each column. The rows contain the segments so that all patients have the same segments. Each segment is a tuple composed of three elements: chromosome, initial region and final region. The values v_{ij} represent gains and losses for segment i and patient j. If the value is positive, or greater than the threshold, it is considered a gain; if it is lower than the value, it is considered a loss.

Table 1 BAC aCGH normalized and segmented

Segment	Patient 1	Patient 2	...	Pantient n
Init-end	v_{11}	v_{12}	...	v_{13}
Init-end	v_{21}	v_{22}	...	v_{23}

The system includes the databases because it extracts the information from genes, proteins and diseases. These databases have different formats but basically there is a tuple of three elements for each row (chromosome, start, end, other information). Altogether, the files downloaded from UCSC included slightly more than 70,000 registries

Figure 1 displays the information for 18 oligo arrays cases. Only the information corresponding to chromosome 5 is shown. The green lines represent gains for the patient in the associated region of the chromosome, while the red lines represent losses. The user can draw the CNVs and use this information to select the relevant information. For example, in figure 3 the CNVs are represented in blue. We can see that there is a gain region with a high incidence in the individuals but this region is not relevant because it belongs to a CNV.

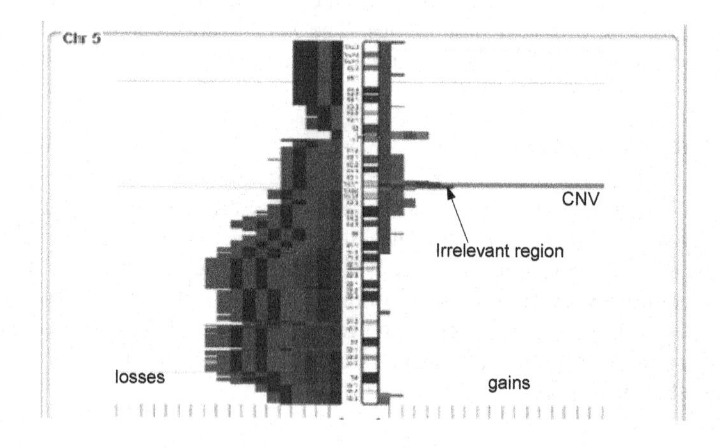

Fig. 1 Selection of segments and genes automatically

When performing the visual analysis, users can retrieve information from a local database or they can browse through UCSC. For example, figure 2 contains the information for the segment belonging to the irrelevant region shown in the previous image.

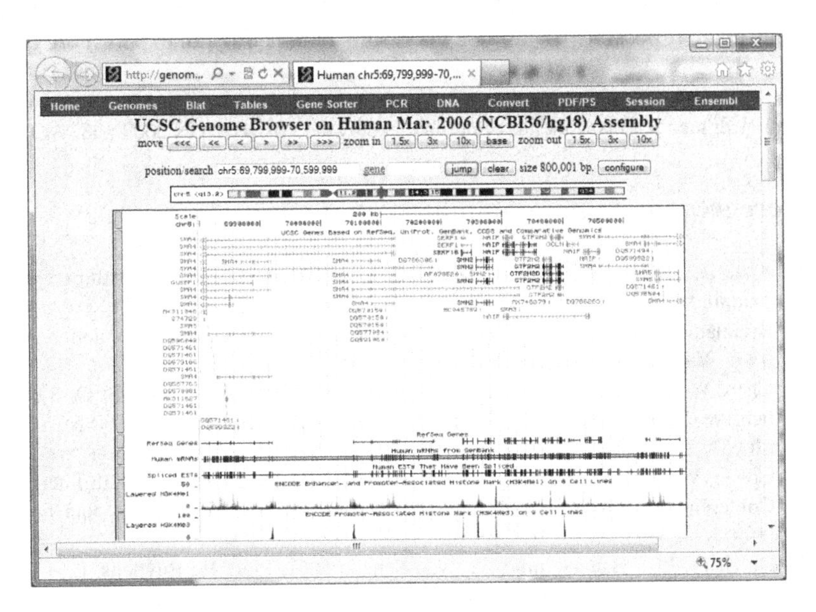

Fig. 2 Browse through UCSC

In order to facilitate the revision and learning phases for the expert, a different visualization of the data is provided. This view helps to verify the results obtained by the hypothesis contrast regarding the significance of the differences between pathologies. Figure 3 shows a bar graph where one bar represents each individual and is divided into different segments with an amplitude proportional to the width of the segment gain (green) or loss (red). We can see that the blue individuals (rectangle over the bars) are not in the range of the green individuals because they remain deactivated when we select the green individuals.

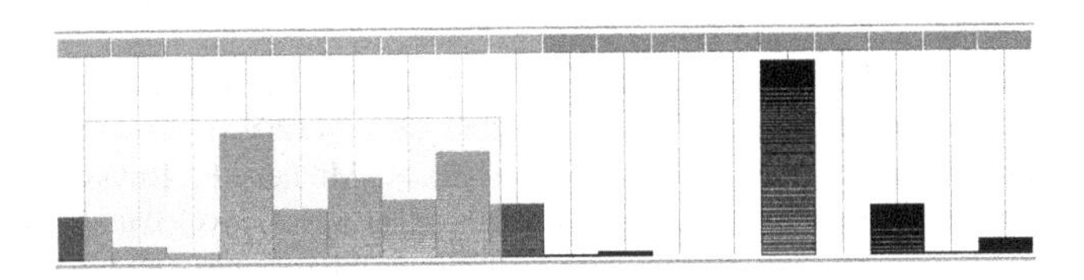

Fig. 3 Selection of segments and genes automatically

The presented system facilitates the use of different sources of information to analyze the relevance in variations located in chromosomic regions. The system is able to select the genes, variants, genomic duplications that characterize pathologies automatically, using several databases. This system allows the management of external sources of information to generate final results. The provided visualizations make it possible to validate the results obtained by an expert more quickly and easily

Acknowledgments. This work has been supported by the MICINN TIN 2009-13839-C03-03.

References

[1] Aha, D., Kibler, D., Albert, M.K.: Instance-based learning algorithms. Machine Learning 6, 37–66 (1991)
[2] Breiman, L., Fried, J.H., Olshen, R.A., Stone, C.J.: Classification and regression trees. Wadsworth International Group. (1984)
[3] Chen, W., Erdogan, F., Ropers, H., Lenzner, S., Ullmann, R.: CGHPRO- a comprehensive data analysis tool for array CGH. BMC Bioinformatics 6(85), 299–303 (2005)
[4] Cohen, W.W.: Fast effective rule induction. In: Proceedings of the 12th International Conference on Machine Learning, pp. 115–123. Morgan Kaufmann, San Francisco (1995)
[5] De Haan, J.R., Bauerschmidt, S., van Schaik, R.C., Piek, E., Buydens, L.M.C., Wehrens, R.: Robust ANOVA for microarray data. Chemometrics and Intelligent Laboratory Systems 98(1), 38–44 (2009)
[6] Duda, R.O., Hart, P.: Pattern classification and Scene Analysis. John Wisley & Sons, New York (1973)
[7] Holmes, G., Hall, M., Prank, E.: Generating Rule Sets from Model Trees. Advanced Topics in Artificial Intelligence 1747(1999), 1–12 (2007)
[8] Holte, R.C.: Very simple classification rules perform well on most commonly used datasets. Machine Learning 11, 63–91 (1993)
[9] Kim, S.Y., Nam, S.W., Lee, S.H., Park, W.S., Yoo, N.J., Lee, J.Y., Chung, Y.J.: ArrayCyGHt, a web application for analysis and visualization of array-CGH data. Bioinformatics 21(10), 2554–2555 (2005)
[10] Lingjaerde, O.C., Baumbush, L.O., Liestol, K., Glad, I.K., Borresen-Dale, A.L.: CGH-explorer, a program for analysis of array-CGH data. Bioinformatics 21(6), 821–822 (2005)
[11] Mantripragada, K.K., Buckley, P.G., Diaz de Stahl, T., Dumanski, J.P.: Genomic microarrays in the spotlight. Trends Genetics 20(2), 87–94 (2004)
[12] Menten, B., Pattyn, F., De Preter, K., Robbrecht, P., Michels, E., Buysse, K., Mortier, G., De Paepe, A., van Vooren, S., Vermeesh, J., et al.: ArrayCGHbase: an analysis platform for comparative genomic hybridization microarrays. BMC Bioinformatics 6(124), 179–187 (2006)
[13] Pinkel, D., Albertson, D.G.: Array comparative genomic hybridization and its applications in cancer. Nature Genetics 37, 11–17 (2005)
[14] Quinlan, J.R.: C4.5: Programs For Machine Learning. Morgan Kaufmann Publishers Inc. (1993)

[15] Rosa, P., Viara, E., Hupé, P., Pierron, G., Liva, S., Neuvial, P., Brito, I., Lair, S., Servant, N., Robine, N., Manié, E., Brennetot, C., Janoueix-Lerosey, I., Raynal, V., Gruel, N., Rouveirol, C., Stransky, N., Stern, M., Delattre, O., Aurias, A., Radvanyi, F., Barillot, E.: Visualization and analysis of array-CGH, transcriptome and other molecular profiles. Bioinformatics 22(17), 2066–2073 (2006)
[16] Wang, P., Young, K., Pollack, J., Narasimham, B., Tibshirani, R.: A method for callong gains and losses in array CGH data. Biostat. 6(1), 45–58 (2005)
[17] Xia, X., McClelland, M., Wang, Y.: WebArray, an online platform for microarray data analysis. BMC Bionformatics 6(306), 1737–1745 (2005)
[18] Ylstra, B., Van den Ijssel, P., Carvalho, B., Meijer, G.: BAC to the future! or oligonucleotides: a perspective for microarray comparative genomic hybridization (array CGH). Nucleic Acids Research 34, 445–450 (2006)
[19] Yue, S., Wang, C.: The influence of serial correlation on the Mann-Whitney test for detecting a shift in median. Advances in Water Resources 25(3), 325–333 (2002)
[20] `http://www.cs.waikato.ac.nz/ml/weka/`
[21] Kruskal, W., Wallis, W.: Use of ranks in one-criterion variance analysis. Journal of American Statistics Association (1952)
[22] Kenney, J.F., Keeping, E.S.: Mathematics of Statistics, Pt. 2, 2nd edn. Van Nostrand, Princeton (1951)
[23] Van de Wiel, M.A., Kim, K.I., Vosse, S.J., Van Wieringen, W.N., Wilting, S.M., Ylstra, B.: CGHcall: calling aberrations for array CGH tumor profiles. Bioinformatics 23(7), 892–894 (2007)

From Networks to Trees

Marco Alves, Joãd Alves, Rui Camacho, Pedro Soares, and Luísa Pereira

Abstract. Phylogenetic networks are a useful way of displaying relationships between nucleotide or protein sequences. They diverge from phylogenetic trees as networks present cycles, several possible evolutionary histories of the sequences analysed, while a tree presents a single evolutionary relationship. Networks are especially useful in studying markers with a high level of homoplasy (same mutation happening more than once during evolution) like the control region of mitochondrial DNA (mtDNA), where the researcher does not need to compromise with a single explanation for the evolution suggested by the data. However in many instances, trees are required. One case where this happens is in the founder analysis methodology that aims at estimating migration times of human populations along

Marco Alves
DEI & Faculdade de Engenharia & IPATIMUP (Instituto de Patologia e Imunologia Molecular da Universidade do Porto), Universidade do Porto, Portugal
e-mail: ei05099@fe.up.pt

João Alves
DEI & Faculdade de Engenharia, Universidade do Porto, Portugal
e-mail: ei08083@fe.up.pt

Rui Camacho
DEI & Faculdade de Engenharia & LIAAD-INESC Porto L.A. (Laboratory of Artificial Intelligence and Decision Support), Universidade do Porto, Portugal
e-mail: rcamacho@fe.up.pt

Pedro Soares
IPATIMUP (Instituto de Patologia e Imunologia Molecular da Universidade do Porto), Universidade do Porto, Portugal
e-mail: pedroa@ipatimup.pt

Luísa Pereira
IPATIMUP (Instituto de Patologia e Imunologia Molecular da Universidade do Porto) & Faculdade de Medicina, Universidade do Porto, Portugal
e-mail: lpereira@ipatimup.pt

M.P. Rocha et al. (Eds.): 6th International Conference on PACBB, AISC 154, pp. 129–136.
springerlink.com

history and prehistory. Currently, the founder analysis methodology implicates the creation of networks, from where a probable tree will be extracted by hand by the researcher, a time-consuming process, prone to errors and to the ambiguous decisions of the researcher. In order to automate the founder analysis methodology an algorithm that extracts a single probable tree from a network in a fast, systematic way is presented here.

1 Introduction

Phylogenetic networks are widely used in population and evolutionary genetics. In the case of mtDNA, which has a considerably faster mutation rate than the nuclear genome and a higher level of homoplasy, phylogenetic networks are particularly appropriate. The evolutionary history of a group of mtDNA sequences can be dubious and a network that presents several possible ways to resolve the relationship between sequences and not a single solution (like in a phylogenetic tree) is more appropriate. This is even more problematic if the researcher is studying the hypervariable segment I (HVS-I) of the mtDNA which has even a higher mutation rate and presents a series of positions that mutate incredibly fast [9] and are recurrently homoplasic in the networks. Two algorithms for constructing networks are commonly used in research of mtDNA, the median joining [3] and the reduced median [2] algorithms, both implemented in the network software package (www.fluxus-engeneering.com). Although networks are the ideal output in many situations, several methodologies require a phylogenetic tree, meaning a single evolutionary history of the sequences. Founder analysis [8], a methodology that aims at estimating the time of occurrence of human migrations, is one of those examples. Despite the original paper that describes this methodology being highly cited (more than 400 times), the founder analysis has not been applied because of the considerable amount of manual work required which can take over several months. The first step of the founder analysis is to consider a single probable phylogenetic tree from each of the various previously constructed networks of the data. This step is usually done by hand, it is time consuming and quite prone to errors and more important is defined by the subjective choices of the researcher. In this work we developed a tool that uses algorithms previously employed in graph theory to make the extraction of probable trees from the network in a semi-automatic way. We have tested different types of algorithms in order to check which one best satisfies the requirements.

The remainder of the paper is organised as follows. Section 2 provides a resume of relevant aspects of phylogenetic analysis and sequence patterns useful for this study. Section 3 describes the proposed algorithm. In Section 4 we present an empirical evaluation of the proposed technique. Finally, in Section 5, we draw some conclusions and propose further work.

2 Basic Concepts of Phylogenetic Analysis

Through the evolutionary history of a species, individuals diverge by accumulating mutations in their DNA called polymorphisms. In mitochondrial DNA, with absence of recombination the accumulated polymorphisms are transmitted in block between generations and new mutations will occur in a specific background of polymorphisms. The sequence considering all the polymorphisms in an individual is called its haplotype and a group of haplotypes that are related and share a common ancestor is named a haplogroup (specific name in mtDNA), a clade or a monophyletic group. Using a phylogenetic approach it is possible to access what are the mutations differentiating haplotypes or clades. These mutations can be homoplasic, meaning that they occurred more than once during evolution. Another problematic issue in phylogenetic reconstruction is that a polymorphism can revert back to an ancestor state in some haplotypes. This complicates the phylogenetic reconstruction since several phylogenetic histories can explain the data. In mtDNA this is particularly problematic due to the high mutation rate and the high recurrence and reversion rate of several mutations, mainly in the HVS-I [9]. In this context it became common practice the use of networks that consider several evolutionary paths in the data for the study of mtDNA. A phylogenetic tree differs from a phylogenetic network in the sense that: a) the tree presents a single evolution suggested by the data, b) a network presents at least one cycle meaning multiple solutions for explaining the data, and c) a network can be considered a combination of various trees, the same way that a tree can be seen as a network without reticulations. In mtDNA research, the network software created and freely distributed by Fluxus Technology Ltd. (currently at version 4.6) [6][1] creates phylogenetic networks from a list of polymorphisms and their state in each sample, using two algorithms based on parsimony [2, 3], meaning that the simplest way of explaining the data or the lower number of required mutations is the best. The network output is a combination of all possible trees with maximum parsimony. One advantage of the network software in relation to the great majority of the available phylogenetic software programs is that it clearly displays which mutations or polymorphisms are represented in each branch and it is possible to further explore which are the most probable trees based on how recurrent are the mutations observed. Currently the major drawback of using Network is that there is no software that automatically suggests plausible trees. In this paper we report on the development of such software [1].

3 Algorithms to Process Network's Output

Basically the main objective of this work is to generate a tree that should be similar to a minimum spanning tree, a tree with the minimum cost, obtained from the elimination of edges in the cycles. The eliminated edges will be the ones with lower cost. Two criteria were implemented: a) more common mutations should appear more times in the tree, and, b) the connection from the root to a given node should

[1] `www.fluxus-engeneering.com`

contain the higher possible number of individual sequences. Regarding the first criterion, the information on how fast a type of mutation is, was based on the published list of occurrence of mutations in a complete sequence tree with 2196 sequences [9]. This provides the weight for each edge. The more frequent the mutation in the tree the lower cost it will have in the graph. In a network output, the information regarding the number of individuals with a given haplotype is indicated on the edges, a value varying between 0 to any integer. In a graphical representation of a network (like the one in Figure 2) the vertices are represented by circles with sizes proportional to the number of individuals with the haplotype represented by that vertex. The vertices's containing a frequency of 0 are called median vectors, representing hypothetical sequences, not observed in any individual but possibly occurring during evolution. This information (frequency of a given haplotype) is the base of the second criterion. After considering several minimum spanning tree algorithms, we selected the Prim's algorithm [5, 7] that is a greedy algorithm that achieves an optimal solution. One important factor in the pre-selection of the Prim algorithm was the facility of implementation. Prim's algorithm will continuously add new edges and vertices's in order to connect all the vertices's of the original graph and form a tree, minimising the total weight of the edges, and it is easy to control for the formation of cycles. The Dijkstra's algorithm [4] was also implemented in order to compare with the results obtained with the Prim's algorithm. One possible advantage of the Dijkstra's algorithm is that it retains the shortest path from the root to every vertex of the graph, which does not necessarily happen in a minimum spanning tree. Both the Prim and Dijkstra's algorithms were adapted to produce several alternative trees within a given overall cost.

We now describe in more detail the algorithmic approach. The steps of the global algorithm of the software tool we have developed is depicted in Figure 1.

The algorithm in Figure 1 is applicable to both Prim and Dijkstra algorithms acting as the "base algorithm". The overall process has the following parts. First the output file of Network is loaded and encoded in a graph (step 1). A minimum tree is then computed providing a cost value that will be taken as the starting cost value. If we use Prim the minimum cost will be the cost of a minimum spanning tree. If we use Dijkstra the minimum cost will be the sum of the shortest paths from a specified root to all other nodes (with no cycles). This first tree is stored and the cost taken as the"current minimum cost". The algorithm enters a loop (steps 5 to 13) where, at each iteration of the loop, a new tree is generated by backtracking on the previously generated tree. Backtracking is achieved looking at previous commitments in edge choice and choosing alternative edges. A new tree is accepted only if its global cost is equal to or less than the "current minimum cost" otherwise is discarded. If there are no alternative trees, at the current cost value, the minimum cost is increased and the process of tree generation continues until the maximum number of trees specified by the user is achieved. Whenever a new tree is accepted it is subject to a post-processing "cleaning" step where "useless" median vectors are merged (step 10). After the main loop completion the resulting set of trees is ranked according to criteria defined by the expert researchers of the field. In order to rank the trees we used the criteria previously defined. In a cycle, nodes that are median vectors

```
Netwwok2Tree(File, N) :
  Given:
    Network 4.6 output file (File)
  Output:
    Ranked set of N phylogenetic trees

1.  Net = readFile(File)
2.  Tree = getMinimumCost(Net)
3.  Cost = cost(Tree)
4.  S = { Tree }
5.  while ||S|| < N do
6.    T = backtrackNewTree(Net, Cost)
7.    if T is empty then
8.      Cost = nextCostValue(Cost, Net)
9.    else
10.     T' = removeMeadianVectores(T)
11.     S = S ∪ { T' }
12.   endif
13. endwhile

14. S = rankTrees(S)
15. return S
```

Fig. 1 Constructing phylogenetic trees from the output of the Network software.

(frequency of 0) were avoided and presented a higher cost than the ones were samples are present. Evolutionary ways that contain a higher number of samples in the nodes have a lower cost and lead to trees with a higher weight. Finally and of most importance edges containing fast mutations have a much lower cost than edges defined by rare mutations.

Finally, the developed software, allows the expert to inspect the constructed trees and decide which is the best one. The expert can also interactively make corrections to that tree in order to achieve the tree he thinks is the correct one.

4 Empirical Testing

We have used Network 4.6 [6] to produce four networks of four different data sets (haplogroups). Networks were constructed using a database of more than 15,000 HVS-I sequences of Asian origin. Case 1 corresponds to haplogroup Y, case 2 to haplogroup C4a2, case 3 to haplogroup M7a and case 4 contains samples from haplogroup R9b. Table 1 displays some basic information about the networks.

The top ranked trees resulting from the extraction of a tree from the four networks using both the Prim s and Dijkstra's algorithms were compared with the results suggested by an expert when doing it manually (Table 2).

Table 1 Characterisation of the Network 4.6 output for the four test cases.

Test case	Number of nodes	Number of edges	Number of median vectors	Number of reticulations
1	61	71	8	10
2	45	51	3	7
3	55	60	3	6
4	92	99	10	8

Table 2 Summary of results of the adapted Prim and Dijkstra algorithms in the four test cases. The expert entries are the base reference for nodes and edges in the final tree.

Test Case	Algorithm	Number of nodes	Number of edges	well resolved reticulations	wrongly resolved reticulations	success rate (%)
1	Dijkstra	56	55	0	10	0
	Prim	56	55	10	0	100
	expert	56	55	-	-	-
2	Dijkstra	44	43	1	6	14.3
	Prim	44	43	5	2	71.4
	expert	44	43	-	-	-
3	Dijkstra	54	53	1	5	16.7
	Prim	52	51	5	1	83.3
	expert	52	51	-	-	-
4	Dijkstra	88	87	4	4	50
	Prim	88	87	3	5	37.5
	expert	87	86	-	-	-

In all cases we can see a reduction in the number of nodes in the trees when compared with the network. This is due to the automatic elimination of median vectors. Prims algorithm gives better results when compared with Dijkstra's. The match rate when compared with the extracted tree by an expert was substantially higher for the output of the Prims algorithm. One should note that many of the branches defined by the expert represent monophyletic clades that are supported by phylogenies of complete mtDNA genomes of those haplogroups. As an example we are showing case 1 in Figure 2. In this case the Prims algorithm resolved the network in the same way as the experts. Dijkstra's algorithm did not resolve any of the cycles in the same way as the experts. Although Dijkstra's algorithm chose one of the shortest routes within the cycles it did not take into account the weight of the edges. For example the edge containing the mutation 265, that is a slow mutation, occurred 3 times in the tree generated with the Dijkstra's algorithm and only once in the Prim's algorithm and in the expert choice. The only case where the Dijkstra's algorithm performed better than Prims (case 4) the network required a very fast mutation to happen four times and various solutions are possible. Even in that case the expert considered the obtained tree by the Prims algorithm to be a valid and acceptable tree.

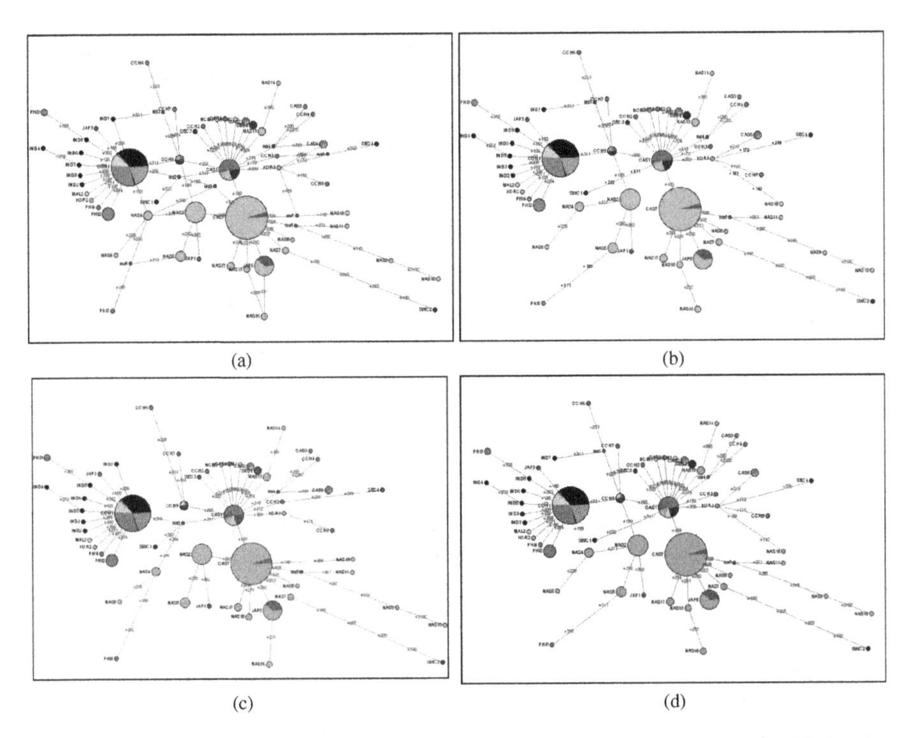

Fig. 2 Test case 1. (a) Output of Network 4.6. (b) Result of applying the modified Prim algorithm. (c) Result of applying the modified Dijkstra algorithm. (d) Result of the domain expert (same as (b)).

5 Conclusions

Phylogenetic trees are constructed automatically from the output of Network 4.6 taking a few seconds. Currently the domain expert constructs the trees manually taking much more time in the process. Dijkstra's algorithm generated trees that did not fit the best parsimonious solution, however Prims algorithm provided a tree that mostly matched the one extracted by the expert and even when it did not match, usually the decision of the modified Prims algorithm is acceptable for the expert. This work is included in a project that aims to automate the founder analysis methodology that can take several months to perform manually, probably a fact partially explaining the high number of citations of the method and the almost complete lack of application in the literature. One of the most time consuming parts was the extraction of a single probable tree from the network output. We will implement the Prims algorithm in an overall package that will also calculate all the required statistics for the founder analysis. The software will allow the validation of the tree obtained and even the possibility for the researcher to perform minor manipulations if required.

Acknowledgements. This work has been partially supported by Reitoria da Universidade do Porto (Project IPG196/2010) and by by Fundação para a Ciência e Tecnologia (FCT). IPATIMUP is an Associate Laboratory of the Portuguese Ministry of Science, Technology and Higher Education and is partially supported by FCT.

References

1. Alves, M.: Biblioteca de programas para análise de genética populacional. Master's thesis, University of Porto (2011)
2. Bandelt, H.J., Forster, P., Sykes, B., Richards, M.: Mitochondrial portraits of human populations using median networks. Genetics 141(2), 743–753 (1995)
3. Bandelt, H.J., Forster, P., Röhl, A.: Median-joining networks for inferring intraspecific phylogenies. Molecular Biology and Evolution 16, 37–48 (1999)
4. Dijkstra, E.W.: A note on two problems in connexion with graphs. Numerische Mathematik 1, 269–271 (1959)
5. Jarník, V.: O jistém problému minimálním [about a certain minimal problem]. Práce Moravské Přírodovědecké Společnosti 6, 57–63 (1930)
6. Ltd FT: Network 4.6 user guide (2011), `http://www.fluxus-engineering.com/Network4600_user_guide.pdf`
7. Prim, R.C.: Shortest connection networks and some generalizations. Bell System Technical Journal 36, 1389–1401 (1957)
8. Richards, M.B., Macaulay, V., Hickey, E., Vega, E., Sykes, B., Guida, V., Rengo, C., Sellitto, D., Cruciani, F., Kivisild, T., Villems, R., Thomas, M., Rychkov, S., Rychkov, O., Rychkov, Y., Gölge, M., Dimitrov, D., Hill, E., Bradley, D., Romano, V., Calì, F., Vona, G., Demaine, A., Papiha, S., Triantaphyllidis, C., Stefanescu, G., Hatina, J., Belledi, M., Rienzo, A.D., Novelletto, A., Oppenheim, A., Ren Nørby, S., Al-Zaheri, N., Santachiara-Benerecetti, S., Scozzari, R., Torroni, A., Bandelt, H.J.: Tracing european founder lineages in the near eastern mtdna pool. The American Journal of Human Genetics 67, 1251–1276 (2000)
9. Soares, P., Ermini, L., Thomson, N., Mormina, M., Rito, T., Röhl, A., Salas, A., Oppenheimer, S., Macaulay, V., Richards, M.: Correcting for purifying selection: an improved human mitochondrial molecular clock. American Journal of Human Genetics 84, 740–759 (2009)

Procedure for Detection of Membranes in Three-Dimensional Subcellular Density Maps

A. Martinez-Sanchez, I. Garcia, and J.J. Fernandez

Abstract. Electron tomography is the leading technique for visualizing the cell environment in molecular detail. Interpretation of the three-dimensional (3D) density maps is however hindered by different factors, such as noise and the crowding at the subcellular level. Although several approaches have been proposed to facilitate segmentation of the 3D structures, none has prevailed as a generic method and thus manual annotation is still a common choice in the field. In this work we introduce a novel procedure to detect membranes. These structures define the natural limits of compartments within biological specimens. Therefore, its detection turns out to be a step towards automated segmentation. Our method is based on local differential structure and on a Gaussian-like membrane model. We have tested our procedure on tomograms obtained under different experimental conditions.

1 Introduction

Electron tomography (ET) is the leading technique for visualizing the molecular organization of the cell environment [6, 11]. Although the stages to derive three-dimensional reconstructions (or tomograms) are well established [11], their interpretation is however not straightforward. This is caused by several factors, including the low signal-to-noise ratio (SNR) and the inherent biological complexity. Significant efforts are thus spent to facilitate their interpretation by (semi-)automated segmentation procedures [17]. Although several such methods have been proposed in the field (reviewed in [15, 17]) none has stood out as a general applicable method yet, and manual segmentation still remains prevalent. Here, the user delineates the features of interest using visualization tools, which is tedious and subjective.

A. Martinez-Sanchez · I. Garcia
Grupo Supercomputacion y Algoritmos. Universidad de Almeria, Spain

J.J. Fernandez
Centro Nacional de Biotecnologia (CSIC), Madrid, Spain
e-mail: JJ.Fernandez@csic.es

M.P. Rocha et al. (Eds.): 6th International Conference on PACBB, AISC 154, pp. 137–145.
springerlink.com

In this work we have focused on detection of membranes. Membranes encompass compartments within biological specimens, define the limits of the intracellular organelles and the cells themselves, etc. Therefore, their detection may play an important role towards automated segmentation of tomograms. Although several segmentation approaches presented in the field are well suited to membrane detection [8, 13, 14, 16], none has shown a good behaviour in the most general case due to several reasons: they are case-specific, or limited performance under low SNR, difficult parameter tuning, user-intervention required, etc.

Here we present a novel procedure intended for detection of membranes. Based on a membrane model, the procedure relies on the characterization of structures at a local scale using differential information. Later, the integration at a global scale yields the definite detection [12]. We show and validate the performance of the algorithm on a number of tomograms under different experimental conditions.

2 Algorithm for Membrane Detection

2.1 Membrane Model

At a local level, a membrane can be considered as a plane-like structure with certain thickness (Fig. 1) [2, 3]. The density along the normal direction progressively decreases as a function of the distance to the centre of the membrane. This density variation across the membrane can be modelled by a Gaussian function (Fig. 1):

$$I(r) = \frac{D_0}{\sqrt{2\pi}\sigma_0} e^{-\frac{r^2}{2\sigma_0^2}} \tag{1}$$

where r runs along the normal to the membrane, D_0 is a constant to set the maximum density value (at the centre of the membrane) and σ_0 is related to its thickness.

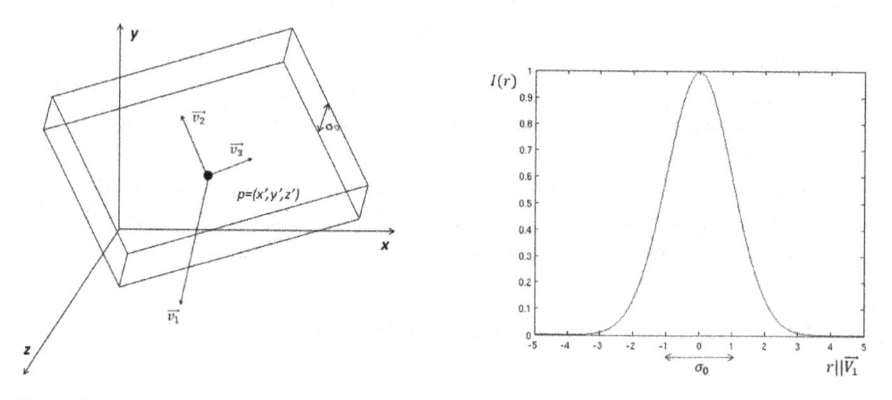

Fig. 1 Membrane model and the density profile across the membrane.

The eigen-analysis of the density function at point $p = (x, y, z)$ of the membrane yields the eigenvectors $\vec{v_1}$, $\vec{v_2}$ and $\vec{v_3}$ with eigenvalues $|\lambda_1| >> |\lambda_2| \approx |\lambda_3|$ (Fig. 1) [2, 3]. This reflects that there are two directions ($\vec{v_2}$, $\vec{v_3}$) with small density variation and the largest variation runs along the direction perpendicular to the membrane ($\vec{v_1}$, parallel to r, i.e. $\vec{v_1} || r$).

Due to its local nature, any detector based on this model can also generate a high response for structures different from membranes. For that reason, it is important to incorporate "global information" to discern true membranes from these others.

2.2 Scale-Space

The scale-space theory was formulated in the 80s [7, 18] and allows isolation of the information according to the spatial scale. At a given scale σ, all the features with a size smaller than the scale are filtered out whereas the others are preserved. Therefore, the scale-space is useful to focus on the structures of a particular size, ignoring other smaller or spurious details. A scale-space for a tomogram f can be generated by convolving f with a set of kernels with size σ [9]. In this work we have used a sampled version of the Gaussian kernel with standard deviation σ, $G(x;\sigma)$.

To analyze the scale-space applied to the membrane model, it can be assumed without loss of generality that r runs along the x direction (i.e. $\vec{v_1} || r || x$), hence reducing the problem to one dimension (along x). Given the Gaussian membrane profile I (Eq. 1), ignoring constants, and taking into account that the convolution of two continuous Gaussian functions yields another Gaussian function whose variance is the sum of the variances [4], the membrane model with thickness σ_0 at a scale σ is:

$$L(x;\sigma) = G(x;\sigma) * I(x) = G(x; \sqrt{\sigma^2 + \sigma_0^2}) = G(x;\sigma_s) \tag{2}$$

2.3 Local Detector and Membrane Strength

Now it is possible to define a detector for the membrane model at a given scale σ (Eq. (2)). This detector is based on differential information, as it has to analyze local structure. In order to make it invariant to the membrane direction, the detector is established along the normal to the membrane (i.e. the direction of the maximum curvature) at the local scale. An eigen-analysis of the Hessian matrix is well suited to determine such direction [5]. At every voxel of the tomogram, the Hessian matrix is defined by:

$$H = \begin{bmatrix} L_{xx} & L_{xy} & L_{xz} \\ L_{xy} & L_{yy} & L_{yz} \\ L_{xz} & L_{yz} & L_{zz} \end{bmatrix} \tag{3}$$

where $L_{ij} = \frac{\partial^2 I}{\partial i \partial j}$ $\forall i, j \in (x, y, z)$. The Hessian matrix provides information about the second order local intensity variation. The first eigenvector $\vec{v_1}$ resulting from the eigen-analysis is the one whose eigenvalue λ_1 exhibits the largest absolute value and points to the direction of the maximum curvature (second derivative).

The Hessian matrix of the membrane model of the previous section (i.e. with maximum curvature along x) at a scale σ has all directional derivatives null, except L_{xx}. As a result, $\lambda_1 = L_{xx}$ and $\overrightarrow{v_1} = (1,0,0)$. Along the direction normal to the membrane, λ_1 turns out to be negative where the membrane has significant values and its absolute value progressively decreases from the centre towards the extremes of the membrane, as shown in Fig. 2(left). Therefore, we propose the use of $|\lambda_1|$ as a local membrane detector (also known as local gauge). In practice, in experimental studies λ_2 and λ_3 are not null. Thus, a more realistic gauge would be:

$$R = \begin{cases} |\lambda_1| - \sqrt{\lambda_2 \lambda_3} & \lambda_1 < 0 \\ 0 & \lambda_1 \geq 0 \end{cases} \tag{4}$$

where $\sqrt{\lambda_2 \lambda_3}$ is the geometrical mean between λ_2 and λ_3.

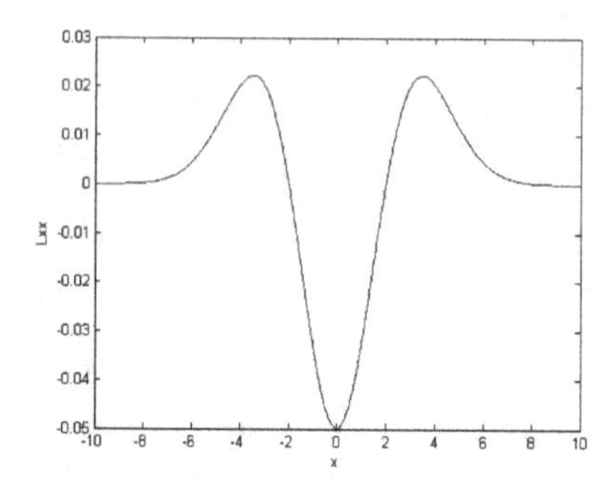

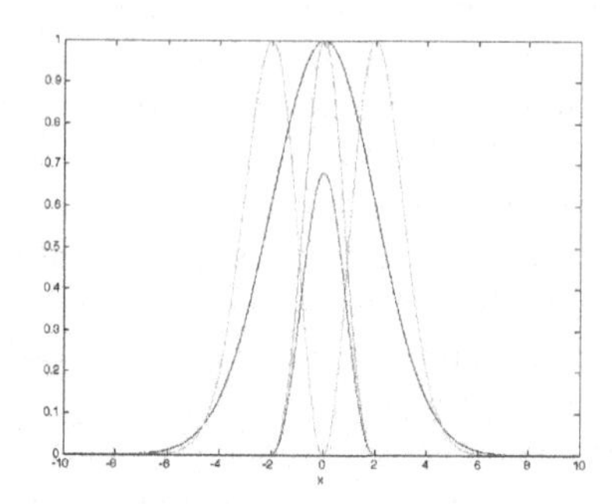

Fig. 2 Second derivative L_{xx} of the membrane model ($\sigma_0 = 1$) at a scale $\sigma = 1$ (left) and Gauges for the density profile of a membrane with $\sigma_0 = 1$ at a scale $\sigma = 1$ (blue): R^2 (red), S (cyan) and membrane strength M (green). The membrane profile, S and M are normalized in the range $[0,1]$. R^2 keeps the scale relative to S.

Unfortunately, R is still sensitive to other local structures that may produce false positives along the maximum curvature direction. To make the gauge robust and more selective, it is necessary to define detectors for these cases. First, the noisy background in the tomogram may generate false positives. However, the background usually has a density level different from that shown by the structures of interest. A strategy based on a density threshold t_l [3] helps to get rid of these false positives.

Local structures resembling 'density steps' in the tomogram also make the gauge R produce a false peak. A suitable detector for a local step is the edge saliency [10]:

$$S = L_x^2 + L_y^2 + L_z^2 \tag{5}$$

where $L_i = \frac{\partial I}{\partial i}\ \forall i \in (x,y,z)$. A membrane exhibits a high value of S at the extremes and a low value at the centre (Fig. 2(right)). Based on their response to a membrane, the ratio between the squared second-order and first-order derivatives (i.e. R^2/S)

quantifies how well the local structure around a voxel fits the membrane model and not a step. We thus define membrane strength as:

$$M = \begin{cases} \frac{R^2}{S} & ,(L > t_l) \text{ and } \left(\text{sign}\left(\frac{\partial R}{\partial r}\right) \neq \text{sign}\left(\frac{\partial S}{\partial r}\right)\right) \\ 0 & , \text{otherwise} \end{cases} \tag{6}$$

The first condition in Eq. (6) denotes the density thresholding described above. The second condition represents the requirement that the slopes of R and S in the gradient direction must have opposite signs. This condition is unique for membranes (see Fig. 2(right)) and it is important to restrict the response of that function for steps. If the local structure approaches the membrane model, M will have high values around the centre of the membrane (high values of R^2, low values of S).

2.4 Hysteresis Thresholding and Global Analysis

Next, M is thresholded to discard voxels unlikely belonging to membranes. Hysteresis thresholding has been shown to outperform the standard thresholding algorithm [15]. Here two thresholds are used, the large value t_u undersegments the tomogram whereas the other t_o oversegments it. Starting from the undersegmented tomogram (seed voxels), adjacent voxels are added to the segmented tomogram by progressively decreasing the threshold until the oversegmenting level t_o is reached.

In this work we have increased the robustness of hysteresis thresholding by constraining the selection of seed voxels to the particular characteristics of membranes, namely the high number of voxels connected. So, we have introduced two additional thresholds so that seed voxels belonging to components with less than t_a pixels planewise, or t_h in 3D, are discarded. This strategy allows isolation of seeds that are most representative of membranes, thereby improving the global performance.

Finally, a global analysis stage intends to identify the segmented components that are actually membranes. A distinctive attribute of membranes is their relatively large dimensions. Therefore, the size (i.e. the number of voxels of the component) can serve as a major global descriptor. A threshold t_v (similar or equal to t_h) is then introduced to set the minimal size for a component to be considered as membrane.

3 Experimental Results

The segmentation algorithm was tested with several tomograms taken under different experimental conditions. The tomograms were rescaled to a common density range of $[0,1]$. The optimal results were obtained using the same basic parameter configuration for hysteresis thresholding, in particular $t_u \in [0.35,1]$, $t_o \in [0.05,0.4]$ and $t_a \in [15,35]$. The values of the parameters σ, t_l, t_h and t_v, however, depend on the specific dataset and were readily set by inspection of the tomogram. σ is the thickness of the membrane.

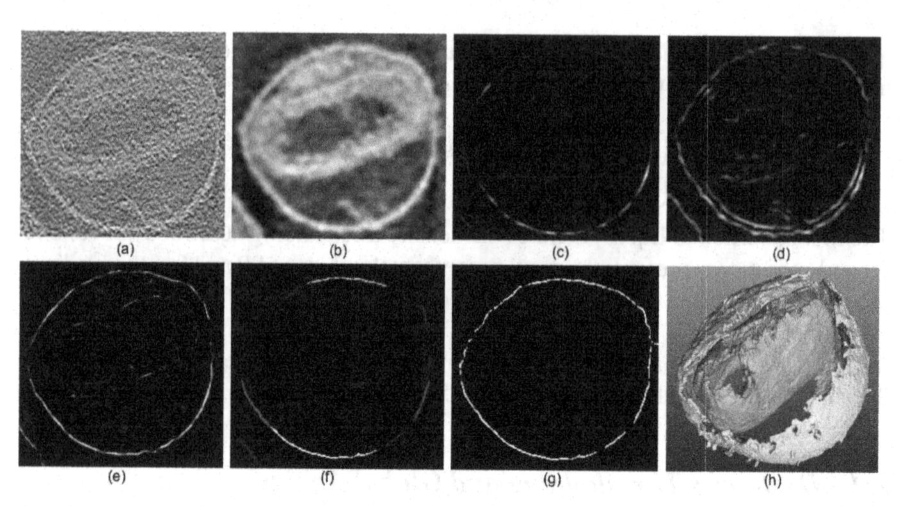

Fig. 3 The procedure applied to Vaccinia virus. (a) slice of the original tomogram. (b) at a scale $\sigma = 3$. (c-e) R, S and M, respectively. (f, g) hysteresis thresholding. seed voxels to extract the outer membrane after the t_a and after t_h thresholding, respectively. brighter colour means larger number of connected voxels. (h) Membranes detected after the global analysis. The membrane of the internal core (in pink) was obtained after running the algorithm at a scale of $\sigma = 6$.

To illustrate the procedure at work, Fig. 3 shows the different stages applied to a tomogram of Vaccinia virus [1], which is particularly noisy. The algorithm succeeds in segmenting both the outer and the internal core membrane by properly tuning the parameter σ. A scale of $\sigma = 3$ was applied to extract the outer membrane. For the core membrane, however, a much higher value was necessary ($\sigma = 6$) because this membrane actually comprises two layers that make it rather thick, thereby needing a higher scale to extract it separately.

Fig. 3(c-e) (which were obtained at $\sigma = 3$, targeting at the outer membrane) clearly shows that, though the gauge R actually quantifies the level of local membrane-ness, it still depends on the density level. Thus, there are some parts of the membrane where R exhibits weak values. On the contrary, M only contains differential information and, therefore, higher strength is shown throughout the membrane regardless of the density value. However, the side effect is that other structures resembling planes at local level also produce a high value of M (for instance, the dense material between the outer membrane and the core seen at the top of (e); or the fiber attached to the internal side of the outer membrane seen at the bottom of (e)). The hysteresis thresholding procedure and the global analysis then manage to extract the true membranes. This behaviour is an inherent feature of the algorithm.

Fig. 4 shows the results of selected stages of the algorithm applied to a tomogram containing a Golgi apparatus. The algorithm was applied at a scale $\sigma = 2.2$ and was capable of segmenting the structure as only one component including all cisternae. The algorithm actually detects other planar structures present in the tomogram such as those of the surrounding mitochondria. By means of the threshold t_v at the global analysis stage, the Golgi apparatus is isolated.

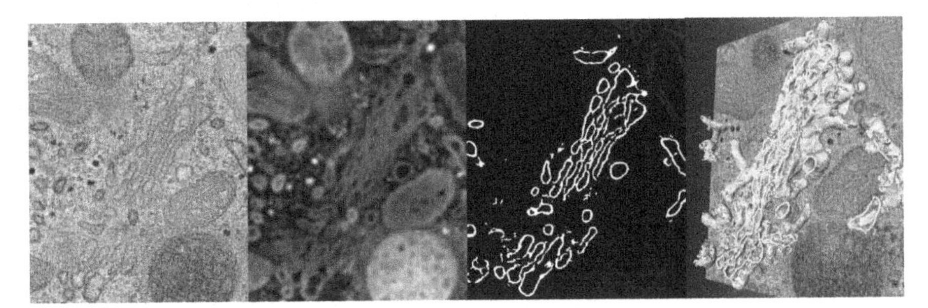

Fig. 4 The procedure applied to a tomogram containing Golgi apparatus. From left to right, a slice of the original tomogram, at scale $\sigma = 2.2$ and reversed contrast, result of the hysteresis thresholding, and 3D visualization of the membranes detected after global analysis.

Finally, Fig. 5 shows a gallery of the membranes detected by the algorithm when applied to tomograms of different specimens, namely vesicles, mitochondrion and chloroplast, respectively. The algorithm was run at a scale σ of 2, 1.5 and 0.1, respectively. As shown, all the membranes present in the tomograms were clearly identified and come out of the background. The datasets shown in Figs. 4 and 5 were taken from the CCDB (Cell-Centered DataBase, http://ccdb.ucsd.edu). They all have a better SNR than the one with the Vaccinia virus (Fig. 3).

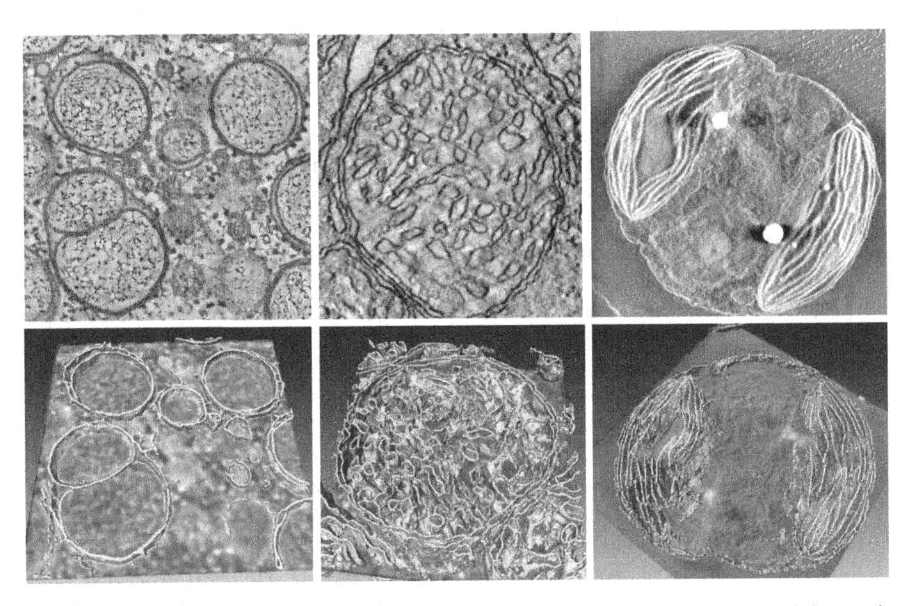

Fig. 5 Membranes detected by the algorithm for three different tomograms containing vesicles, mitochondrion and chloroplast, respectively. Top: a slice of the original tomogram is shown. Bottom: 3D visualization of the membranes.

4 Conclusion

We have presented a procedure to detect membranes in tomograms. It relies on a simple local membrane model and the local differential structure to determine points whose neighbourhood resembles plane-like features. Later stages of the algorithm then intend to definitely determine which of those points do actually constitute the membranes. The performance of algorithm has been shown on a set of representative tomograms. In general, the algorithm has turned out to be effective to detect membranous structures. Therefore, it has potential to be a useful tool for interpretation in electron tomography.

Acknowledgements. Grants MCI-TIN2008-01117 and JA-P10-TIC6002, in part financed by the European Reg. Dev. Fund (ERDF). A.M.S. is a fellow of the Spanish FPI programme.

References

1. Cyrklaff, M., Risco, C., Fernandez, J.J., Jimenez, M.V., Esteban, M., Baumeister, W., Carrascosa, J.L.: Cryo-electron tomography of vaccinia virus. Proc. Natl. Acad. Sci. USA 102, 2772–2777 (2005)
2. Fernandez, J.J., Li, S.: An improved algorithm for anisotropic diffusion for denoising tomograms. J. Struct. Biol. 144, 152–161 (2003)
3. Fernandez, J.J., Li, S.: Anisotropic nonlinear filtering of cellular structures in cryoelectron tomography. Comput. Sci. Eng. 7(5), 54–61 (2005)
4. Florack, L.J., Romeny, B.H., Koenderink, J.J., Viergever, M.A.: Scale and the differential structure of images. Image and Vision Computing 10, 376–388 (1992)
5. Frangi, A.F., Niessen, W.J., Vincken, K.L., Viergever, M.A.: Multiscale Vessel Enhancement Filtering. In: Wells, W.M., Colchester, A.C.F., Delp, S.L. (eds.) MICCAI 1998. LNCS, vol. 1496, pp. 130–137. Springer, Heidelberg (1998)
6. Frank, J. (ed.): Electron tomography. Springer, Heidelberg (2006)
7. Koenderink, J.J.: The structure of images. Biol. Cybern. 50, 363–370 (1984)
8. Lebbink, M.N., Geerts, W.J., Krift, T.P., Bouwhuis, M., Hertzberger, L.O., Verkleij, A.J., Koster, A.J.: Template matching as a tool for annotation of tomograms of stained biological structures. J. Struct. Biol. 158, 327–335 (2007)
9. Lindeberg, T.: Scale-space for discrete signals. IEEE Trans. PAMI 12(3), 234–254 (1990)
10. Lindeberg, T.: Edge detection and ridge detection with automatic scale selection. Int. J. Computer Vision 30, 117–154 (1998)
11. Lucic, V., Forster, F., Baumeister, W.: Structural studies by electron tomography: from cells to molecules. Ann. Rev. Biochem. 74, 833–865 (2005)
12. Martinez-Sanchez, A., Garcia, I., Fernandez, J.J.: A differential structure approach to membrane segmentation in electron tomography. J. Struct. Biol. 175, 372–383 (2011)
13. Moussavi, F., Heitz, G., Amat, F., Comolli, L.R., Koller, D., Horowitz, M.A.: 3D segmentation of cell boundaries from whole cell cryogenic electron tomography volumes. J. Struct. Biol. 170, 134–145 (2010)
14. Nguyen, H., Ji, Q.: Shape-driven three-dimensional watersnake segmentation of biological membranes in electron tomography. IEEE Trans. Med. Imaging 27, 616–628 (2008)

15. Sandberg, K.: Methods for image segmentation in cellular tomography. Methods in Cell Biology 79, 769–798 (2007)
16. Sandberg, K., Brega, M.: Segmentation of thin structures in electron micrographs using orientation fields. J. Struct. Biol. 157, 403–415 (2007)
17. Volkmann, N.: Methods for segmentation and interpretation of electron tomographic reconstructions. Methods Enzymol. 483, 31–46 (2010)
18. Witkin, A.P.: Scale-space filtering. In: Proc. 8th Intl. Conf. Artif. Intell., pp. 1019–1022 (1983)

A Cellular Automaton Model for Tumor Growth Simulation

Ángel Monteagudo and José Santos

Abstract. We used cellular automata for simulating tumor growth in a multicellular system. Cells have a genome associated with different cancer hallmarks, indicating if those are activated as consequence of mutations. The presence of the cancer hallmarks defines cell states and cell mitotic behaviors. These hallmarks are associated with a series of parameters, and depending on their values and the activation of the hallmarks in each of the cells, the system can evolve to different dynamics. We focus here on how the cellular automata simulating tool can provide a model of the tumor growth behavior in different conditions.

1 Introduction and Previous Work

Cancer is a genetic disease, which arises from mutations in single somatic cells. These mutations alter the proliferation control of the cells which leads to uncontrolled cell division. The transformed cells form a neoplastic lesion that may be invasive (carcinoma) or benign (adenoma). The decisive factor between these two behaviors is the growth rate and invasiveness of the cells in the neoplasm. These two properties are in turn driven by what mutations the cells have acquired. In the invasive case the tumor grows in an uncontrolled manner up to a size of approximately 10^6 cells. At this size the diffusion driven nutrient supply of the tumor becomes insufficient and the tumor must initiate new capillary growth (angiogenesis). When the tumor has been vascularized the tumor can grow further and at this stage metastases are often observed.

Hanahan and Weinberg described the phenotypic differences between healthy and cancer cells in an article entitled "The Hallmarks of Cancer" [8]. The six essential alterations in cell physiology that collectively dictate malignant growth are: self-sufficiency in growth signals, insensitivity to growth-inhibitory (antigrowth) signals,

Ángel Monteagudo · José Santos
Computer Science Department, University of A Coruña (Spain)
e-mail: {jose.santos}@udc.es

M.P. Rocha et al. (Eds.): 6th International Conference on PACBB, AISC 154, pp. 147–155.
springerlink.com

evasion of programmed cell death (apoptosis), limitless replicative potential, sustained angiogenesis, and tissue invasion and metastasis.

Tumor growth in cellular systems is an example of emergent behavior, which is present in systems whose elements interact locally, providing global behavior which is not possible to explain from the behavior of a single element, but rather from the "emergent" consequence among the interactions of the group [2]. In this case, it is an emergent consequence of the local interactions between the cells and their environment. Emergent behavior was studied using models like Cellular Automata (CA) and Lyndenmayer Systems [9][10]. CAs have been the focus of attention because of their ability to generate a rich spectrum of complex behavior patterns out of sets of relatively simple underlying rules and they appeared to capture many essential features of complex self-organizing cooperative behavior observed in real systems [9]. In this line, we used CAs to model cell behavior, which will depend on several of the main cancer hallmarks, which define the cell state and cell mitotic behavior.

The traditional approach to model cancer growth was the use of differential equations to describe avascular, and indeed vascular, tumor growth. As Patel and Nagl [11] state, the use of differential equations automatically assumes that the current state of the system is a consequence of the previous global state of the system. In reality, a single cell will behave according to its immediate locality, not according to the state of the tumor as a whole, and local environments are differing. This is not the case of the CA approaches, where the state of each cell is described by its local environment.

Along this line, Abbott et al. [1] investigated the dynamics and interactions of the hallmarks in a CA model the authors called *CancerSim*. The main interest of the authors with their simulation was to describe the likely sequences of precancerous mutations or pathways that end in tumorigenesis. They were interested in the relative frequency of different mutational pathways (what sequences of mutations are most likely), how long the different pathways take, and the dependence of pathways on various parameters associated with the hallmarks.

Adamopoulos [3] developed a simple cellular automaton model, where the simulations of cancer growth were based on a 2-dimensional probabilistic CA with fixed lattice structure. The cells could be in 3 different basic states (normal, immature cancer, matured cancer), being the mechanism of local interactions of the CA (the rules of the CA) established by the designer. As indicated by the author, the results of the simulations were in good agreement with both in vitro experiments of cell cultures and statistical models of cancer growth. Bankhead and Heckendorn [4] used a CA which incorporated a simplified genetic regulatory network simulation to control cell behavior and predict cancer etiology. Their simulation used known histological morphology, cell types, and stochastic behavior to specifically model ductal carcinoma in situ (DCIS), a common form of non-invasive breast cancer.

Ribba et al. [12] used a Hybrid Cellular Automaton (HCA), "hybrid" because the automaton combines discrete and continuous fields, as it incorporates nutrient and drug spatial distribution together with a simple simulation of the vascular system in a 2-dimensional lattice model. The authors presented an application of the model for assessing chemotheraphy treatment for non-Hodgkin's lymphoma (NHL).

In the CA model of Gerlee and Anderson [5] each cell was equipped with a micro-environment response network that determined the behavior of the cell based on the local environment. The response network was modeled using a feed-forward neural network, which was subject to mutations when the cells divided. This implies that cells might react differently to the environment and when space and nutrients are limited only the fittest cells survive. For example, for low oxygen concentration, the authors observed tumors with a fingered morphology, while increasing the matrix density gave rise to more compact tumors with wider fingers.

Gevertz et al. [6] adapted a previous CA model developed by the authors (designed to simulate spherically symmetric tumor growth) to study the impact that organ-imposed physical confinement and heterogeneity have on tumor growth, that is, to incorporate the effects of tissue shape and structure. The results of the authors indicate that the impact is more pronounced when a neoplasm is growing close to, versus far from, the confining boundary.

As indicated by Gerlee and Anderson [5], tumor invasion has successfully been modeled by both continuous and discrete mathematical approaches, but most of such models have failed to capture the evolutionary dynamics of tumor growth, and thus neglected a very important aspect of carcinogenesis. In our proposal the CA rules will be the designed to model the mitotic and apoptotic behaviors in each cell from the information of the cell state and from its surrounding environment. We will define the rules from the main behavior of the cells when the main hallmarks are present in the cells. Opposite to Abbott et al.'s work [1], focused on possible pathways or sequences of hallmark mutations that end in tumorigenesis, our aim is different, as our simulation tries to determine the consequences of the key parameters on the multicellular system behavior, which have implications for cell population dynamics that are difficult to foresee without a model and associated simulating tool.

2 Cellular System Modeling

2.1 Hallmark Definitions

In the simulation each cell resides in a point in a cubic lattice and has a "genome" associated with different cancer hallmarks. The essential alterations in cell physiology that collectively dictate malignant growth are [7][8]:

H1. Growth even in the absence of normal "go" signals: Most normal cells wait for an external message (growth signals from other cells) before dividing. Cancer cells often counterfeit their own pro-growth messages.

H2. Growth despite antigrowth signals issued by neighboring cells: As the tumor expands, it squeezes adjacent tissue, which sends out chemical messages that would normally bring cell division to a halt. Malignant cells ignore the commands.

H3. Evasion of programmed cell death (apoptosis): In healthy cells, genetic damage above a critical level usually activates a suicide program. Cancer cells bypass this mechanism.

Table 1 Definition of the parameters associated with the hallmarks

Parameter name	*Default value*	*Description*
Telomere length (t)	100	Initial telomere length in each cell. Every time a cell divides, there is a shortening of one unit of the length. When it reaches 0, the cell dies, unless the "Effective immortality" hallmark (H5) is ON.
Evade apoptosis (e)	10	A cell with n hallmarks mutated has an extra n/e likelihood of dying each cell cycle, unless the "Evade apoptosis" hallmark (H3) is ON.
Base mutation rate (m)	100000	Each gene (hallmark) is mutated (when the cell divides) with a $1/m$ chance of mutation.
Genetic instability (i)	100	There is an increase of the base mutation rate by a factor of i for cells with this mutation (H7).
Ignore growth inhibit (g)	10	Cells with the hallmark "Ignore growth inhibit" (H2) activated have a probability $1/g$ of killing off a neighbor to make room for mitosis.
Random apoptosis (a)	1000	In each cell cycle every cell has a $1/a$ chance of death from several causes.

H4. Ability to stimulate blood vessel construction (angiogenesis): Tumors need oxygen and nutrients to survive. They obtain them by co-opting nearby blood vessels to form new branches that run throughout the growing mass.

H5. Effective immortality: Healthy cells can divide no more than several times (< 100). The limited replicative potential arises from the inability of DNA polymerase to replicate chromosomes completely. With the duplication there is a loss of base pairs in the telomeres (chromosomes ends which protect the bases), so when the DNA is unprotected, the cell dies. Malignant cells overproduce the telomerase enzyme, appending molecules to the telomeres, so they overcome the reproductive limit.

H6. Power to invade other tissues and spread to other organs: Cancers usually become life-threatening only after they somehow disable the cellular circuitry that confines them to a specific part of the particular organ in which they arose. New growths appear and eventually interfere with vital systems.

In our modeling, each cell genome indicates if any of these hallmarks are activated as consequence of mutations. Additionally, we considered a "*Genetic instability*" hallmark or factor (H7), that accounts for the high incidence of mutations in cancer cells, allowing rapid accumulation of genetic damage. The simulation implies that the cells with this factor will increase their mutation rate. Depending on the activation of the hallmarks in each of the cells, the system can evolve to different dynamics.

Algorithm 2.1: EVENT MODEL FOR CANCER SIMULATION()

```
t ← 0   // Simulation time. Initial cell at the center of the grid.
SCHEDULE MITOTIC EVENT(5, 10) // Schedule mitotic event with a random time
       // (ts:) between 5 and 10 time instants in the future (t+ts). The events
       // are stored in an event queue. The events are ordered on event time.
while event in the event queue
   do {
      POP EVENT( ) // Pop event with the highest priority (the nearest in time).
      t ← t + ts of popped event
      RANDOM APOPTOSIS TEST( ) // The cell can die with a given probability.
      GENETIC DAMAGE TEST( ) // The larger the number of hallmark mutations,
                       // the greater the probability of cell death. If
                     // "Evade apoptosis" (H3) is ON, death is not applied.
      MITOSIS TESTS( ) :
        TISSUE EXTENT CHECKING( ) // A cell must have a number of
          // neighbors greater than a parameter (2), which simulates the growth
          // factor, to be able to perform a division. If the hallmark "Self-
          // growth" (H1) is mutated (ON), this checking is not applied.
        IGNORE GROWTH INHIBIT CHECKING( ) // If there are not empty cells in
          // the neighborhood, the cell cannot perform a mitotic division. If the
          // "Ignore growth inhibit" hallmark (H2) is ON, then the cell can perform
          // the division with a given probability, killing one of the neighbor cells.
        LIMITLESS REPLICATIVE POTENTIAL CHECKING( )// If the telomere length
          // is 0, the cell dies, unless the hallmark "Effective immortality"
          // (Limitless replicative potential, H5) is ON.
      if the three tests indicate possibility of mitosis
        then PUSP EVENTS( ) // If these tests determine a mitotic division,
          // then schedule mitotic events (push in event queue) for both cells:
          // Mother and daughter, with the random times in the future.
          // Add mutations to the daughter cell according the base mutation rate (1/m)
          // Increase the base mutation rate if genetic instability (H7) is ON.
          // Decrease telomere length in both cells.
        else PUSP EVENT( ) // Schedule mitotic event (in queue) for mother cell
   }
```

2.2 Event Model

Every cell has its genome which consists in five hallmarks plus some parameter particular to each cell. Metastasis and angiogenesis are not considered (hallmarks H6 and H4), as we are interested in this work in the first avascular phases of tumorigenesis. All the parameters are commented in Table 1. The parameters *telomere length* and *base mutation rate* can change their values in a particular cell over time, as explained in the table. The cell's genome is inherited to the daughter cells when a mitotic division occurs.

In the simulation of the cell life cycle, most elements do not change observably each time step. The only observable changes to cells are apoptosis and mitosis. In an actual tissue, only a fraction of all cells are undergoing such transitions at any given time. We used an event model, similar to the one used by Abbott et al. [1], where a mitosis event is scheduled several times in the future. Such time is a random variable distributed uniformly between 5 and 10 time steps, simulating the variable duration of the cell life cycle.

The simulation begins by initializing all elements of the grid to represent empty space. Then, the element at the center of the grid is changed to represent a single normal cell (no mutations). Mitosis is scheduled for this initial cell, and this mitotic division is the next event. After the new daughter cells are created, mitosis is scheduled for each of them, and so on. Each mitotic division is carried out by copying the genetic information (the hallmark status and associated parameters) of the cell to an unoccupied adjacent space in the grid. Random errors occur in this copying process, so some hallmarks can be activated, taking into account that once a hallmark is activated in a cell, it will be never repaired by another mutation. Frequently, cells are unable to replicate because of some limitation, such as contact inhibition or insufficient growth signal. Cells overcome these limitations through mutations in the hallmarks.

Moreover, cells undergo random apoptosis with low probability. On each cell cycle, each cell is subjected to a $1/a$ chance of death, where a is a tunable parameter. This might be due to mechanical, chemical or radiological damage, aging, or the immune system. Algorithm 2.1 summarizes the simulation, which takes into account the main aspects of the cell cycle from the application point of view.

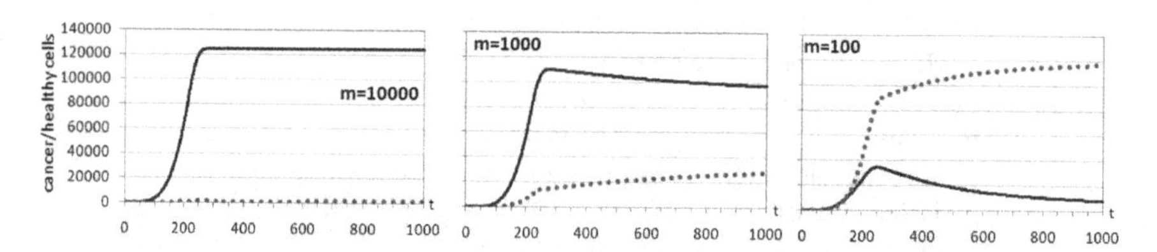

Fig. 1 Evolution through time iterations of the number of healthy cells (continuous lines) and cancer cells (dashed lines) for different base mutation rates ($1/m$).

3 Results

We show here the capabilities of the simulation tool with some representative examples. First, Figure 1 shows the evolution over time of the number of healthy and cancer cells for different values of the parameter m, which defines the base mutation rate, maintaining the rest of the parameters in their default values and using the same grid size (125000) employed in [1]. The number of time iterations was 1000 in the different runs. Given the stochastic nature of the problem, the graphs are always an average of 5 different runs. A cell was considered as cancerous if any of the hallmarks was present. As expected, with increasing values of the base mutation rate ($1/m$), the faster is the increase of cancer cells. These tests serve to check the difficulty of appearance of a cancer growth with such default parameters, requiring high values of the base mutation rate. Note that most of the m values imply a fixed dynamics in terms of complex systems, as the dynamics ends with most of the cells in healthy or cancerous states.

In Figure 2 we repeated the simulations but using a parameter set that facilitates the appearance of cancer cells. The values chosen ($m = 100000$, $t = 40$, $e = 20$, $i = 100$, $g = 4$, $a = 400$ and a grid size of 125000) were the same used by Abbott et al. [1] for the determination of possible mutational pathways, that is, the sequence of appearance of hallmarks that end in a tumor growth. For example, the lower value of t implies fewer mitoses in healthy cells, and the lower value of a facilitates that more vacant sites are available for cancer cells to propagate, in connection with the higher probability of killing neighbors to make room for mitosis (lower value of g). The right part of the Figure shows snapshots at different time states of the multicellular system in a run with such parameters. In this case, we used a grid size of 10^6, for a better visualization of the tumor progression. The cancer cells are shown in different colors (depending on which hallmarks are activated) and the healthy cells in gray color. The dominant hallmark in such tumor growth is $H5$ (Effective immortality, green color cells), allowing the progression of the cells with such mutation even when the telomere length reaches its limit, which represents and advantage with respect to healthy (non-mutated) cells.

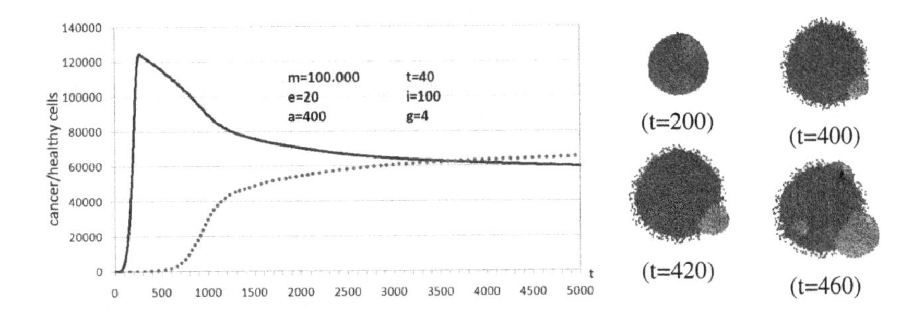

Fig. 2 Time evolution of the number of healthy cells (continuous line) and cancer cells (dashed line) with a parameter set which facilitates cancer growth (grid size=125000). The right part shows snapshots of the cellular system at different time steps using such parameter set (grid size=10^6).

One of our aims is to inspect the behavior in possible transitions from states with predominance of one type of cells. Within this objective, Figure 3 shows a test of the decrease of cancer cells when a possible target acts against such cancer cells, killing them. The x axis indicates the probability of eliminating cancer cells in each iteration or simulation time, whereas the y axis indicates the number of final cancer cells. Fig. 3a and Fig. 3b include two tests using two values of the parameter m, whereas the rest of parameters were set to their default values. In both cases, the Figures show the final number of cancer cells after 1000 iterations. As the Figures denote, it is easy to maintain the number of cancer cells to a minimum. In fact, there is a rapid decrease in the number of cancer cells (with probabilities < 0.2), which is similar to critical phase transitions of many emergent and complex systems [13], which present sudden shifts in behavior arising from small changes in circumstances.

We repeated the experiment with the parameter set used in Figure 2, which facilitates the appearance of cancer cells. Figure 3c shows again the final number of cancer cells after the simulation run, taking into account that the cancer cells in each simulation iteration are eliminated with a given probability (x axis value). In this case the y axis indicates the number of final cancer cells after 3000 time iterations, since with this parameter set more iterations are needed to obtain a stable number in both types of cells (as Figure 2 denotes). With this parameter set, the decrease is smoother with respect to the previous cases, indicating the difficulty to maintain the cancer cells under a low threshold when the conditions make easier their appearance, even with this simulation that only kills the cancer cells.

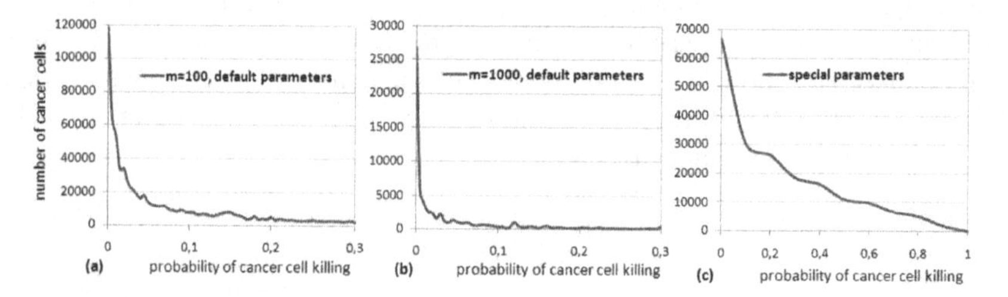

Fig. 3 Effect of killing cancer cells during tumor growth for different killing probabilities and using three parameter sets.

4 Conclusions

We used the CA tool in this bioinformatics application not only as a method of simulation of the real dynamic process, but also as a model that will lead to a better understanding and characterization of the underlying biological processes involved.

Because the hallmarks specify cell behavior, they have implications for cell population dynamics that are difficult to foresee without a simulation model. Our focus was on the dependences of the first phases of cancer growth on the hallmark parameters, showing how the simulation can determine the emergent behavior of the cancer growth dynamics in different conditions and anti-cancer scenarios. Along this line, when we consider the elimination of cancer cells during the tumor progression, the simulation tool shows the greater difficulty to maintain the cancer cells under a low threshold when the conditions or hallmark parameters make easier their appearance.

Acknowledgments. This paper has been funded by the Ministry of Science and Innovation of Spain (project TIN2011-27294).

References

1. Abbott, R.G., Forrest, S., Pienta, K.J.: Simulating the hallmarks of cancer. Artificial Life 12(4), 617–634 (2006)
2. Adami, C.: Introduction to artificial life. Telos-Springer Verlag (1998)

3. Adamopoulos, A.V.: CA2: In silico simulations of cancer growth using cellular automata. WSEAS Transactions on Systems 4(9), 1467–1473 (2005)
4. Bankhead, A., Heckendorn, R.B.: Using evolvable genetic cellular automata to model breast cancer. Genet. Program Evolvable Mach. 8, 381–393 (2007)
5. Gerlee, P., Anderson, A.R.A.: An evolutionary hybrid cellular automaton model of solid tumour growth. Journal of Theoretical Biology 246(4), 583–603 (2007)
6. Gevertz, J.L., Gillies, G.T., Torquato, S.: Simulating tumor growth in confined heterogeneous environments. Phys. Biol. 5 (2008)
7. Gibbs, W.W.: Untangling the roots of cancer. Scientific American 289, 56–65 (2003)
8. Hanahan, D., Weinberg, R.A.: The hallmarks of cancer. Cell 100, 57–70 (2000)
9. Ilachinski, A.: Cellular automata. A discrete universe. World Scientific (2001)
10. Langton, C.G.: Life at the edge of chaos. In: Langton, C.G., Taylor, C., Farmer, J.D., Rasmussen, S. (eds.) Artificial Life II, pp. 41–49. Addison-Wesley (1992)
11. Patel, M., Nagl, S.: The role of model integration in complex systems. An example from cancer biology. Springer, Heidelberg (2010)
12. Ribba, B., Alarcón, T., Marron, K., Maini, P.K., Agur, Z.: The Use of Hybrid Cellular Automaton Models for Improving Cancer Therapy. In: Sloot, P.M.A., Chopard, B., Hoekstra, A.G. (eds.) ACRI 2004. LNCS, vol. 3305, pp. 444–453. Springer, Heidelberg (2004)
13. Solé, R., Goodwin, B.: Signs of life. How complexity pervades biology. Basic Books (2000)

Ectopic Foci Study on the Crest Terminalis in 3D Computer Model of Human Atrial

Carlos A. Ruiz-Villa, Andrés P. Castaño, Andrés Castillo, and Elvio Heidenreich

Abstract. Atrial fibrillation (AF) is the most common arrhythmia in clinical practice. Epidemiological studies show that AF tends to persist over time, creating electrophysiological and anatomical changes called remodeled atrial. It has been shown that these changes result in variations in conduction velocity (CV) in the atrial tissue. The changes caused by electrical remodeling in a model of action potential (AP) of atrial myocytes have been incorpotated in this study, coupled with an anatomically realistic three-dimensional model of human dilated atrium. Simulations of the spread of AP in terms of anatomical and electrical remodeling and remodeling of gap junctions were measured vulnerable windows of reentry generation on the crest terminalis of the atrium. The results obtained indicate that

---

Carlos A. Ruiz-Villa · Andrés P. Castaño
Center for research and innovation in engineering, University of Caldas, Street 65 # 26-10, Manizales, Colombia
e-mail: {carv,andres.castano}@ucaldas.edu.co

Andrés Castillo
Language, Informatics Systems and Software Engineer Department Pontifical University of Salamanca, Campus Madrid, Paseo Juan XXIII, 3, Madrid, Spain
e-mail: andres.castillo@upsam.net

Elvio Heidenreich
Engineering Faculty, University of Lomas of Zamora, Ruta Provincial 4, Km 2, Buenos Aires, Argentina
e-mail: elvioh@gmail.com

Carlos A. Ruiz-Villa
Informatic and Computation Department, National University of Colombia, Campus La Nubia, Manizales, Colombia
e-mail: caruizvi@unal.edu.co

M.P. Rocha et al. (Eds.): 6th International Conference on PACBB, AISC 154, pp. 157–164.
springerlink.com

vulnerable window in the remodeling of gap junctions shifted 38 ms with respect to the model dilated, which shows the impact of structural remodeling Several types of permanent reentry of figures in form of eight and in form of rotor, favored by the underlying anatomy of the atrium were obtained.

Keywords: Atrial fibrillation, electrical remodeling, action potential, cresta terminalis, reentry, 3D model.

1 Introduction

Atrial fibrillation (AF) is the most common of sustained cardiac arrhythmias, and it contributes greatly to cardiovascular morbidity and mortality. It is also the most frequent supraventricular arrhythmia, with an increased prevalence of 0.5% in people over 50 years, and almost 10% in people over 80 years[1]. Thus it has a tendency to become chronic and resistant to conversion to sinus rhythm with increased duration of illness [2,3]. Although the onset of atrial fibrillation has been intensively investigated [4-6], theories that try to determine the onset of AF did not explain the persistent tendency of the arrhythmia. From this point of view, current research has focused on electrophysiological remodeling in chronic AF. The AF causes shortening of the refractory period due to specific reductions in current densities [7,8], the sinus node dysfunction and reduced speed intra-atrial conduction [7]. Various mechanisms such as rapid ectopic focal activity, a single reentry circuit, or multiple reentry circuits can cause AF. The reentry need a suitable substrate in which the firing of an ectopic focus is sufficient to initiate reentry. Therefore it is necessary to determine the anatomical regions in which these sources are more vulnerable to reentry and thus identify both the most critical anatomical regions under pathological conditions facilitate atrial fibrillation and the morphology of the reentrant found. Experimental studies establish that the crest terminalis region is prone to the occurrence of reentrant events caused by ectopic foci in terms of electrical and structural remodeling[9].

The geometry, anisotropy and fiber orientation of atrial tissue plays an important role during the spread of AP, therefore anatomically realistic deployment models can accurately reproduce the electrical behavior of the tissue, both in normal physiological condition as remodeling, allowing better analysis for understanding cardiac pathologies. However, the complex atrial anatomy and the high computational costs caused most of the studies on the spread of AP in human atrial models of a simplified geometry to have been implemented[10,11]. Our group has been developing 3D models improved[12], that take into account anatomical structures such as large beams and structural remodeling.

The aim of this paper is to study the behavior and appearance of ectopic foci in the region of the crest terminalis in a highly realistic model of a dilated human atrial.

2 Methods

2.1 3D Model of Human Atrial

The overall anatomy was developed in previous studies [13]. The model was fitted to the geometric specifications [14] and [15], thus, includes among others the left and right atria, the union interatrial, the pectineus, the fossa ovalis and its ring, the bundle of Bachmann, terminal crest, left and right appendages, the coronary sinus, the union interatrial of the coronary sinus and holes to the right and left pulmonary veins and for upper and lower vena cava, and tricuspid and mitral valves. Additionally, we defined an area close to the superior vena cava which defined the sinoatrial node.

The model consists of 52,906 hexahedral elements and 100,554 nodes. Areas of high (crest terminalis, Bachmann's bundle and ring the oval fossa), low (right atrial isthmus and fossa ovalis) and half (the remaining atrial tissue) conductivity were identified. The conductivities were adjusted to obtain conduction velocities consistent with experimental data [16,17]. The values assigned to the conductivities were 0.25, 0.40 and 0.10 S / cm for medium, high and low conductivity zones respectively.

The model is divided into 42 areas, in order to assign a realistic fiber direction on the basis of histological data from human atrial [18]. The longitudinal direction of the fiber is considered parallel to the direction of the large bundles.

2.2 Atrial Remodeling

Experimental studies conducted by [19] and [20] have shown that episodes of AF induce changes in conductance and kinetics of some ionic channels of atrial myocytes. These changes were incorporated into a model of AP human atrial cell, developed by Nygren [21], to reproduce the electrical behavior of atrial tissue of patients with chronic AF.

The parameters adjusted in the model were: Ik1 channel conductance increased by 250%, ICal was decreased by 74%, Ito was decreased by 85%, rapid inactivation kinetics of ICa was decreased by 62%, the activation curve of Ito was shifted +16 mV and INa inactivation curve was shifted +1.6 mV in the depolarizing direction.

Some studies show the importance of dilation in determining the propagation properties of transverse and changes in gap junctions [22]. The results suggest that in a pathological arrhythmogenic substrate, the size of the cell during the remodeling of gap junctions is important in maintaining a maximum rate of depolarization. Thus it is reasonable to model the remodeling effect of gap junctions in view of the atrial dilatation, atrial remodeling, tissue conductivity and anisotropy. In this work we assumed a simplified gap junction remodeling considering: Situations of AF, electrical and structured (dilation) remodeling, a reduction of the conductivity of tissue by 40% [23], a anisotropic ratio of 10:1 in the crest terminalis [17] and reduced conductivity in the SAN [24] and [25].

2.3 Action Potential Propagation and Stimulation Protocol

For the calculation of propagation of the AP a single-domain solution approach is considered. AP propagation extended to three-dimensional tissue, is described as

$$\nabla \cdot (D_i \nabla V_m) = C_m \frac{dV_m}{dt} + I_{ion} \qquad \text{in } \Omega \qquad (1),$$

Where Vm is the potential in the intracellular space, Di is the anisotropic conductivity tensor, Cm capacitance of the medium and Iion is the set or currents that describe the ionic state cells in the tissue as a function of time and ionic concentrations. Extracellular space is assumed with infinite resistance.

With the following boundary conditions:

$$(D_i \nabla V_m) \cdot n = 0 \qquad in\ \Gamma_n \qquad (2),$$

Where n is the normal vector to surface.

To solve the equation (1) of diffusion reaction, a code parallel using the finite element method (FEM) was implemented. In this discretization a system of linear equations with nonlinear reactive term represented by Iion appears. The term reactive is explicitly resolved while the temporal equation is solved implicitly. The system to be solved at each time step has a number of unknowns that is the number of nodes (100554).

In order to simulate sinus rhythm of the heart, a train of 10 pulses in the region of the sinoatrial node was applied. The basic cycle length (BCL) was 300 ms. An ectopic focus was modeled by a suprathreshold pulse applied to a small group of cells in the vicinity of the inferior vena cava and the region of the crest terminalis. The focal activity was applied during the repolarization phase of the tenth sinus rhythm.

The vulnerable window for reentry corresponds to the time frame during which you can find at least one re-entry and is measured in ms. The initial time window is obtained by applying stimuli following the protocol established in an intermediate zone of the crest terminalis near the inferior vena cava, to find the first generation of reentry. The process is followed by observation until more reentry is not generated; this time defines the upper limit of the window.

3 Results and Discussion

We approach the study of anatomical reentry induced by applying an ectopic focus near the inferior vena cava and in the region of the crest terminalis, following established protocol. As shown in Figure 1, The crest terminalis region, noted for being the one with the largest vulnerable window in all the models studied. For the model of dilated atria and simplified remodeled of gap junctions, the vulnerable window was 46% higher in relation to the atrium dilated and 58% in relation to atrial electrical remodeling just being the vulnerable window of 60 ms defined between coupling intervals of 82 ms and 141 ms.

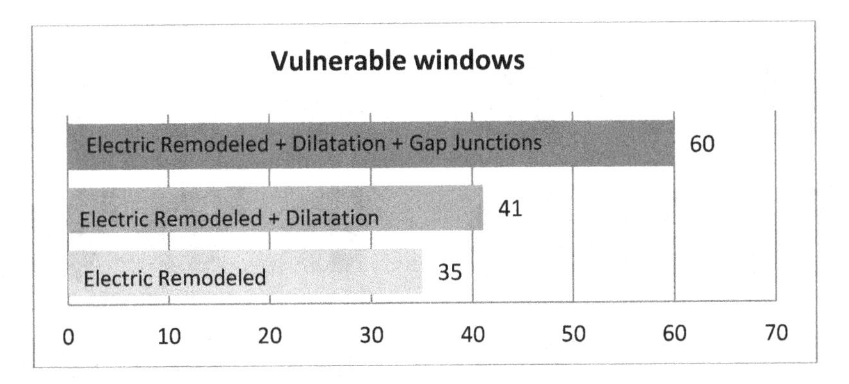

Fig. 1 Vulnerability to reentry with focus on near the inferior vena cava and in the region of the crest terminalis in the different models to study human´s atrial.

Across the range of the vulnerable window, the wave fronts generated by the unidirectional block formed reentry. In this region the focal activity is characterized by the initial appearance of functional reentry in figure eight to which then is followed by a rotor in the free wall of the right atrium whose origin moves. The front moves in the direction of the superior vena cava and the ends rotate in the direction of its own tail refractory, against each other forming a figure eight and generating functional reentry, which propagates back into the superior vena cava, turn right and creating at this time at one end of a rotor spreads through the free wall of the right atrium, in turn generating another reentry. Reentrant activity is maintained indefinitely. To illustrate this situation we observe in Figure 2, here is the sequence of activation caused by a focus fired at 85 ms coupling interval in the lower region of the crest terminalis. At 105 ms the front generated by the crest terminalis is directed towards the SAN. At 165 ms two higher speed branches are formed, approaching each other by the refractory tail that generates the spread. At 205 ms the two fronts collide and generate a reentry forming a figure eight. At 365 ms fronts collide again in the beam region of the right atrium intercaval and a new front reentrant is generated to 385 ms. At the same time a front passes through the interatrial bundles into the left atrium spread again by the septum and reappears at the 445 ms. At 485 ms the front thus formed constitutes a new reentry. The process is repeated indefinitely observed the formation of rotors, figure-eight figures and macro-reentry. It was observed a 70% decrease in action potential duration and a decrease in conduction velocities between 14. 6 and 26%, measures in different regions of the atrium dilated.

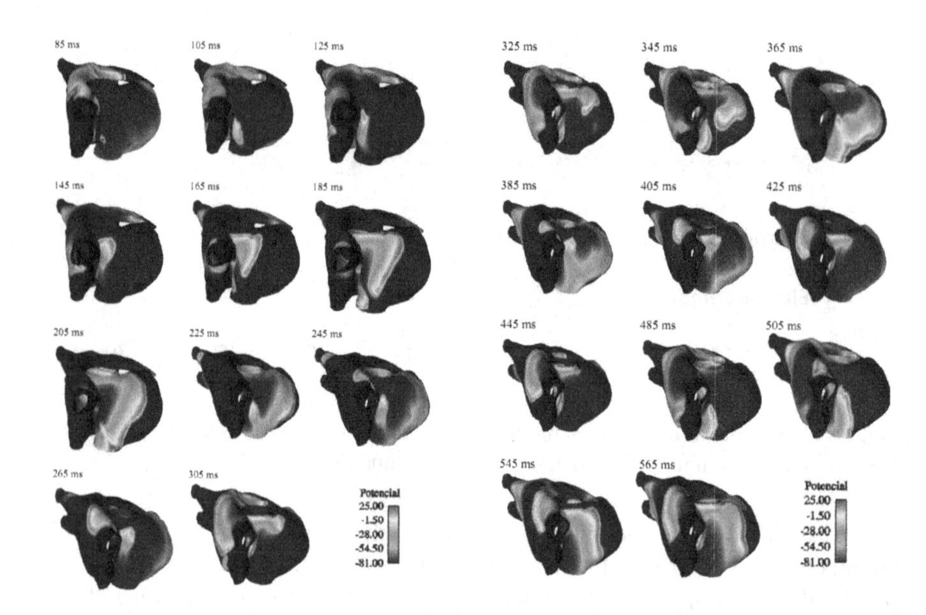

Fig. 2 Action potencial propagation as result of a focus on the crest terminalis region for the atrium dilated with remodeled gap junctions (Potential in mv).

4 Conclusions

By implementing the 3D model highly realistic human atrial, the condition atrial electrical remodeling caused by chronic AF episodes was reproduced.

The electrical remodeling produced a 70% decrease in action potential duration and decreased conduction velocities between 14.6 and 26%, measures in different regions of the atrium dilated. As shown in Figure 1, increases vulnerability to re-entry in dilated atrial and is even higher in dilated atrial with remodeling of gap junctions.

The focus shot in the neighborhood of the inferior vena cava on the crest terminalis, wavefront generated a regular activity reentrant with the appearance of rotors, and figures in eight form who shifted their center and kept indefinitely their activity.

Clearly the impact of the underlying anatomy of the region to facilitate the re-entry process as in the free wall of the right atrium and pulmonary veins into the left atrium.

The values obtained are qualitatively similar to those obtained in other experimental studies and simulation.

Acknowledgments. Finally a special acknowledgement for the support of the Institute for Research and Innovation in Bioengineering from the Polytechnic University of Valencia, Spain and the group of structures and materials modeling at the University of Zaragoza, Spain.

References

1. Nattel, S.: New ideas about atrial fibrillation 50 years on. Nature 415(6868), 219–226 (2002), doi:10.1038/415219a415219a
2. Suttorp, M.J., Kingma, J.H., Jessurun, E.R., Lie, A.H.L., van Hemel, N.M., Lie, K.I.: The value of class IC antiarrhythmic drugs for acute conversion of paroxysmal atrial fibrillation or flutter to sinus rhythm. J. Am. Coll. Cardiol. 16(7), 1722–1727 (1990); doi:0735-1097(90)90326-K
3. Crijns, H.J., Van Gelder, I.C., Van Gilst, W.H., Hillege, H., Gosselink, A.M., Lie, K.I.: Serial antiarrhythmic drug treatment to maintain sinus rhythm after electrical cardioversion for chronic atrial fibrillation or atrial flutter. Am. J. Cardiol. 68(4), 335–341 (1991)
4. Allessie, M., Ausma, J., Schotten, U.: Electrical, contractile and structural remodeling during atrial fibrillation. Cardiovasc. Res. 54(2), 230–246 (2002); doi:S0008636302002584
5. Kirchhof, C., Chorro, F., Scheffer, G.J., Brugada, J., Konings, K., Zetelaki, Z., Allessie, M.: Regional entrainment of atrial fibrillation studied by high-resolution mapping in open-chest dogs. Circulation 88(2), 736–749 (1993)
6. Konings, K.T., Kirchhof, C.J., Smeets, J.R., Wellens, H.J., Penn, O.C., Allessie, M.A.: High-density mapping of electrically induced atrial fibrillation in humans. Circulation 89(4), 1665–1680 (1994)
7. Elvan, A., Wylie, K., Zipes, D.P.: Pacing-induced chronic atrial fibrillation impairs sinus node function in dogs. Electrophysiological Remodeling. Circulation 94(11), 2953–2960 (1996)
8. Tieleman, R.G., De Langen, C., Van Gelder, I.C., de Kam, P.J., Grandjean, J., Bel, K.J., Wijffels, M.C., Allessie, M.A., Crijns, H.J.: Verapamil reduces tachycardia-induced electrical remodeling of the atria. Circulation 95(7), 1945–1953 (1997)
9. Sanchez-Quintana, D., Anderson, R.H., Cabrera, J.A., Climent, V., Martin, R., Farre, J., Ho, S.Y.: The terminal crest: morphological features relevant to electrophysiology. Heart 88(4), 406–411 (2002)
10. Virag, N., Jacquemet, V., Henriquez, C.S., Zozor, S., Blanc, O., Vesin, J.M., Pruvot, E., Kappenberger, L.: Study of atrial arrhythmias in a computer model based on magnetic resonance images of human atria. Chaos 12(3), 754–763 (2002); doi:10.1063/1.1483935
11. Jacquemet, V., Virag, N., Ihara, Z., Dang, L., Blanc, O., Zozor, S., Vesin, J.M., Kappenberger, L., Henriquez, C.: Study of unipolar electrogram morphology in a computer model of atrial fibrillation. J. Cardiovasc. Electrophysiol. 14(10 suppl.), S172–S179 (2003)
12. Ruiz-Villa, C., Tobón, C., Heidenreich, E., Hornero, F.: Propagación de potencial de acción en un modelo 3D realista de Aurícula Humana. In: Congreso Anual de la Sociedad Española de Ingeniería Biomédica, Pamplona (2006)
13. Ruiz-Villa, C., Tobón, C., Rodriguez, F.J., Heidenreich, E.: Efecto de la dilatación auricular sobre la vulnerabilidad a reentradas. In: Congreso Anual de la Sociedad española de Ingeniería Biomédica, Valladolid, pp. 205–208 (2008)
14. Wang, K., Ho, S.Y., Gibson, D.G., Anderson, R.H.: Architecture of atrial musculature in humans. Br. Heart J. 73(6), 559–565 (1995)

15. Cohen, G.I., White, M., Sochowski, R.A., Klein, A.L., Bridge, P.D., Stewart, W.J., Chan, K.L.: Reference values for normal adult transesophageal echocardiographic measurements. J. Am. Soc. Echocardiogr 8(3), 221–230 (1995)
16. Feld, G.K., Mollerus, M., Birgersdotter-Green, U., Fujimura, O., Bahnson, T.D., Boyce, K., Rahme, M.: Conduction velocity in the tricuspid valve-inferior vena cava isthmus is slower in patients with type I atrial flutter compared to those without a history of atrial flutter. J. Cardiovasc. Electrophysiol. 8(12), 1338–1348 (1997)
17. Hansson, A., Holm, M., Blomstrom, P., Johansson, R., Luhrs, C., Brandt, J., Olsson, S.B.: Right atrial free wall conduction velocity and degree of anisotropy in patients with stable sinus rhythm studied during open heart surgery. Eur. Heart J. 19(2), 293–300 (1998); doi:S0195668X97907429
18. Ho, S.Y., Sanchez-Quintana, D., Anderson, R.H.: Can anatomy define electric pathways? In: Computer Simulation and Experimental Assessment of Electrical Cardiac Function, Lausanne, Switzerland, pp. 77–86 (1998)
19. Bosch, R.F., Zeng, X., Grammer, J.B., Popovic, K., Mewis, C., Kuhlkamp, V.: Ionic mechanisms of electrical remodeling in human atrial fibrillation. Cardiovasc. Res. 44(1), 121–131 (1999); doi:S0008-6363(99)00178-9
20. Workman, A.J., Kane, K.A., Rankin, A.C.: The contribution of ionic currents to changes in refractoriness of human atrial myocytes associated with chronic atrial fibrillation. Cardiovasc. Res. 52(2), 226–235 (2001); doi:S0008636301003807
21. Nygren, A., Fiset, C., Firek, L., Clark, J.W., Lindblad, D.S., Clark, R.B., Giles, W.R.: Mathematical model of an adult human atrial cell: the role of K+ currents in repolarization. Circ. Res. 82(1), 63–81 (1998)
22. Spach, M.S., Heidlage, J.F., Dolber, P.C., Barr, R.C.: Electrophysiological effects of remodeling cardiac gap junctions and cell size: experimental and model studies of normal cardiac growth. Circ. Res. 86(3), 302–311 (2000)
23. Jongsma, H.J., Wilders, R.: Gap junctions in cardiovascular disease. Circ. Res. 86(12), 1193–1197 (2000)
24. Boyett, M.R., Honjo, H., Kodama, I.: The sinoatrial node, a heterogeneous pacemaker structure. Cardiovasc. Res. 47(4), 658–687 (2000); doi:S0008-6363(00)00135-8
25. Wilders, R.: Computer modelling of the sinoatrial node. Med. Biol. Eng. Comput. 45(2), 189–207 (2007); doi:10.1007/s11517-006-0127-0

SAD_BaSe: A Blood Bank Data Analysis Software

Augusto Ramoa, Salomé Maia, and Anália Lourenço

Abstract. The main goal of this project was to build a Web-based information system – SAD_BaSe – that monitors blood donations and the blood production chain in a user-friendly way. In particular, the system keeps track of several data indicators and supports their analysis, enabling the definition of collection and production strategies and, the measurement of quality indicators required by the Quality Management System of blood establishments. Data mining supports the analysis of donor eligibility criteria.

Keywords: Decision support system, blood donations, process monitoring, quality management.

1 Introduction

The first assignment of blood establishments is to educate society for the need to donate blood. Therefore, they set up marketing campaigns so people constantly hear about it and schedule regular collection sessions not only in their facilities, but also through mobile collection sessions around their geographic area.

In the collection of blood from the donors, several safety measures must be implemented to protect the donor and the receiver. On the donor's side, a set of requirements has been established to ensure a safe procedure, accounting for the considerable amount of blood that is to be drawn (approximately 0.45 liters).

Augusto Ramoa · Salomé Maia
Instituto Português do Sangue,
Centro Regional de Sangue do Porto,
Rua de Bolama, 133, 4200-139 Porto, Portugal
e-mail: {augustoramoa,maia.salome.m}@gmail.com

Anália Lourenço
Institute for Biotechnology and Bioengineering,
Centre of Biological Engineering,
Campus de Gualtar, University of Minho, 4710-057 Braga, Portugal
e-mail: analia@deb.uminho.pt

M.P. Rocha et al. (Eds.): 6th International Conference on PACBB, AISC 154, pp. 165–171.
springerlink.com

On the recipient's side, the donor's behavior regarding drug use and sexual activity is assessed during a clinical interview to minimize the risk of infecting the patient. In parallel, laboratory tests are performed to determine the blood group and to check for a number of blood-borne diseases.

Most blood units are collected from donors as Whole Blood units (WB) and then must be divided in the laboratory. There are different therapeutic products such as Red Cell Concentrates (RCC), Fresh Frozen Plasma (FFP) and Pooled Platelet Concentrates (PPC) [2]. The blood components remain in quarantine until all lab tests are validated and the components are considered safe for use. Then, the blood establishment distributes blood components balancing hospital requests and stock thresholds. Due to the small shelf life of some blood components (e.g. only 5 days for platelets) a good planning is required to avoid blood wastage and, more importantly, blood shortage.

This work proposes a blood bank data analysis software that provides a novel contribution to the monitoring of blood collection and component production and distribution. The remainder of the paper describes the work as follows. In Section 2, current software systems in use are addressed. In Section 3, SAD_BaSe is presented and results presented. Some conclusions are provided in section 4.

2 Blood Bank Software Systems

From the moment of collection to the administration of the blood to a patient, a complex process takes place, mostly based on computer-aided procedures, to ensure the safety of the transfusion and the traceability of all steps. Indeed, the Council of Europe's Guide to the preparation, use and quality assurance of blood components, one of the most important guides of a blood bank, dedicates a subchapter ("Data processing systems") to the design, implementation, testing and maintenance of adequate software.

Currently, blood bank software systems are almost exclusively focused in assuring safety (Does this unit meet the necessary requirements to be issued?) and traceability (To whom the unit was collected? Who collected it? To which hospital was it issued?), however process monitoring is also very important.

The variability associated with the source product (e.g. weight and the values of hemoglobin and platelet), collection (e.g. you cannot have a good collection if the donor does not have a good vein or if the nursing staff is not well trained to proceed with the collection) and production (e.g. the considerable human manipulation inherent to the whole process) leads inevitably to wastes that need to be minimized. A unit of blood cannot be reprocessed if something goes wrong. So, by learning where the process can fail we have the opportunity to improve the process and minimize blood losses.

3 SAD_BaSe Blood Bank Data Analysis Software

According to a survey on the Portuguese Blood Establishments, one of the major user complaints to existing blood bank software systems concerns report capabilities [3]. To assist blood banks in process monitoring, we have developed a complementary information system, named SAD_BaSe. This software suite includes the following modules: a data processing module that gathers and manipulates data generated throughout the collection, processing and distribution stages; a report module that enables query customization (e.g. detailed reports on the different processes for time periods, hospitals, collection sessions or blood components of interest); and a data mining module that enables advanced data analysis, namely to monitor and control donor rejection.

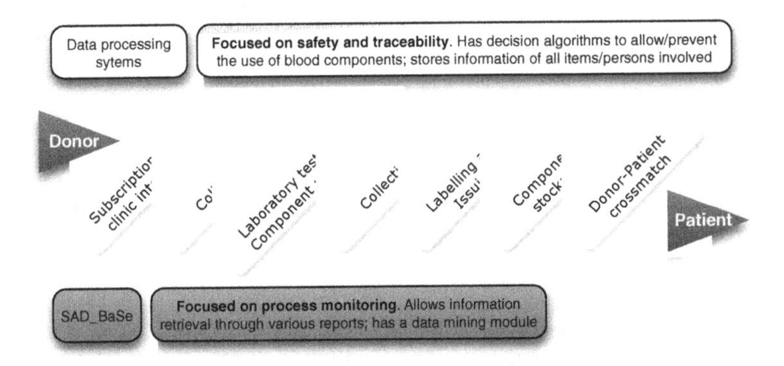

Fig. 1 Blood chain from donor to patient and the differences between existing data processing systems and SAD_BaSe

3.1 Data Modeling and Database Population

The information flow from the blood establishment system to SAD_BaSe database is fully implemented (Fig. 2). Currently, the database covers for information on the collection sessions, donors, lab analysis results and hospitals.

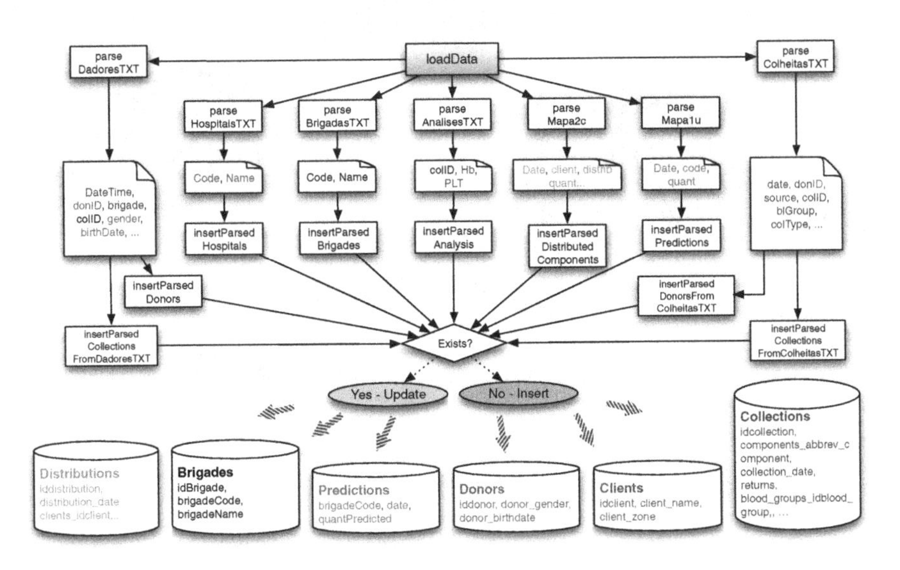

Fig. 2 SAD_BaSe information flow.

SAD_BaSe was implemented and tested in a Portuguese blood establishment. A snapshot was retrieved from its information system encompassing data from January 2009 to December 2011.

For validation purposes, the data related to 2009 was compared with the reports obtained from the blood establishment software. Some data incoherencies were detected. However, they were related to the way data is being used in the two systems. For example, the blood establishment software reports the number of units discarded in a specific time interval, whereas SAD_BaSe reports the number of units discarded that were collected in that time interval.

3.2 Reporting

The tool supports on-demand reporting and querying facilities to enable searches by component, time series, discard cause and collection site.

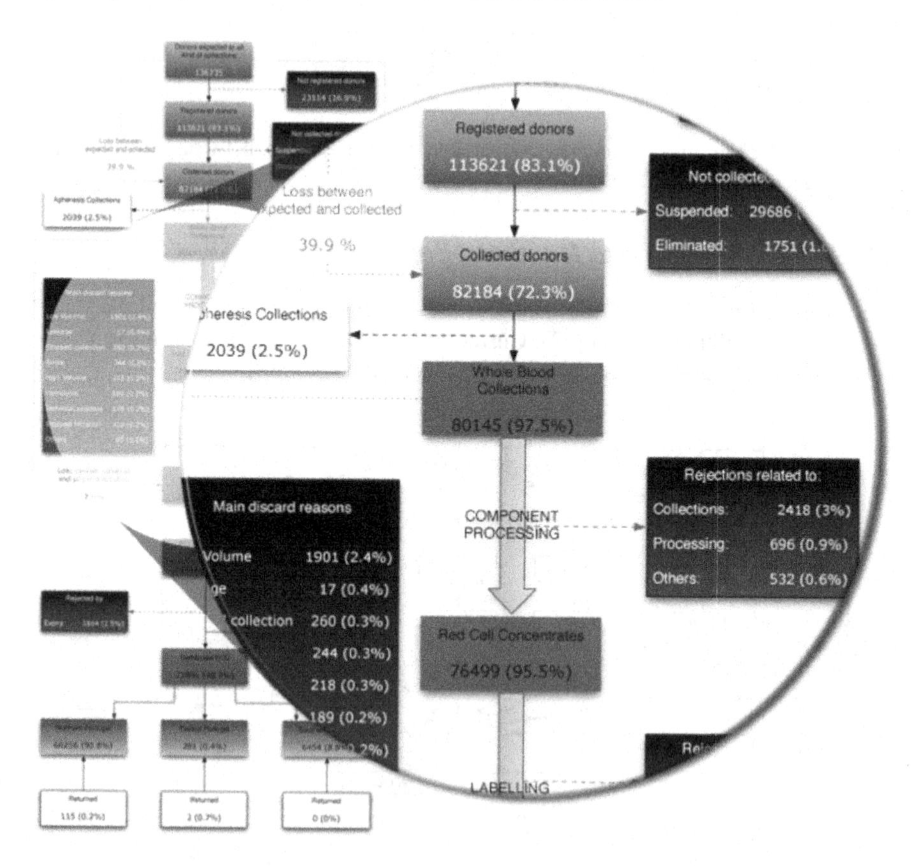

Fig. 3 SAD_BaSe Flowchart

Reports are primarily intended to provide Board members and the Head of departments a summary of the blood establishment activity. The flowchart generated (an execution example is provided in Fig. 3) highlights a set of quality indicators to be monitored, such as: predicted number of donors for all collection sessions; losses by absent donors, suspended and deferred donors, discards related to collection, component processing, serological markers, immuno-hematology problems and expired components; components available for distribution and components received from other blood establishments; and components issued to the hospitals by geographic region.

3.3 Data Mining

Analysis was set on the examination of potential donors and the alert for eventual donation discard. It must be mentioned beforehand that we are aiming to analyze a process where it is expected that most donors are considered suitable to undergo collection. However, young, first time donors are often suspected to be less predisposed to go through the donation without any problems and, large collection sessions (e.g. in universities) are suspected to have more donors suspended. Also, first time donors unaware of donation requirements are presumed to be more ineligible for donation. Despite this "common knowledge", neither the blood establishment had a study to corroborate these suspicions nor other databases stored data on such features.

So, a dataset was prepared containing all donor interviews in the first semester of 2011 (56666 records). The description of the selected attributes is presented below:

- Age of the donor (age) – arranges donations in three groups, donors with less than 24 years old, donors aged between 24 and 60, and donors above 60;
- Gender (gender) – generally speaking, women have less body blood volume than men, but the volume drawn is the same in both cases;
- Previous donations (pdon) – two groups were created according to the number of previous donations, first time and recurrent;
- Collection site (brigade) – the collection sites where the donations take place are quite different. The collection site that is operated in the blood establishment facility is the only one that runs daily. Others run weekly and many run less frequently. This attribute considers three categories: daily collections (PP), weekly collections (regular) and the remaining collections (others);
- Predicted collections (predCol)– before each collection session an estimate of the number of donors is necessary for staff and materials setup. This attribute divides the dataset by the size of the collection session in small sessions (less than 50 donors), medium sessions (between 51 and 100 donors), large sessions (101 to 199 donors) and larger sessions (200 or more donors);
- Conclusion – if the donor is somewhere along the blood chain suspended or permanently deferred (eliminated) it is labeled as S+E, if the donor is considered apt, it is labeled as A.

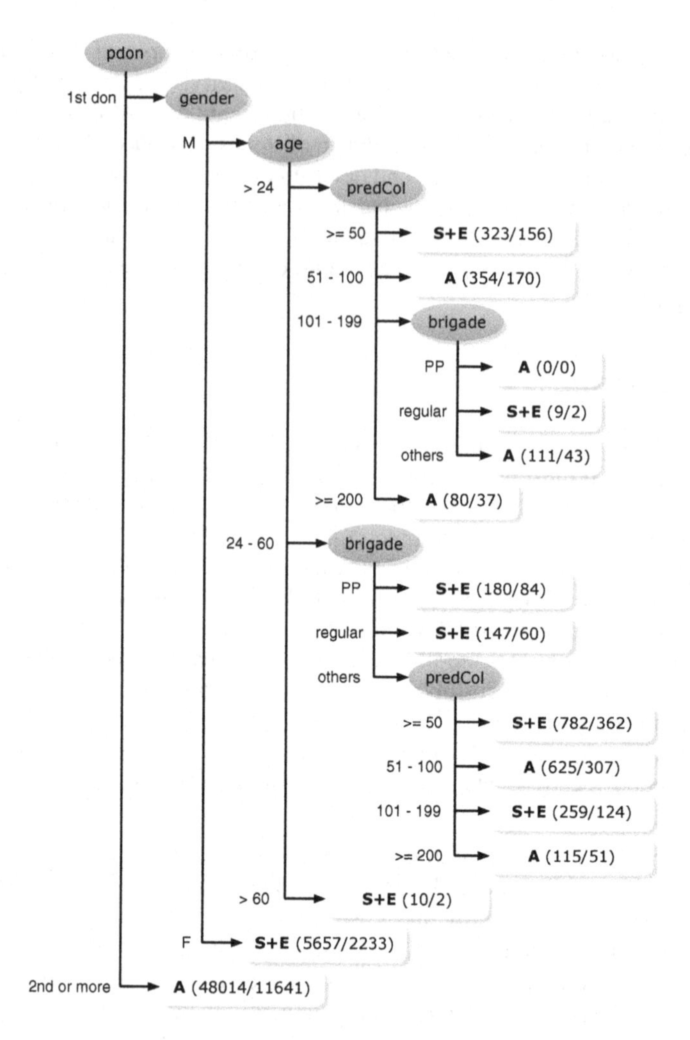

Fig. 4 Decision tree obtained using Weka j48

To perform data mining we have used the software workbench, Waikato Environment for Knowledge Analysis (WEKA 3.6.0.) [4]. The dataset was classified using the decision tree algorithm C4.5 [5] through its Java implementation, j48.

The obtained results point to 72.89% correctly classified instances and the obtained tree illustrates the importance of the previous donations (Fig. 4) which is mainly explained by the fact that returning donors already know that some medical conditions and behaviors are incompatible with blood donations. The gender is also a very important attribute. When looking at the data stored in SAD_BaSe of first time donors, low hemoglobin is the main causes of donor deferral in the female population with 626 deferrals whereas in the males only 51 donors were deferred by the same cause. The main cause of men deferrals was the existence of a new sexual partner in the last six months accounting for 204 deferrals, whereas in the females this accounted for a similar number of deferrals, 177.

4 Conclusions

Blood transfusions are nowadays a routine procedure in hospitals. Blood establishments are equipped with suitable operational information systems but monitoring procedures is still a requirement.

SAD_BaSe complements blood establishment software in two ways: it monitors mandatory quality indicators and supports production planning, maximizing production and minimizing process costs/losses. Its monitoring and analysis abilities respond not only to current requirements, but also to upcoming ones, making new points of monitoring available, with less intervention of the users.

References

1. Council of Europe – Health Policy web page, `http://www.coe.int/t/dg3/health/themes_en.asp`
2. Guide to the Preparation, Use and Quality Assurance of Blood Components, European Directorate for the Quality of Medicines & HealthCare - Council of Europe, Strasbourg (2011)
3. Leal, J.: Validação de Sistemas Informáticos de Serviços de Sangue e de Medicina Transfusional. Faculdade de Medicina da Universidade do Porto (2010)
4. Hall, M.: The WEKA Data Mining Software: An Update. SIGKDD Explorations 11(1) (2009)
5. Quinlan, R.: C4.5: Programs for Machine Learning. Morgan Kaufmann Publishers, San Mateo (1993)

A Rare Disease Patient Manager

Pedro Lopes, Rafael Mendonça, Hugo Rocha, Jorge Oliveira,
Laura Vilarinho, Rosário Santos, and José Luís Oliveira

Abstract. The personal health implications behind rare diseases are seldom considered in widespread medical care. The low incidence rate and complex treatment process makes rare disease research an underrated field in the life sciences. However, it is in these particular conditions that the strongest relations between genotypes and phenotypes are identified. The rare disease patient manager, detailed in this manuscript, presents an innovative perspective for a patient-centric portal integrating genetic and medical data. With this strategy, patient's digital records are transparently integrated and connected to wet-lab genetics research in a seamless working environment. The resulting knowledge base offers multiple data views, geared towards medical staff, with patient treatment and monitoring data; genetics researchers, through a custom locus-specific database; and patients, who for once play an active role in their treatment and rare diseases research.

Keywords: Rare Diseases, Genetic Mutations, LSDB, Personalized Medicine.

1 Introduction

Biomedical research states that a rare disease is a particular condition affecting at most 1 in 2000 patients. The European Organization for Rare Diseases (EURORDIS) estimates that there are approximately 6000 to 8000 rare diseases, affecting about 6% to 8% of the population. Within these, about 80% are caused by

Pedro Lopes · Rafael Mendonça · José Luís Oliveira
DETI/IEETA, Universidade de Aveiro,
3880-193 Aveiro, Portugal
e-mail: {pedrolopes,rmendonca,jlo}@ua.pt

Hugo Rocha · Jorge Oliveira · Laura Vilarinho · Rosário Santos
Centro de Genética Médica Jacinto Magalhães, INSA,
4099-028 Porto, Portugal
e-mail: {hugo.rocha,jorge.oliveira,laura.vilarinho,
rosario.santos}@insa.min-saude.pt

M.P. Rocha et al. (Eds.): 6th International Conference on PACBB, AISC 154, pp. 173–180.
springerlink.com

genetic disorders. Due to the reduced incidence of each individual disease, it is difficult for patients to find support, both at clinical and psychological level [1]. The existence of a small number of patients for each rare disease also delays the creation of adequate research studies as it is difficult to identify and coordinate a correct cohort [2,3]. Despite the low statistic impact regarding these diseases, the combined amount of patients suffering from one of these rare diseases is considerably high.

The Portuguese National Programme for Rare Disorders, approved in 2008, strives towards a concerted system for rare disease tracking and patient monitoring. This endeavor relies heavily on a platform designed to gather information from the different stakeholders involved in rare disease patient care, ranging from diagnostic labs to health providers up to policy-makers. The set of requirements behind such a large-scale project touches miscellaneous research areas. The combination of heterogeneous data from medical reports with patient records and genetic datasets triggers problems in the areas of privacy and security, service composition and data integration.

The solution we envisage relies on edge-of-breed software engineering strategies to provide a seamless working environment for everyone involved in rare disease research. The platform, a rare disease patient manager, collects and connects data for patients during their entire stay in the system. This widespread information collection, including medical reports with the initial diagnosis, patient genetic test results or the patient treatment program, is fused into a single knowledge base. From there, distinct data views are provided to distinct user roles, each accessing its own private and independent web information system.

This Portuguese case study revolves around the creation of reference centres. These are more than a single physical institution, enabling the cooperation and networking amongst several public entities in the Portuguese health system and detaining a wealth of data on patients diagnosed with known rare diseases. This organizational structure, involving efforts from multidisciplinary teams, permits the best procedures to be available, simplifying processes and minimizing the resources needed. With the innovative rare disease patient manager proposed in this manuscript, the reference centres will be endowed with the required set of software tools to study the underlying rare disease causes and adverse effects and, above all, vastly improve patient care.

2 Overview

The importance of a patient's genetic records in diagnosis and resolution of rare diseases has taken medicine to a level where wet-lab research is crucial to unravel disease causes and consequences. Hence, databases emerging with information about human genome, such as the HGMD (Human Gene Mutation Database) [4] or the 1000 Genomes Project [5] retain growing relevance. It is important to collect their public data in novel biomedical software, enabling its usage on the daily medical workflow. Despite this growing, the absence of public medical reports, specially regarding rare disease patients, indicates the vulnerability of these patient groups [6].

Personalized medicine requires the combination of data from both these fields along with individual patient information. Data from sequence mutations, genes and drug interactions – the genotype – must be combined with data from electronic health records (EHR), DICOM imaging and disease-specific patient data – the phenotype. Henceforth, the patient-centric rare disease patient manager offers a unique holistic view promoting the collaboration of multiple entities towards the study of rare diseases and assessment of patients' evolution – Fig. 1.

The *de facto* standard in rare disease software is Orphanet [7], a web platform directed to the general public, health professionals and patients, to inform about orphan drugs and rare diseases. It also displays information on specialized consultations, diagnostics, research projects, clinical trials and support groups. Diseasecard [8] is another portal aggregating information regarding rare diseases and pointing to key elements for both the education and the biomedical research field.

Along with these applications, we must take in account EHRs, gathered from miscellaneous health information systems, containing clinically relevant patient information. Only by mining these reports and collecting them for future analysis we will be able to provide the adequate knowledge set for clinicians involved in the reference center.

Locus-specific databases (LSDB) are gene-centric applications, focused on providing variant-related features, such as store and search, to gene curators. They are usually used by a group of researchers in collaboration, with expertise in a particular gene or genotype, and provide a valuable tool for analysis of gene expression and phenotype, both in normal conditions and disease [9].

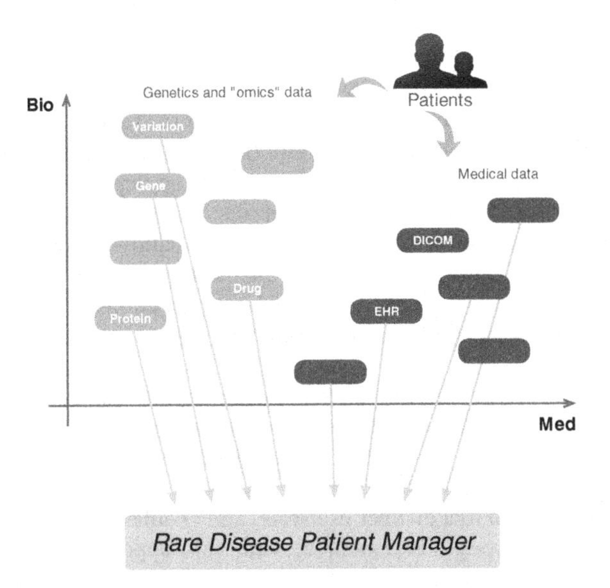

Fig. 1 The rare disease patient manager seamlessly integrates a collection of patient-centric data, combining the quadrants of genetics and "omics" information with miscellaneous medical data in a centralized knowledge base.

Concomitant with the systematic, widespread collection of patient data, LSDBs ultimate goal is a contribution to the global effort of the Human Variome Project (HVP), which aims to collect, curate and distribute all human genetic variation affecting health [10-12].

The goal behind the creation of reference centres is to connect individuals from disparate research areas and specialized in a particular rare disease. Hence, a reference centre is composed of multidisciplinary teams with access to resources for the diagnosis, treatment and medical monitoring of patients.

When a patient, whose clinical conditions signal the presence of a known rare disease, arrives for medical screening, he is directed to the reference centre, initiating his presence in the rare disease patient manager. Initial contact with the clinicians permits the analysis of symptoms and the evaluation of future treatment scenarios. Along with the clinical analysis, patients are sent for genetic testing labs, where their genetic sequence will be sampled and explored, searching for well-known mutation and other biomarkers. The treatment of rare diseases at this level requires a deep collaboration between clinicians and genetics researchers.

3 Information Workflow

The rare disease patient manager provides a unified set of tools ranging from the integration of patient genetic and clinical data to the various independent data access views – Fig. 2.

Rare disease information in the patient manager knowledge base is built from OMIM, *Online Mendelian Inheritance in Man* [13], and HGNC, *HUGO Gene Nomenclature Committee* [14]. To cope with the increasing data in this field, new tools have emerged, namely MUTbase [15], Universal Mutation Database (UMD) [16] and Leiden Open-source Variation Database (LOVD) [17]. LOVD innovates with the "LSDB-in-a-box" approach: it is offered as a downloadable software package containing the full set of tools required for the deployment of a local locus-specific database. Using LOVD within the rare disease patient manager enables the quick arrangement of an in-house LSDB, allowing the registration of genetic mutations, linking them to patients in the rare disease patient manager. The resulting knowledge network, with direct connections from mutations to patients up to diseases represents a true complete view over a rare disease influence area.

When dealing with rare disease patients, the anonymity and privacy of data are regarded as imperative. To cope with the caveat of aggregating all data in a single knowledge base, the patient manager provides distinct "data views" over acquired information.

Each view is assigned to a distinct user-role, and enables/disables access to certain features or data-types. For example, with this approach, reference centre coordinators are differentiated from diagnostic lab technicians: the former are able to manage patient monitoring forms whereas the latter can only access the system's LSDB.

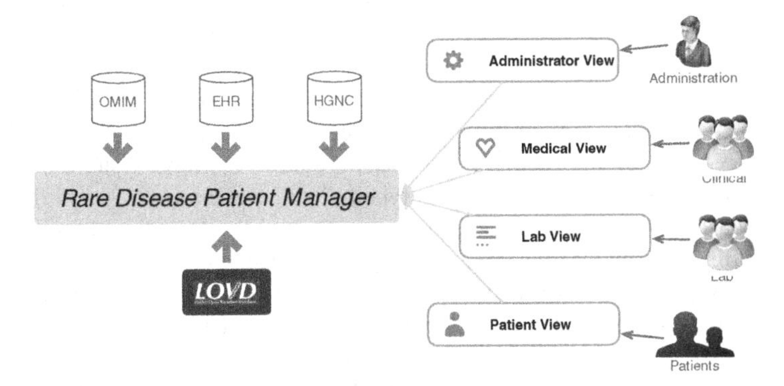

Fig. 2 Data aggregated in the rare disease patient manager is provided to various stakeholders, each assigned a distinct role and with its respective data views.

Furthermore, by placing the patient as the central element of this system, he is given the power to explore and control his data. Given the personal nature of collected data, patients have the option of not sharing their data to statistical analysis or, going the opposite way, and liberate their personal records to whomever they want within the reference centre team. Moreover, with the foreseen advances in personal electronic health records, patients will be able to get their data out of the system and share it with family, physicians or other information systems.

4 Architecture

The rare disease patient manager acts as a centralized, patient-centric, information portal. To reduce data distribution and increase the overall coherency, the system is deployed as a portal of websites. Leveraging on modern web information technologies, each reference centre is able to launch any number of rare disease-oriented websites, each with its own customized rare disease patient management infrastructure. Despite being a part of the same physical architecture, each rare disease patient manager application is virtually independent and unaware of its sibling systems.

In parallel with the main rare disease patient manager system there is a LOVD server. This server enables the creation of new customized locus-specific databases that can be used independently. Hence, genetic lab teams within the reference centre can use their own LSDBs without interfering with the overall rare disease patient manager architecture.

Privacy and security are major requirements inherent to the rare disease patient manager model. Consequently, distinct types of users are given distinct roles. Role assignment is a responsibility of the reference centre coordinator. This role-based strategy allows for a transparent data access division: users assigned with a given role can only access features and data that are available to that role.

To complement the LSDB independent usage with the rare disease patient manager, the transition between systems needs to be transparent. For this matter, an innovative data sharing strategy enables cross authentication between the rare disease patient manager and the independent LOVD instances.

5 Discussion

The rare disease patient manager, overviewed in Fig. 3, provides a solid breakthrough in the seamless integration of genetic, medical and patient data. The patient registry, centralizing medical and genetic data in a single system, delivers a fresh and unique perspective to various types of researchers involved in the study of rare diseases. This provides comprehensive analysis and medical monitoring of patients from the moment they are registered in the system.

In addition to the aforementioned personalized genetic testing datasets, disease-specific templates can be created to monitor a broad number of items. This permits the creation of custom forms at each rare disease patient manager, enabling both the tracking of individual rare disease markers, e.g. neuromuscular disorder level, or general phenotypic traits, e.g. blood type or pressure. Each tracking form contains the title, a short description and a set of fields. The fields can be of different types, mimicking the final user interaction: *text*, *true/false*, *number*, *multiple selectable options* or *enumerations*. With these fields, the system aims to respond to the overall needs of form personalization.

The innovative integration of gene-centric LSDBs with patient provides a tighter connection between a patient's genetic profile and his medical condition. With this, information collected in medical studies is complemented with genetic analysis carried out by wet-lab experts. As a result, this distinctive combination of data and custom disease tracking supports the creation of modern standardized

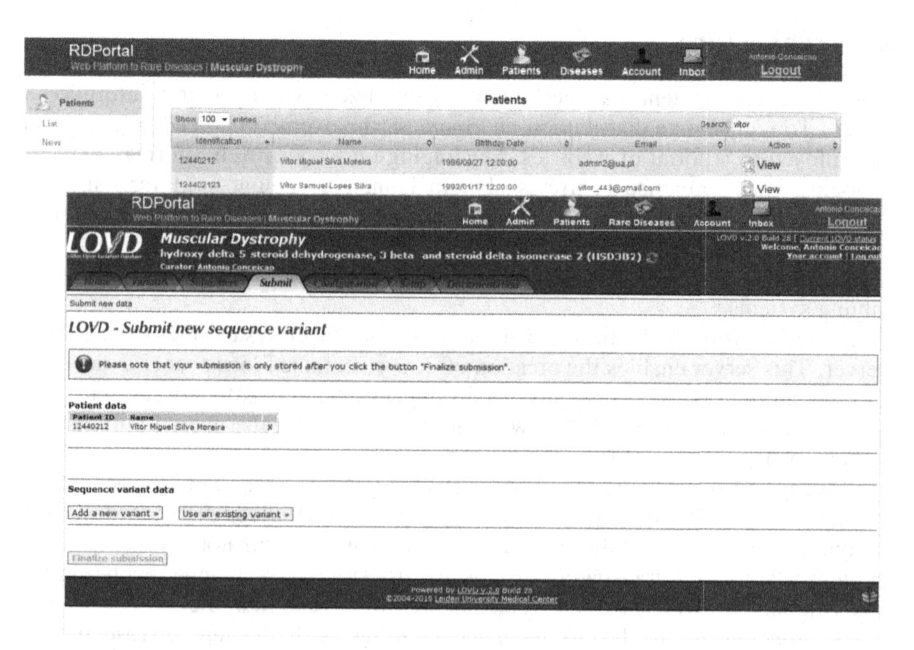

Fig. 3 Rare disease patient manager prototype interface highlighting the patient browser, on top, and the independent LOVD application, at the bottom.

rare disease monitoring procedures, aiding in the achievement of a more accurate diagnosis and of an improved patient care.

One obvious advantage of this system is its web connectivity, which enables the different participants to submit and update the clinical and mutational information in a central database using secure protocols. Simultaneously this information can be accessed and retrieved in real-time by its registered and authorized users. The general purpose solutions currently available are usually focused on intra-institutional use and do not incorporate such features. In addition, due to the diseases' specificities, the system design/architecture should be as flexible as possible. This platform is actually a "core-system" where the different reference centers can customize the structure according to their specific needs and workflow. Finally, it should be taken into account that genetic information management modules are not usually included in general laboratory or patient management systems. The mutational data integrated in this centralized platform can also be easily transferred to an international LSDB or central database.

6 Conclusions

Rare diseases are seldom qualified as mainstream life sciences research. Their specific features require novel research strategies involving the cooperation of researchers from the clinical, genetics and pharmaceutical domain. Despite successful advances on the last two decades, adequate information structure to support these multidisciplinary teams is severely absent. Information from diseases, symptoms, diagnostics, drugs and underlying genetic impact is missing or poorly available. The rare disease patient manager platform described in this manuscript aims to overcome these difficulties. This fusion of data from disparate research fields is of paramount importance to physicians, genetics researchers and, consequently, indispensable to patient care.

For clinicians, this system represents a breakthrough in rare disease patient data management. Clinicians can now track and manage patient data, treatments, diagnostics, procedures and genetic profiles in an integrated fashion. As for gene curators and wet-lab researchers, the system enables the usage of locus-specific databases in an independent yet complementary way, thus enriching patient's medical profiles. At last, for patients, being actively included in their disease-monitoring workflow, in the research for better treatments and with total control over personal data, represents a new standard in personalized medicine.

Acknowledgments. The research leading to these results has received funding from the European Community's Seventh Framework Programme (FP7/2007-2013) under grant agreement no. 200754 - the GEN2PHEN project.

References

1. Nabarette, H., Oziel, D., Urbero, B., Maxime, N., Aymé, S.: Use of a directory of specialized services and guidance in the healthcare system: the example of the Orphanet database for rare diseases. Revue d'épidémiologie et de santé publique 54(1), 41 (2006)

2. Schieppati, A., Henter, J.I., Daina, E., Aperia, A.: Why rare diseases are an important medical and social issue. The Lancet 371(9629), 2039–2041 (2008)
3. Burke, W.: Genetic testing. N. Engl. J. Med. 347(23), 1867–1875 (2002)
4. Krawczak, M., Ball, E.V., Fenton, I., Stenson, P.D., Abeysinghe, S., Thomas, N., Cooper, D.N.: Human gene mutation database—a biomedical information and research resource. Human Mutation 15(1), 45–51 (2000)
5. Via, M., Gignoux, C., Burchard, E.G.: The 1000 Genomes Project: new opportunities for research and social challenges. Genome Medicine 2(1), 1–3 (2010)
6. Schneeweiss, S., Avorn, J.: A review of uses of health care utilization databases for epidemiologic research on therapeutics. Journal of Clinical Epidemiology 58(4), 323–337 (2005)
7. Seoane-Vazquez, E., Rodriguez-Monguio, R., Szeinbach, S.L., Visaria, J.: Orphanet Journal of Rare Diseases. Orphanet Journal of Rare Diseases 3, 33 (2008)
8. Oliveira, J.L., Dias, G., Oliveira, I., Rocha, P., Hermosilla, I., Vicente, J., Spiteri, I., Martín-Sánchez, F., Pereira, A.S.: DiseaseCard: A Web-Based Tool for the Collaborative Integration of Genetic and Medical Information. In: Barreiro, J.M., Martín-Sánchez, F., Maojo, V., Sanz, F. (eds.) ISBMDA 2004. LNCS, vol. 3337, pp. 409–417. Springer, Heidelberg (2004)
9. Claustres, M., Horaitis, O., Vanevski, M., Cotton, R.G.H.: Time for a unified system of mutation description and reporting: a review of locus-specific mutation databases. Genome Research 12(5), 680 (2002)
10. Kohonen-Corish, M.R., Al-Aama, J.Y., Auerbach, A.D., Axton, M., Barash, C.I., Bernstein, I., Beroud, C., Burn, J., Cunningham, F., Cutting, G.R., den Dunnen, J.T., Greenblatt, M.S., Kaput, J., Katz, M., Lindblom, A., Macrae, F., Maglott, D., Moslein, G., Povey, S., Ramesar, R., Richards, S., Seminara, D., Sobrido, M.J., Tavtigian, S., Taylor, G., Vihinen, M., Winship, I., Cotton, R.G.: How to catch all those mutations–the report of the third Human Variome Project Meeting, UNESCO Paris. Hum. Mutat. 31(12), 1374–1381 (2010); doi:10.1002/humu.21379
11. Patrinos, G.P., Al Aama, J., Al Aqeel, A., Al-Mulla, F., Borg, J., Devereux, A., Felice, A.E., Macrae, F., Marafie, M.J., Petersen, M.B., Qi, M., Ramesar, R.S., Zlotogora, J., Cotton, R.G.: Recommendations for genetic variation data capture in developing countries to ensure a comprehensive worldwide data collection. Hum. Mutat. 32(1), 2–9 (2011); doi:10.1002/humu.21397
12. Lopes, P., Dalgleish, R., Oliveira, J.L.: WAVe: Web Analysis of the Variome. Human Mutation 32 (2011); doi:10.1002/humu.21499
13. Hamosh, A., Scott, A.F., Amberger, J.S., Bocchini, C.A., McKusick, V.A.: Online Mendelian Inheritance in Man (OMIM), a knowledgebase of human genes and genetic disorders. Nucleic Acids Research 33(suppl. 1), D514–D517 (2005)
14. Bruford, E.A., Lush, M.J., Wright, M.W., Sneddon, T.P., Povey, S., Birney, E.: The HGNC Database in 2008: a resource for the human genome. Nucleic Acids Research 38(suppl. 1), D445–D448 (2008)
15. Riikonen, P., Vihinen, M.: MUTbase: maintenance and analysis of distributed mutation databases. Bioinformatics 15(10), 852–859 (1999); doi:10.1093/bioinformatics
16. Béroud, C., Collod-Béroud, G., Boileau, C., Soussi, T., Junien, C.: UMD (Universal Mutation Database): A generic software to build and analyze locus-specific databases. Human Mutation 15(1), 86–94 (2000)
17. Fokkema, I.F.A.C., den Dunnen, J.T., Taschner, P.E.M.: Taschner LOVD: Easy creation of a locus-specific sequence variation database using an ldquoLSDB-in-a-boxrdquo approach. Human Mutation 26(2), 63–68 (2005)

MorphoCol: A Powerful Tool for the Clinical Profiling of Pathogenic Bacteria

Ana Margarida Sousa, Anália Lourenço, and Maria Olívia Pereira*

Abstract. Pathogenicity, virulence and resistance of infection-causing bacteria are noteworthy problems in clinical settings, even after disinfection practices and antibiotic courses. Although it is common knowledge that these traits are associated to phenotypic and genetic variations, recent studies indicate that colony morphology variations are a sign of increased bacterial resistance to antimicrobial agents (i.e. antibiotics and disinfectants) and altered virulence and persistence.

The ability to search for and compare similar phenotypic appearances within and across species is believed to have vast potential in medical diagnose and clinical decision making. Therefore, we are developing a novel phenotypic ontology, the Colony Morphology Ontology (CMO), to share knowledge on the colony morphology variations of infection-causing bacteria. A study on the morphological variations of Pseudomonas aeruginosa and Staphylococcus aureus strains, two pathogenic bacteria associated with nosocomial infections, supported the development of CMO. We are also developing a new Web-based framework for the modelling and analysis of biofilm phenotypic signatures, supported by the CMO. This framework, named MorphoCol, will enable data integration and interoperability across research groups and other biological databases.

Keywords: infection-causing bacteria, biofilms, colony morphology, phenotype ontology, microbial database.

1 Introduction

In the clinical context, bacteria can be isolated from various sources such as patients, indwelling medical devices (e.g. catheters, prostheses, pacemakers,

---

Ana Margarida Sousa · Anália Lourenço · Maria Olívia Pereira
IBB - Institute for Biotechnology and Bioengineering,
Centre of Biological Engineering, University of Minho,
Campus de Gualtar, 4710-057 Braga, Portugal
e-mail: {anamargaridasousa,analia,mopereira}@deb.uminho.pt

* Corresponding author.

M.P. Rocha et al. (Eds.): 6th International Conference on PACBB, AISC 154, pp. 181–188.
springerlink.com

endotracheal tubes), biological fluids (e.g. blood, urine and sputum) and abiotic surfaces (e.g. clinical and medical equipment and air). In addition, bacteria are usually isolated from biofilms, i.e. organised microbial communities entrapped on protective polymeric matrices, which are formed in a variety of biotic [1] and abiotic surfaces [2].

Bacterial persistence due to biofilm formation is a noteworthy downside in clinical settings, challenging disinfection and antibiotic procedures. Recent studies have pointed out colony morphology variations as a sign of increased resistance to antimicrobial agents (i.e. antibiotics and disinfectants) as well as of altered virulence traits and persistence of infection-causing microorganisms. Small colony variants (SCV) [3-5] and mucoid variants [6, 7] are examples of this phenomenon – e.g. the reduced size and the mucoidity of colony variants appear to be linked to a higher resistance to some significant antibiotics of medical use, such as vancomicin, kanamycin and ciprofloxacin.

Colony observation has thus become a crucial asset to medical diagnose and clinical decision making. However, the characterization has to go well beyond the simple identification of the microorganisms involved and some qualitative or quantitative description of the altered colony morphology. Despite of the simplicity of colony observation procedures, they often diverge on the experimental settings (namely, the conditions of cultivation) and colony morphology is likely to be altered under different conditions [8, 9].

The development of computational tools in assistance of phenotypic data deposition and further comparison is the key to integrate phenotypic observations into clinical routines. The challenge lays on establishing standard laboratorial guidelines and consensual criteria to identify the morphological characteristics of the colonies and then, creating a suitable and unambiguous computational representation for both the experiment profile and microscopic observations.

This work proposes the Colony Morphology Ontology (CMO), a novel ontology in support of the comprehensive annotation of colony phenotypic studies. This ontology supports MorphoCol, a Web-based framework suitable for the modelling and analysis of biofilm phenotypic signatures. The framework is being submitted to in-house testing and will be soon made publicly available at http://stardust.deb.uminho.pt/morphocol/.

2 The Colony Morphology Ontology

One of the signs of bacterial adaptability to aggressive environments is the change between distinct phenotypes similar to a mechanism On/Off. This changeability is augmented when bacteria face antimicrobial pressures or switch their mode of growth to the biofilm state [10]. A rapid tool to notice this variability is the spread of bacterial cultures on solid media and the observation of the morphology of the colonies adopted by bacteria. But, in order to take full advantage of these outputs, researchers are in need of controlled vocabulary to annotate the experimental settings and the parameters used to characterise the morphotypes unequivocally.

The rationale of the CMO is to unify the vocabulary commonly used to describe colony morphologies within and between research and clinician groups and therefore being able to perform large-scale study comparison. This ontology is a module of a more comprehensive ontology on biofilm studies that describes the overall concept of biofilm signature as being a set of information on the microorganisms, the laboratory procedures, the antimicrobial susceptibility and virulence profiles (Fig. 1 – the top frame).

The hierarchical structure of the ontology is represented by a directed acyclic graph and the terms hold relationships of the type "part_of" – which represent part-whole relationships - and the type "is_a" – defining subtypes. The "microorganism" sub-tree (Fig. 1-a) embraces the identification of the infection-causing microorganism regarding the reference collection and the source from which it has been isolated (patient, indwelling devices, medical surfaces, etc.). Microbiologists need a bacteria identification to be able to compare studies and extrapolate the main phenotypic conclusions, i.e. it is not reasonable to consider the comparisons between different bacterial species.

The "experimental procedure" sub-tree (Fig. 1-b) encompasses the subclasses "growth assay", "plating assay" and "stress assay", "molecular assay", "omics assay" and "*in vivo* assay". A growth assay characterises a set of parameters associated to bacterial growth, such as growth media, temperature, pH, time of growth, aeration, cell mode of growth, etc. Stress assays study the antimicrobial action (biocide, bacteriostatic and surfactant) of chemical and physical stressors on biofilms. Plating assay include the operational parameters associated to the spread of the bacterial cultures onto solid media and the development of the colonies. Molecular, 'omics' and *in vivo* assays address the use of high-throughput methods, such as polymerase chain reaction (PCR), expressing cloning, DNA arrays, MALDI-MS, gel electrophoresis, and *in vivo* techniques and conditions that may be used to confirm the preliminary identification and performs the comprehensive characterisation of the morphotypes. Again, results are comparable if and only if they were produced under the same culture and plating conditions.

The CMO sub-tree contains one root term "colony morphology" with two direct descendants: the terms "qualitative measure" and "quantitative measure" (Fig. 1-c). Colony morphology is firstly analysed in terms of qualitative parameters such as form (circular, irregular, filamentous, etc.), margin (entire, undulated, curled, etc.), surface (homogeneous, heterogeneous), texture (smooth, rough), elevation, size and colour. Then, it is quantified in terms of frequency, i.e. the number of colonies of a given morphotype variant per the total number of colonies in the plate.

The "antimicrobial susceptibility" sub-tree reports the identification of the morphotype according its tolerance to a range (and concentrations) of antimicrobials (antibiotics, disinfectants and biocides). The "virulence" sub-tree profiles the pathogenicity of colony variants in terms of virulence factors such as motility, enzymes production, tissue invasion ability and so on.

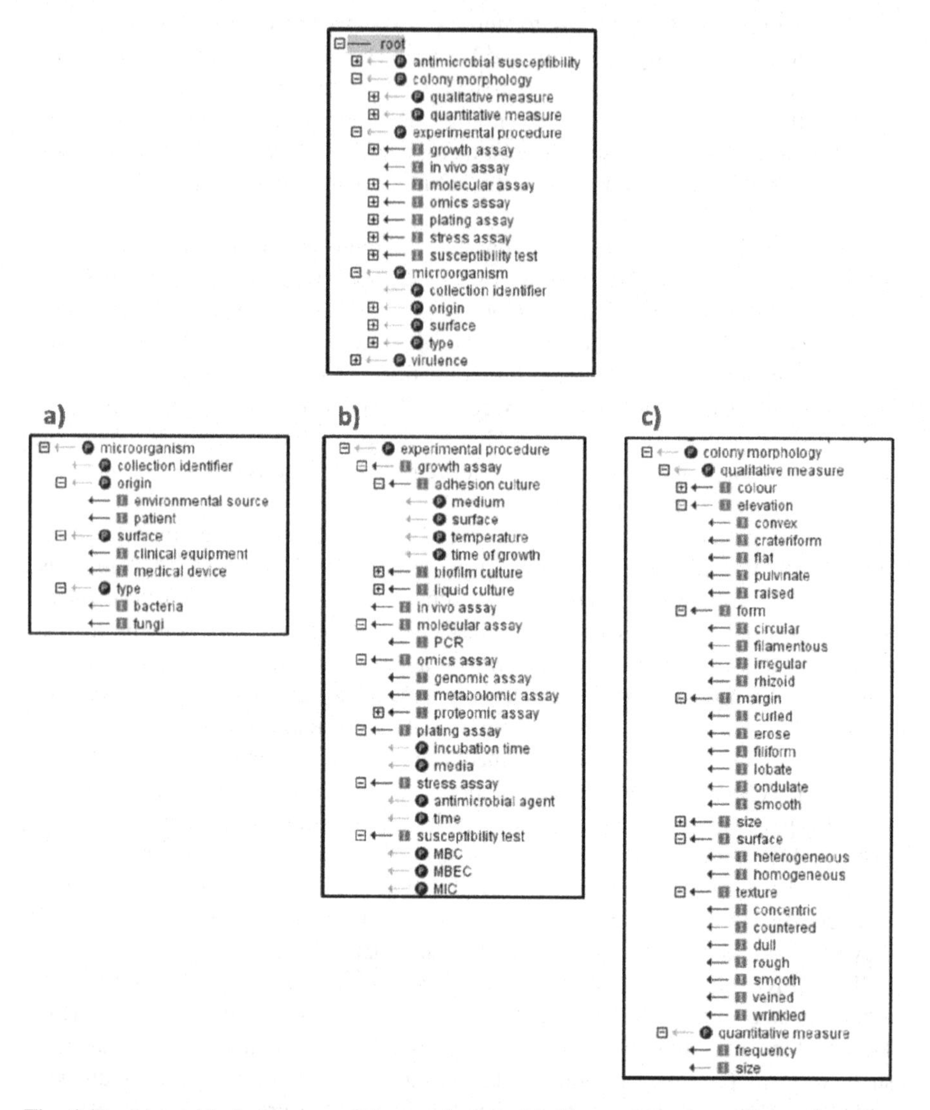

Fig. 1 The hierarchical structure of the ontology on biofilm studies. A profile on the morphology of the colony encompasses information on the microorganisms (a), the experimental assays performed (b) and the morphological characterisation (c).

3 Results and Discussion

The demonstration of the CMO and the development of the modelling and analysis framework are supported by an in-house study on the morphological variations of three pathogenic strains - *P. aeruginosa* ATCC 10145, a clinical isolated from medical equipment (PAI), and *S. aureus* ATCC 25923. The aim of this study is to evaluate the impact of experimental settings (in particular the solid media growth and incubation time) and the susceptibility profile of the colonies.

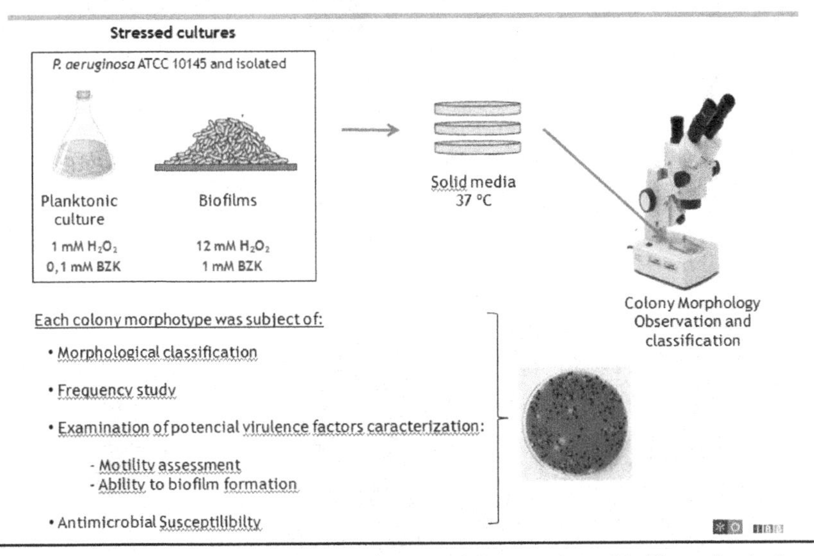

To include in CMO annotation: persistence of *P. aeruginosa* biofilms after hydrogen peroxide stress; colony diversity of *S. aureus* on different solid growth media; antibiotic susceptibility of different types of SCV.

Fig. 2 The experimental scheme to obtain, identify and characterize colonies of unstressed and stressed cultures. Data obtained on virulence and antimicrobial resistance about each morphotype was attached to experiment signature.

As illustrated in Fig. 2, bacteria are cultured on tryptic soy broth (TSB) and agar (TSA) media at 37 °C. During a period of 24 hours, bacteria are cultured in suspension (planktonic state) and in biofilm. Next, planktonic and biofilm cells are exposed to different antimicrobial agents of clinical use, namely to peroxide hydrogen and benzalkonium chloride. Then, non-stressed and stressed planktonic and biofilm cells are spread onto different solid media to observe and characterise the colonies at different incubation times. The observation of colony morphology was supported by directly placing petri plates on magnifying glass and photographed.

For our use case, it was possible to observe that the morphology of the colonies was strongly dependent of the experimental conditions. Data obtained revealed that the morphology of colonies was strongly dependent of the strain, cell mode of growth, type of solid media and incubation time. For only three bacterial strains it was observed more than one hundred distinct colonies. Also, the stressed cultures of all the strains have exhibited colony diversity. Fig. 3 presents some examples of the colonies observed.

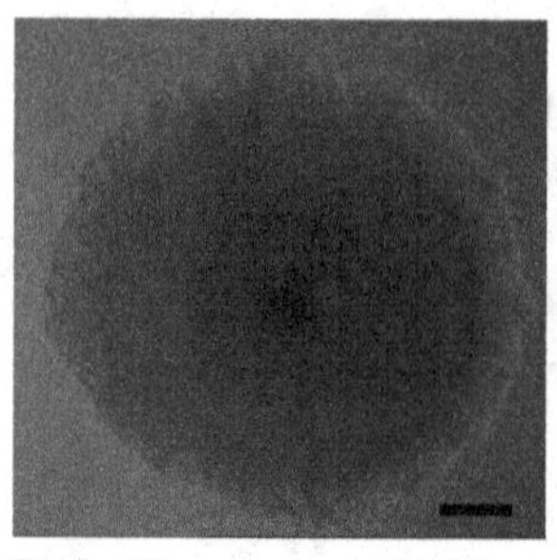

Strain: *P. aeruginosa* ATCC
Diameter: 1,5 cm
Colour: yellow
Colony classification: LIRC
(a)

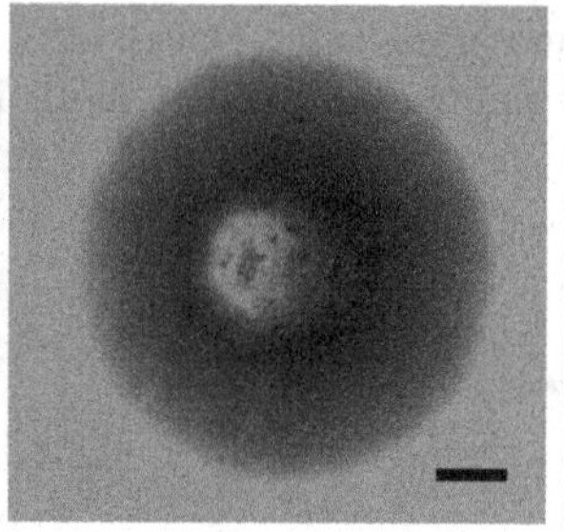

Strain: Isolated *P. aeruginosa*
Diameter: 1,2 cm
Colour: light brown
Colony classification: LRSC
(b)

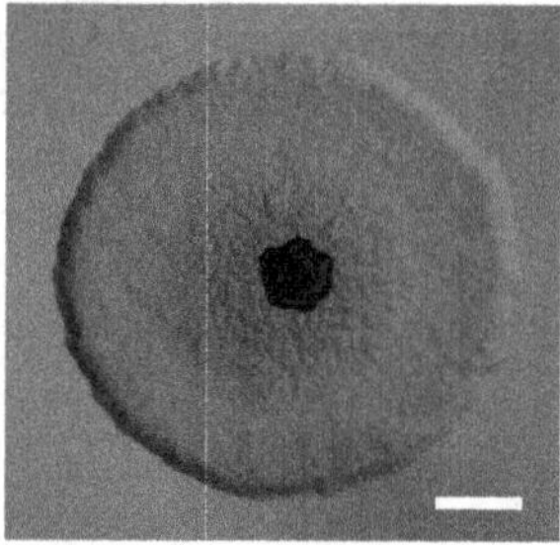

Strain: Isolated *P. aeruginosa*
Diameter: 0,2 cm
Colour: light brown
Colony classification: RSCV
(c)

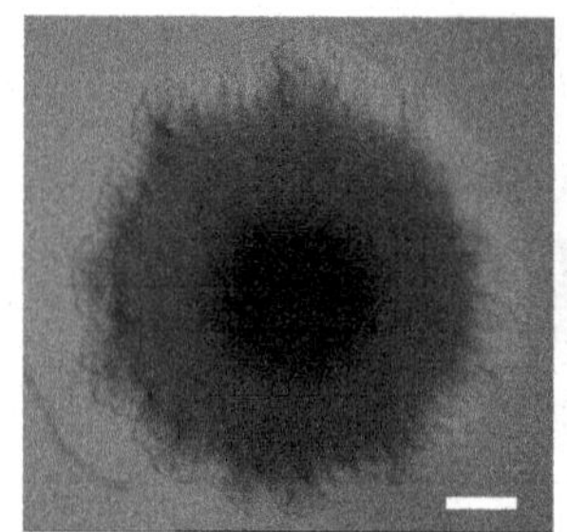

Strain: Isolated *P. aeruginosa*
Diameter: 0,2 cm
Colour: light brown
Colony classification: RSCV
(d)

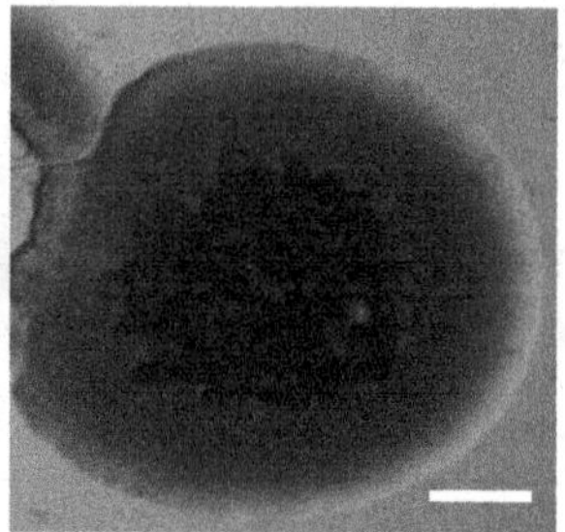

Strain: Isolated *P. aeruginosa*
Diameter: 0,2 cm
Colour: light brown
Colony classification: SwSCV
(e)

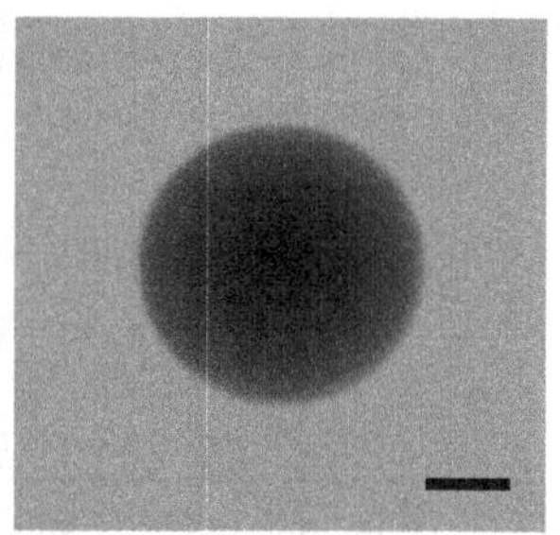

Strain: *S. aureus* ATCC
Diameter: 0,25 cm
Colour: light brown
Colony classification: SCV
(f)

Fig. 3 Examples of colony morphology diversity and the corresponding CMO annotations: (a) larger (more than 3 mm of diameter), irregular and rough colonies (LIRC) formed by *P. aeruginosa* ATCC; (b) larger, regular and smooth colonies (LRSC) with craters formed by the isolated *P. aeruginosa*; (c), (d) and (e) are SCV (less than 3 mm of diameter) formed by isolated *P. aeruginosa* under antimicrobial stress and (f) formed by *S. aureus* ATCC. The SCV exhibited distinct surfaces: (c) and (d) present a rough surface (RSCV), (e) has a smooth and wrinkled surface (SwSCV) and (f) has smooth surface (SSCV). The coloured bar indicates the scale of the image: black bars = 1 mm, white bars = 0,5 mm.

These data provide novel insights into the proteomic alterations that occur during the complex process of morphotype switching, and lend support to the idea that this is associated with a fitness advantage *in vivo*.

Table 1 The ciprofloxacin susceptibility profile of the SCV of *P. aeruginosa* ATCC and isolated *P. aeruginosa*. Analytical assessment: minimal inhibitory concentration (MIC) and minimal bactericidal concentration (MBC).

Microorganism	Morphotype	Ciprofloxacin (mg/L)	
		MIC	MBC
P. aeruginosa ATCC			
	LIRC	0,1563	0,625
Isolated *P. aeruginosa*			
	LISC	0,0391	0,625
	SrSCV	0,0781	1,25
	RSCV	0,0781	0,625
	SwSCV	0,0781	1,25

4 Conclusions and Future Work

Rapid identification of pathogenic microorganisms is crucial in medical diagnosis and clinical decision-making. Specifically, the ability to search for and compare similar phenotypic appearances within and across species is believed to have vast clinical potential. In this work, we propose a novel ontology on colony morphology – the CMO – as a step to the development of a fully integrated modelling and analysis framework for biofilm signatures. This approach has the potential to facilitate to efforts to link details on colony morphologies to specific environmental stressors and make our understanding of adaptive evolution more comprehensive.

Further development and refinement of the CMO will occur continuously and in parallel with phenotype annotation; thus, the evolution of the ontology will reflect the complexity with which phenotypes are described in biofilm research. The creation of terms will be based on need, specifically by curation of primary literature, by annotation of in-house phenotypic studies and by expert input from members of the biofilm research community.

Based on CMO, a public platform – MorphoCol – is under construction for deposition, description and characterization of colony morphologies. Thus, MorphoCol will help the consistent representation of colony morphology data.

Acknowledgements. The financial support from IBB-CEB and Fundação para a Ciência e Tecnologia (FCT) and European Community fund FEDER, through Program COMPETE, in the ambit of the FCT project "PTDC/SAU-SAP/113196/2009/ FCOMP-01-0124-FEDER-016012" and Ana Margarida Sousa PhD Grant (SFRH/BD/72551/2010) are gratefully acknowledged.

References

1. Burmolle, M., Thomsen, T.R., Fazli, M., Dige, I., Christensen, L., Homoe, P., Tvede, M., Nyvad, B., Tolker-Nielsen, T., Givskov, M., Moser, C., Kirketerp-Moller, K., Johansen, H.K., Hoiby, N., Jensen, P.O., Sorensen, S.J., Bjarnsholt, T.: Biofilms in chronic infections - a matter of opportunity - monospecies biofilms in multispecies infections. FEMS Immunol. Med. Microbiol. 59(3), 324–336 (2010)
2. Donlan, R.: Biofilms and device-associated infections. Emerging Infectious Diseases 7(2), 277–281 (2001)
3. Kirisits, M.J., Prost, L., Starkey, M., Parsek, M.: Characterization of colony morphology variants isolated from Pseudomonas aeruginosa biofilms. Appl. Environ. Microbiol. 71(8), 4809–4821 (2005)
4. Proctor, R.A., Peters, G.: Small colony variants in staphylococcal infections: diagnostic and therapeutic implications. Clin. Infect. Dis. 27(3), 419–422 (1998)
5. Haussler, S., Tummler, B., Weissbrodt, H., Rohde, M., Steinmetz, I.: Small-colony variants of Pseudomonas aeruginosa in cystic fibrosis. Clin. Infect. Dis. 29(3), 621–625 (1999)
6. Hogardt, M., Heesemann, J.: Adaptation of Pseudomonas aeruginosa during persistence in the cystic fibrosis lung. Int. J. Med. Microbiol. 300(8), 557–562 (2010)
7. Lyczak, J.B., Cannon, C.L., Pier, G.B.: Lung infections associated with cystic fibrosis. Clin. Micro-Biol. Rev. 15(2), 194–222 (2002)
8. Sousa, A.M., Loureiro, J., Machado, I., Pereira, M.O.: The role of antimicrobial stress on Pseudo-monas aeruginosa colony morphology diversity, tolerance and virulence. In: International Conference on Antimicrobial Research, Valladolid, Spain (2010)
9. Sousa, A.M., Loureiro, J., Machado, I., Pereira, M.: In vitro adaptation of P. aeruginosa: colony morphology variants selection and virulence characterization. In: International Conference on BIOFILMS IV, Winchester, England (2010)
10. Sousa, A.M., Machado, I., Pereira, M.O.: Phenotypic switching: an opportunity to bacteria thrive. In: Mendez-Vilas, A. (ed.) Science Against Microbial Pathogens: Communicating Current Research and Technological Advances (2011) (accepted)
11. Starkey, M., Hickman, J.H., Ma, L., Zhang, N., De Long, S., Hinz, A., Palacios, S., Manoil, C., Kirisits, M.J., Starner, T.D., Wozniak, D.J., Harwood, C.S., Parsek, M.R.: Pseudomonas aeruginosa rugose small-colony variants have adaptations that likely promote persistence in the cystic fibrosis lung. J. Bacteriol. 191(11), 3492–3503 (2009)
12. Seaman, P.F., Ochs, D., Day, M.J.: Small-colony variants: a novel mechanism for triclosan resistance in methicillin-resistant Staphylococcus aureus. J. Antimicrob. Chemother. 59(1), 43–50 (2007)

Applying AIBench Framework to Develop Rich User Interfaces in NGS Studies

Hugo López-Fernández, Daniel Glez-Peña, Miguel Reboiro-Jato,
Gonzalo Gómez-López, David G. Pisano, and Florentino Fdez-Riverola

Abstract. AIBench is a Java application framework for building translational software in Biomedicine. In this paper, we show how to use this framework to develop rich user interfaces in next-generation sequencing (NGS) experiments. In particular, we present PileLineGUI, a desktop environment for handling genome position files in next-generation studies based on the PileLine command-line toolbox.

Keywords: next-generation sequencing, pileup, vcf, single nucleotide variants, PileLine, AIBench framework.

1 Introduction

Nowadays, research experiments in the field of next-generation sequencing (NGS) technologies are generating genome sequence data on an unprecedented scale. This fact is due to: *(i)* the increasing resolution of these techniques and *(ii)* the remarkable reduction in per-base sequencing costs. These factors have produced a significant increase of NGS-based studies. Consequently the NGS data analysis is being more and more demanded by wet-labs and new efficient, practical and adaptable bioinformatics tools are needed in order to get a best performance setting up NGS data analysis workflows [1].

Hugo López-Fernández · Daniel Glez-Peña ·
Miguel Reboiro-Jato · Florentino Fdez-Riverola
ESEI: Escuela Superior de Ingeniería Informática, University of Vigo,
Edificio Politécnico, Campus Universitario As Lagoas s/n, 32004, Ourense, Spain
e-mail: {hlfernandez,dgpena,mrjato,riverola}@uvigo.es

Gonzalo Gómez-López · David G. Pisano
Bioinformatics Unit (UBio), Biología Estructural y Biocomputación,
Centro Nacional de Investigaciones Oncológicas (CNIO), Madrid, Spain
e-mail: {ggomez,dgonzalez}@cnio.es

M.P. Rocha et al. (Eds.): 6th International Conference on PACBB, AISC 154, pp. 189–196.
springerlink.com

PileLine [2] is a novel command-line toolbox for efficient handling, filtering and comparison of standard genome position files (GPFs) employed in NGS studies , such as SAMTools pileup, UCSC's BED (Browser Extensible data), UCSC's GFF (General Feature Format) and the 1000-Genomes Project VCF (Variant Call Format). However, some researchers may not be familiarized with the command-line and they demand user-friendly applications which allow performing their NGS studies using rich user interfaces.

Developing applications in a scientific context presents a large number of specific requisites [3], like the extensive use of logging messages to monitor the progress of long processes, setting values for a high and variable number of parameters before running the experiments, the ability to repeat the same workflow but changing a few parameters or input data, taking the maximum advantage of multi-threading, the need to integrate third party (or previously developed) software or cross-platform compatibility. The development of applications meeting these requirements is not easy. Furthermore, if we take into account that this software must include a sophisticated user interface, the time invested and the difficulty of the development increase significantly.

In this context, we have developed AIBench [3], an application framework for building translational software in Biomedicine. In this work, we describe how AI-Bench has been used to develop PileLineGUI, a PileLine front–end and genome browser featuring an intuitive desktop environment.

The paper is structured as follows. Section 2 contains an overview of the AIBench architecture and its key design concepts. Section 3 contains an overview of the Pile-Line toolbox. Section 4 describes the main functionalities and design the of PileLine-GUI application. Finally, Section 5 includes the conclusions and future work.

2 AIBench Framework

AIBench was specially designed for applications based on the input-process-output (IPO) model, in which the output of one task can be the input of another one. For example, one application could first load some sample genomic data from a file taking the path as input, then execute a loading procedure and finally, generate an in-memory data representation of its results as output, which could be rendered and displayed to the final user at any time after being produced. After that procedure, the loaded data may be used as input of another process (e.g. an analysis) producing a new in-memory representation of its output.

As we see, the IPO model is particularly well suited to the scientific domain. The framework acts as the glue between each executed task so the developer only needs to concentrate on the internals of the processes, data and viewers. In order to provide the basis for application development, AIBench manages three key concepts that are present in every AIBench application: *operations*, implementing the algorithms and data processing routines, *data-types*, storing relevant problem-related information and *views*, rendering data-types obtained from executed operations. From an abstract point of view, an AIBench application can be seen as a collection of operations, data-types and views, reusable in more than one final application.

From an architectonic perspective, AIBench is structured in several layers, as it is shown in Fig. 1. The AIBench framework runs over a plug-in engine able to define a straightforward reusable component model where both the framework native components and the application-specific functionalities are divided and packaged into plug-ins. AIBench plug-ins are isolated by default, increasing the modularity and ensuring that accidental coupling is not introduced, but they can also interact by establishing dependencies (i.e. one plug-in to require other plug-ins to be present at run-time) or extension points.

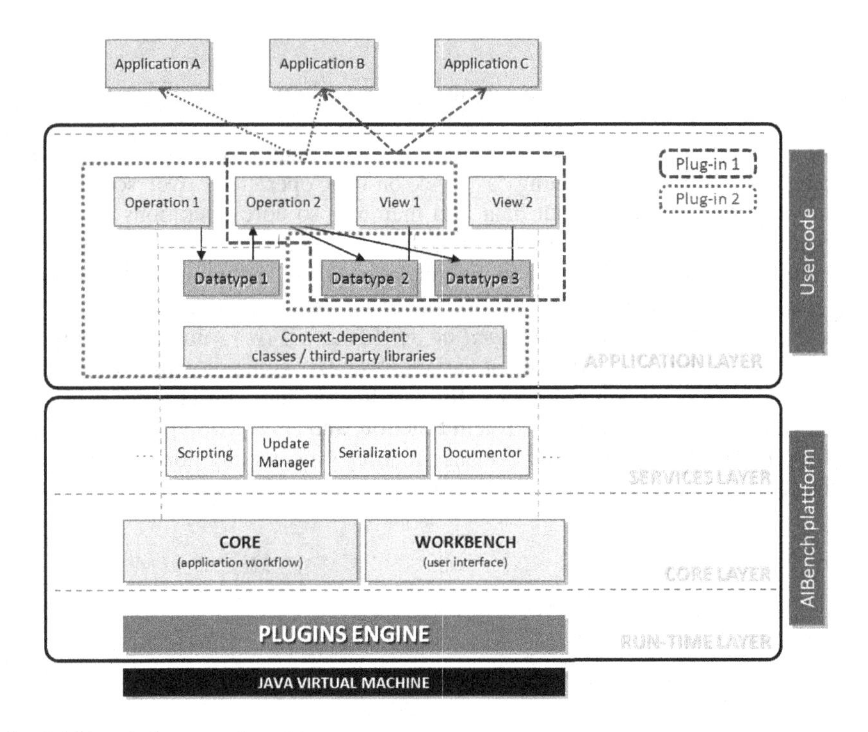

Fig. 1 AIBench framework architecture.

The Core layer contains two native plug-ins: the Core and the Workbench. The AIBench Core detects and registers the application-specific operations, executes them upon request, keeps the results in the Clipboard structure and stores the session workflow in the History. The graphical user interface aspects are implemented in the Workbench plug-in, which creates the main application window, composes a menu bar with all the implemented operations, generates input dialogs when some operation is requested for execution, instantiates the registered results viewers, etc. All additional services bundled with AIBench belong to the Services layer and are also implemented via independent plug-ins that can be easily removed to meet the application specific needs. The Core and Services layers are maintained by the AIBench team and constitute all the code built-in and distributed with the framework, being the starting point of every development.

The application layer is placed on the top of the architecture and contains the application specific code (operations, data-types and views) provided by applications developers (AIBench users). Operations, data-types and views can (and should) be shared among applications related to the same area, especially when they are developed inside the same team. These higher-level components, along with other third-party libraries are also packaged in one or more plug-ins. For example, Fig. 1 shows two plug-ins containing and sharing some operations, views and data-types.

3 PileLine Toolkit

PileLine is a novel and versatile command-line toolbox that provides a catalogue of functions to analyze, compare, filter and annotate genome-position files giving support to common NGS analysis workflows. The main design principle is to be memory efficient by performing fast seek on-disk operations over sorted GPFs. This way avoids loading input data into memory, so core operations operate directly on disk without using auxiliary indexes.

PileLine provides several functionalities over GPFs, including (*i*) full standard annotation with human dbSNP, HGNC Gene Symbol and Ensembl IDs, (*ii*) custom annotation through standard *.bed* or *.gff* files, (*iii*) two sample (i.e.: case VS control) and *n* sample comparison at variant level, (*iv*) generation of SIFT, Firestar and PolyPhen compatible outputs for predicting the consequences of non-synonymous coding variants on protein function, and (*v*) a genotyping quality control (QC) test for estimating performance metrics on detecting homo/heterozygote variants against a given gold standard genotype.

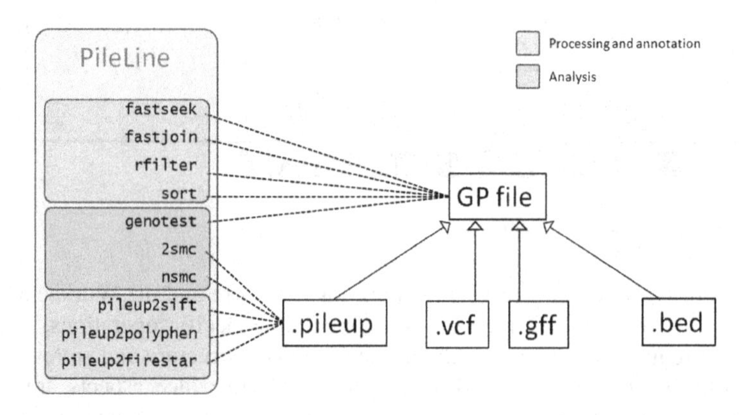

Fig. 2 PileLine commands and accepted input files.

The tools comprising PileLine are focused on two different but complementary activities: *(i)* processing and annotation, implementing simple but reusable operations over input GPFs and *(ii)* analysis, giving support to more advanced and specific requirements. Regarding to the file formats, while *2smc*, *nsmc*, *pileup2sift*, *pileup2polyphen* and *pileup2firestar* work with specific SAMTools .pileup file format, *fastseek*, *fastjoin*, *rfilter* and *genotest* work with generic GPFs (i.e.: .pileup, .vcf, .gff, .bed, etc.) (Fig. 2).

4 Results: PileLineGUI

PileLineGUI is a novel desktop application specifically intended for effectively handling and browsing GPFs generated in DNA-seq experiments [4]. The application includes all the core functionalities of the PileLine command-line toolbox and it has been designed to be intuitively used by biomedical researchers. By also including an interactive genome browser to explore DNA variants, this tool is useful for wet-lab users demanding fast and easy-to-use local applications able to analyze huge amounts of data coming from next-generation sequencing studies. PileLineGUI is implemented using the Java programming language on top of the PileLine tools and was developed using the AIBench application framework.

The first step to develop a user interface for the PileLine tools using AIBench is to divide and structure the problem-specific code into objects of the three AIBench entities: *operations*, *data-types* and *views*.

Data-types store relevant problem-related information, so every file format supported by the application (e.g.: generic GPFs, .pileup, etc.) must be mapped into this entity. In this sense, a *data-type* consists in a Java class that stores: *(i)* path to the file, *(ii)* a PileLine *Fastseek* object in order to retrieve data from the file and *(iii)* some specific attributes like sequence column or position column, among others. FASTA files are also supported to be used by the genome browser for reference genome visualization. These files are not genome position files and they have their own *data-type*: *Genome*, which contains a PileLine *GenomeIndex* object to access the data.

Once the data structures are available, there are necessary some AIBench *operations* in order to load and manage the NGS data. AIBench *operations* define high-level problem-oriented processes. Loading operations (i.e. *Load file* GPFs and *Load genome* for FASTA files) take as input a file and some parameters if needed (e.g.: genome position column, sequence column, etc.), construct the in-memory representation of the data and return this result. Rest of operations (*Sort file*, *Join files*, *Filter file*, *Annotate file*, *2smc*, *nsmc* and *Export*) involve one or more PileLine commands. Each operation is implemented trough only one Java class (which can delegate its internal behavior to other classes).

Generally speaking, one operation is a unit of logic with a well-defined input and output specified via a set of ports. For example, the PileLine sort command (*pileline-sort*), which sorts a GPF by position coordinate, takes an input file to sort, an output file to store the result and has two parameters that specify the sequence column and the position column. In the corresponding operation (*Sort file*), the input data and the two columns parameters must be defined as *INPUT* ports and the results file as *OUTPUT* port.

Every time an operation is executed, one instance of its class is allocated in memory and all the methods associated to the ports are invoked in a predefined order, which the parameters already retrieved from the user. The framework is responsible for dynamically generating complex dialogs for all the input ports, giving a valuable aid in a time-consuming task present in every desktop application development cycle. However, it is possible to provide custom dialogs. In this case, five operations require custom dialogs in order to provide a best user-interaction. Anyway, the framework is the responsible for rendering the dialog layout.

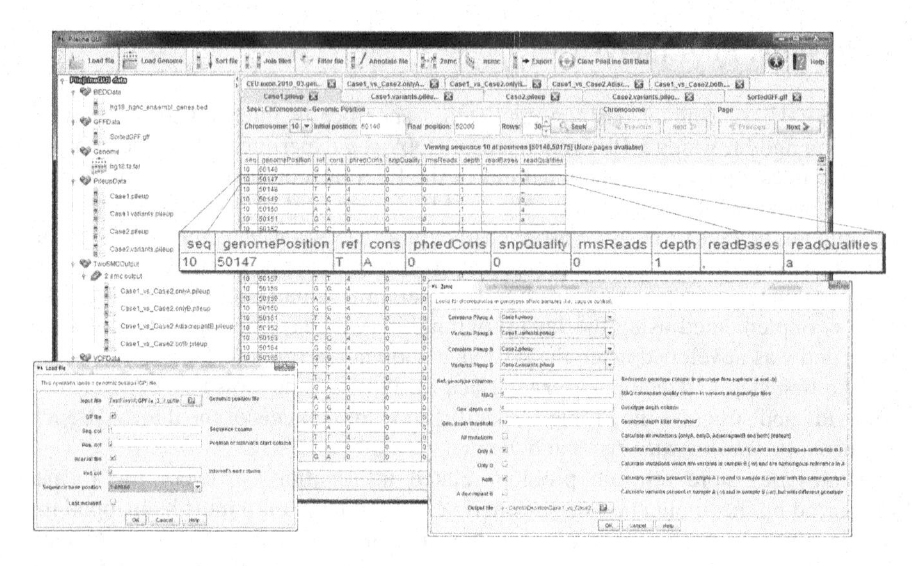

Fig. 3 PileLineGUI *GPF browser* interface.

The third AIBench entities are *views*. The function of a *view* is to render data-types obtained from executed operations. In order to visualize the results operations in a friendly way, every *data-type* may be rendered with a *view*. Considering that GPFs have usually a considerable size and it cannot be opened with standard file editors, a view called *GPF browser* is provided to allow the user to explore the content of NGS files, taking advantage of the PileLine FastSeek which allows performing very fast searches without loading input data into memory. Fig. 3 shows a screenshot of the GPF browser.

In addition, there is also another view called *Genome browser*. This *view* allows the user to render *Genome* objects and consists in a powerful genome browser for GPFs including multiple track support, fully interactive zoom and image export facilities. The user may add several GPFs as individual tracks to the browser. Single-nucleotide GPFs (.pileup and .vcf) are rendered in the genome by displaying the value of a custom column at the corresponding genome position. Moreover, pileup files can be dynamically filtered by depth, SNP quality and consensus quality. If the zoom level is low (showing a very large of range of bases), the browser plots a configurable histogram to see the distribution of occurrences in the GPF. Intervals GPFs (.bed and .gff) are rendered as segments. If there are overlapping segments, they are plotted at different heights in the track. An example of the genome browser can be seen in Fig. 4.

The AIBench components used to build the PileLineGUI application are packaged in two plug-ins: the 'PileLineGUI' plug-in and the 'Genome browser' plug-in. The genome browser is implemented as a separate plug-in ('Genome browser') to facilitate its later inclusion and reuse in other projects. These components are summarized in Table 1.

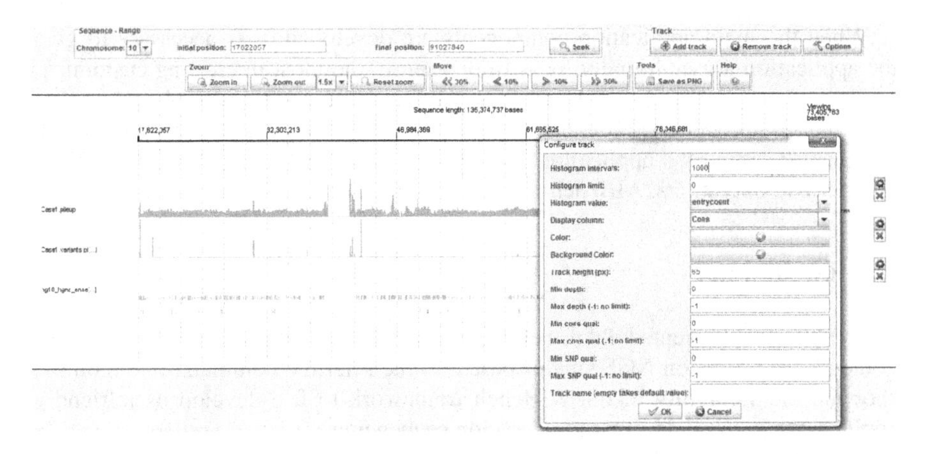

Fig. 4 PileLineGUI *Genome Browser* interface.

Table 1 AIBench components of the PileLineGUI application.

AIBench component	Description
Operations	
Load file	Loads a genomic position file.
Load genome	Loads a FASTA file.
Sort file	Sorts a GPF by position coordinate.
Join files	Joins two sorted GPFs.
Filter file	Filters a positional file with range-based annotations (in bed, gff or custom formats).
Annotate file	Annotates a positional file with range-based annotations (in bed, gff or custom formats).
2smc	Looks for discrepancies in genotypes of two samples (i.e.: case vs control) in pileup format files.
nsmc	Takes the output of several 2smc comparisons commands to reports where variants are reproduced.
Export	Generates a Polyhen/SIFT-compatible change column for each variant line in pileup files.
Datatypes	
GPData	A generic genome position file.
IntervalsGPData	A generic intervals genome position file.
*{file_type}*Data	Represents each type of supported formats (i.e .pileup, .bed, .gff and .vcf).
Genome	A FASTA file.
Views	
GP files browser	Allows the user to browse genomic position files.
Genomic browser	Allows the use to render entire genomes and add multiple GPFs.

When the main application components are developed, it is necessary to give the application a user-friendly look. In this sense, the most interesting customizations are: *(i)* reorganizing of the toolbar in other menu positions to obtain a coherent and comprehensive application handling, *(ii)* changing the start-up splash screen, *(iii)* setting the application icon and *(iv)* setting specific icons for both data-types and operations. AIBench provides several customization facilities.

5 Conclusions

This paper has presented PileLineGUI, a desktop environment for handling genome position files on NGS studies based on the PileLine command-line toolbox, showing the suitability of our AIBench framework to fast-develop user friendly applications in the field of next-generation sequencing.

The future work is divided in two objectives. By one hand, the presented application will be improved in future versions. Somatic mutation functions (*2smc* and *nsmc*) will be extended to handle VCF (Variant Call Format) files, since the SAMTools pileup format is now deprecated and there will be included more standard file formats to the genome browser, such as BAM. In the other hand, the AIBench framework is in continuous development, including new features and bug fixes frequently.

Acknowledgements. This work is supported in part by the project TIN2009-14057-C03-02 from Ministerio de Ciencia e Innovación (Spain), and by the project 10TIC305014PR from Xunta de Galicia (Spain).

References

1. The International Cancer Genome Consortium. Nature 464(7291), 993–998 (2010)
2. Glez-Peña, D., Gómez-López, G., Reboiro-Jato, M., Fdez-Riverola, F., Gómez-Pisano, D.: PileLine: a toolbox to handle genome position information in next-generation sequencing studies. BMC Bioinformatics 12, 31 (2011)
3. Glez-Peña, D., López-Fernández, H., Reboiro-Jato, M., Méndez, J.R., Fdez-Riverola, F.: A Java application framework for scientific software development. John Wiley & Sons, Ltd. (2011)
4. López-Fernández, H., Glez-Peña, D., Reboiro-Jato, M., Gómez-López, G., Gómez Pisano, D., Fdez-Riverola, F.: PileLineGUI: a desktop environment for handling genome position files in next-generation sequencing studies. Nucleic Acids Research Web Server (1-5) (2011)

Comparing Bowtie and BWA to Align Short Reads from a RNA-Seq Experiment

N. Medina-Medina, A. Broka, S. Lacey, H. Lin, E.S. Klings, C.T. Baldwin, M.H. Steinberg, and P. Sebastiani

Abstract. High-throughput sequencing technologies are a significant innovation that can contribute to important advances in genetic research. In recent years, many algorithms have been developed to align the large number of short nucleotide sequences generated by these technologies. Choosing within the available alignment algorithms is difficult; to assist this decision we evaluate several algorithms for the mapping of RNA-Seq data. The comparison was completed in two phases. An initial phase narrowed down the comparison to the three algorithms implemented in the tools: ELAND, Bowtie and BWA. A second phase compared the tools in terms of runtime, alignment coverage and process control.

Keywords. RNA-Seq, high-throughput sequencing, short reads alignment, ELAND, BWA and Bowtie.

N. Medina-Medina
Department L.S.I,
Technical School of Computer and Telecommunications Engineering,
University of Granada, Spain
e-mail: nmedina@ugr.es

A. Broka
Boston University LinGA Computing Resource
e-mail: brokaa@bu.edu

S. Lacey · P. Sebastiani
Department of Biostatistics,
Boston University School of Public Health, USA
e-mail: seanmlacey@gmail.com, sebas@bu.edu

H. Lin · E.S. Klings · C.T. Baldwin · M.H. Steinberg
Department of Medicine,
Boston University School of Medicine, USA
e-mail: {hhlin,klingon,cbaldwin,mhsteinb}@bu.edu

M.P. Rocha et al. (Eds.): 6th International Conference on PACBB, AISC 154, pp. 197–207.
springerlink.com

1 Introduction

Decoding the human genome is important to understand human evolution, the genetic causes of common diseases, the relationship between genes and the environment, and the personalized medicine [1]. The Human Genome Project (HGP) was initially proposed in 1985 with the goal of determining the complete nucleotide sequence of the human genome [2]. The HGP started in 1990 and it took approximately 10 years to complete a draft of the human genome using a "shotgun approach" [3]. Since Sanger's methods for DNA sequencing was introduced in 1977 using chain terminators [4], advances in technology have made the sequencing methods less tedious and more efficient. Nowadays, several "next generation sequencing" technologies are available [5 and 6], which by parallelizing the sequencing process can produce millions of sequences in a matter of days. To do this, next generation sequencing technologies produce short reads ---i.e. short sequence of nucleotide bases --- which must be subsequently mapped to a known reference genome or "de-novo" assembled to constitute a new genome.

The task of assembling genomic sequences is daunting, considering the complexity of genomes and the amount of short reads required [7]. In contrast, the alignment of short reads to a known reference genome is relatively easier [8], although various types of genomic variations, such as single nucleotide polymorphisms (SNPs) or short insertions/deletions (indels) can make the alignment very challenging. Multiple local alignments must be performed to align each short sequence within the complete genome sequence and different algorithms have been proposed, most of them based on dynamic programming [9 and 10]. Numerous programs have been developed to work with sequence data and they provide tools for visualization, assembly and alignment. In addition, programs differ in terms of applicable platforms, functionality and user interface. For example, Geneious [11] or CLC Main Workbench 5.5 [12] are Window-based tools with a friendly graphic interface to visualize sequences, perform multiple sequence alignment and generate phylogenetic trees. Bowtie [13] and BWA [14] are ultra-fast tools designed to efficiently align a large amount of short sequences in different environments, such as Mac Os X, Windows/MinGW and Linux; and other tools, such as Galaxy [15] (a platform for interactive large-scale genome analysis) are also available on line.

Due to the large number of alignment tools, before conducting any genetic analysis it is necessary to make a careful choice of the program to be used for the alignment. This paper compares some of the mainstream programs for alignment of short reads produced by next generation sequencing technology. The paper is structured as follow: Section 2 describes the use of three different software tools used to align the short reads generated from seven human samples to the reference human genome hg18 (NCBI Build 36, 2006); Section 3 presents the comparison of results using as benchmarking criteria: time, sensitivity and specificity; and Section 4 provides some conclusions and suggestions for future work.

2 Short Reads Alignment Using ELAND, Bowtie and BWA

Following an initial evaluation, in which more than 40 candidate programs were considered, we chose only three to conduct a more thorough comparison: ELAND [16] a mapper which indexes and hashes the reads before scanning the reference genome; and Bowtie [13] and BWA [14] two mappers which index the reference genome with the Burrows-Wheeler transform [17]. The three programs were chosen for several reasons including popularity, capability to operate on large data sets such as whole genome transcriptome or RNA-Seq, efficiency in time and space, coverage and accuracy. An additional advantage of BWA and Bowtie is that they do not require server-class hardware but can run in a desktop machine (Win/Mac/Linux) with a relatively small memory footprint.

The data used in this evaluation are whole genome RNA-Seq data from 7 human samples generated with the Illumina GAII sequencer. RNA was extracted from sickle cell (n=3) and control (n=4) blood outgrowth endothelial cells and sequenced using recommended protocol. Image analysis and base calling were performed using the Firecrest and Bustard modules, and the short reads were saved in FASTQ format (sequence files). On average, 3,000,000 single end reads per sample were obtained and each short read has a size of 40 bases. The results obtained with each tool are briefly described in the sections 2.1 (Illumina/ELAND), 2.2 (Bowtie) and 2.3 (BWA) respectively.

2.1 Alignment in Illumina

The reads were aligned to the human genome (hg18) using the alignment algorithm ELANDv2 in the module GERALD of the Illumina pipeline software (export files) with default parameters. As shown in Table 1, an average of 80% of filtered reads could be aligned to the human genome.

Table 1 ELAND alignments.

Sample	s1	s2	s3	s4	s5	s6	s7
Percentage of alignments	81%	82%	83%	83%	83%	78%	79%

Since there is variation within the human population's coding sequences, for example in polymorphic bases, it is expected that some reads cannot be aligned exactly to the reference genome. However, when ELAND is used with default parameters, there is no constraint on the maximum number of mismatches that are allowed for a read to be mapped to the reference genome and certain alignments generated by ELAND can have a large number of mismatches. To obtain information about the mismatches of the aligned reads, we analysed the match descriptors of the alignments generated with ELAND. The match descriptor is a string of numbers and letters, with numbers that represent run of matching bases, and letters that denote substitutions of bases relative to the referent sequence. The match descriptor “40” means a perfect match of the whole read of size 40 in the reference genome while, for example, the match descriptor “10G29” indicates that the bases 1-10 and 12-29 match the referent, while the nucleotide in position 11 is replaced by “G”. The match descriptor **“10G18C2TTAG3T”** (obtained in an alignment of the sample 1) indicates that the original nucleotides in the positions 1-10, 12-29, 31-32, 37-39 are perfect matches, while the nucleotides in position 11, 30, 33, 34, 35, 36 and 40 are base substitution. Therefore, the length of the match descriptor inform about the number of base substitution. In sample 1 we found 4,211 alignments whose match descriptor was longer than 13, and only two match descriptors of length 18. A match descriptor of length 13 usually has 7 or 8 nucleotides replaced, and a match of length 18 has 10 or 11 substitutions. However, the number of matches with a length less or equal to 6 was 16,304,860 for sample 1. As a match of size 6 usually has two or three replaces, this means that most of the alignments in ELAND have less than four mismatches.

2.2 Alignment in Bowtie

Bowtie is a very fast and memory-efficient short reads aligner. It is capable of aligning short DNA sequences to the human genome at a rate of over 25 million 35bp reads per hour. In addition, Bowtie indexes the genome with a Burrows-Wheeler index to reduce the memory in use, typically about 2.2 GB for the human genome. Thus, the first phase to Bowtie requires creating the indexes for the reference genome. In a second phase, these indexes are used to align the short reads stored in the data file on the reference genome.

To make the results comparable with the results generated with ELAND, we had to choose among different parameters to execute Bowtie. Two important parameters are the number of mismatches allowed per read, and the number of multiple alignments allowed per read. For example, the export files generated with ELAND include only one alignment for each aligned read, while in Bowtie, by default, all the possible alignments for each read are reported. So, we used the option “--best --strata” combined with the option “-m 1” in order to report only the best alignment for each read. This choice of parameters "-m 1 --best --strata" reports only reads with one alignment in best stratum (alignments in the best stratum are those having the least number of mismatches). As a consequence, reads with several best stratum alignments will not be reported. However, having multiple alignments in best stratum is not common, so the number of alignments that are lost in Bowtie respect to ELAND for this circumstance is low. Note that ignoring multiple alignments implies a loss of sensitivity. Another important parameter is the number of allowed mismatches, that is, the number of gaps (bases deleted) or substitutions (bases mutated, for example, a base “Cytosine” is replaced by the base “Guanine”) in each nucleotide sequence (short read) that are allowed to map a read to a certain position in the reference genome. In ELAND, the number of errors was not limited, so, in Bowtie we chose the largest allowable value, i.e., the value 3 (using for that the parameter “-v 3”). Using the parameters outlined above and detailed in the Table 2, the alignments reported in Bowtie for each sample are summarized in Table 3. In addition to the alignment percentages for 3 mismatches, the table includes the percentages for 1 and 2 mismatches. Clearly, the number of allowed mismatches positively correlates with the number of aligned reads (sensitivity) at the price of data quality (specificity).

Table 2 Bowtie parameters.

Phase 1 (Create the index): **bowtie-build hg18 index**	
hg18	Reference genome in FASTA format.
index	The base name of the index files to write.
Phase 2 (Create and report the alignments): **bowtie -p 8 -t -q -a -m 1 --best --strata -v Num –sam --solexa1.3-quals hg18 File**	
-q	The query input files are FASTQ files.
-t	Print the amount of wall-clock time taken by each phase.
-p 8	The -p option launches the specified number of parallel search threads. In our case, we use 8 threads.
--sam	The output in SAM format.
--solexa1.3-quals	Option for use with (unconverted) reads emitted by GA Pipeline version 1.3 or later.
-v Num	Report alignments with at most Num mismatches (maximum Num=3). We use Num=1 to 3 in different executions of Bowtie.
--best --strata	If many valid alignments exist and they fall into more than one alignment "stratum", report only those alignments that fall into the best stratum.
-a	Report all valid alignments per read. If more than one valid alignment exists and the --best and --strata option is specified, then only the best alignment "stratum" are reported.
-m 1	Suppress all alignments for a particular read if more than 1 reportable alignment exists for it. Reportable alignments are those that would be reported given the options: -v, -a, and --best --strata.
File	Sequence file obtained with the Illumina pipeline. We use 7 samples in different executions, File=s_1_sequence.fq to s_7_sequence.fq

Table 3 Percentage of aligned reads using Bowtie (the rows show the percentage of alignments for samples s1 to s7 with one, two and three allowed mismatches respectively).

Mismatches	s1	s2	s3	s4	s5	s6	s7
1	66%	71%	73%	73%	72%	65%	67%
2	70%	74%	75%	75%	75%	69%	72%
3	74%	76%	77%	77%	77%	72%	75%

2.3 Alignment in BWA

Burrows-Wheeler Aligner (BWA) is another popular and very efficient program to align short nucleotide sequences against a reference genome. It implements two algorithms, bwa-short, designed for short reads, and BWA-SW, designed for longer reads. bwa-short is designed to work with less errors than BWA-SW and performs gapped global alignment while BWA-SW performs a heuristic Smith-Waterman-like alignment to find high-scoring local hits. Because our reads have a length of 40 bases, we used the algorithm bwa-short that is recommended to align reads shorter than 200bp and an error rate less than 3%. The algorithm bwa-short is implemented with the "aln" command, which produces the suffix array (SA) coordinates of good hits of each individual read. Then, the "samse" command converts SA coordinates to chromosomal coordinates. An index of the reference genome must be generated with "bwa index" before "bwa aln" and "bwa samse".

By default BWA generates output in SAM format [18] (Bowtie was executed using the option "--sam" with the intention that the output format of both tools was the same) and finds all equally best hits only; that is, the reported alignments are in best stratum. Regarding the number of mismatches, the allowed maximum is established using the parameters "–n" and "–k" during the execution of the "aln" command. The option "–n" indicates the maximum number of differences allowed in the whole sequence. We used values 1, 2 and 3 for "–n", in a similar way to Bowtie (with "–v"). The option "–k" indicates the maximum number of differences allowed in the seed. The seed are the first bases of the sequence (first 32bp by default). In order to compare BWA with Bowtie, we disabled seeding in BWA using the option "-k 41". 41 is a number of errors higher than the length of the seed, which means that any number of errors is permitted in the seed. This implies that the errors allowed with the option –n can occur in any position of the read. Finally, the number of alignments reported for each read is limited to 1 (as in Bowtie), using the parameter "-n 1" in the execution of the "samse" command (similar to –m 1 in Bowtie). So, in identical circumstances of execution to Bowtie (with the parameters shown in Table 4), the tool BWA generates the alignments shown in Table 5.

Table 4 BWA parameters.

Phase 1 (Create the index): **bwa index -a bwtsw hg18**	
-a bwtsw	The option –a specifies the algorithm to construct the BWT index. In our case, we use the bwtsw method to be able to work with the whole human genome.
hg18	Reference genome in FASTA format.
Phase 2 (Create the alignments): **bwa aln -t 8 -n Num -k 41 hg18 File**	
-t 8	The `-t` 8 option permits to launch 8 parallel search threads.
`-n Num`	The option –n indicates the maximum number of differences allowed in the whole sequence. We use Num=1 to 3.
-k 41	The option –k establishes the maximum edit distance in the seed. Using a number greater than the size of the sequences in the data file, that is 40, we disable the seeding. So, the allowed mismatches (with –n) can occur in any position of the sequence.
File	Sequence file obtained with the Illumina pipeline. We use File=s_1_sequence.fq to s_7_sequence.fq
Phase 3 (Report the alignments): **bwa samse -n 1 hg18 aln_sa.sai File**	
aln_sa.sai	File with the SA coordinates obtained as result of the previous phase.
-n 1	This option is used to limit the maximum of reported alignments for a read. So, if a read has more than 1 hit (by default in best stratum), the alignment will not be written.

Table 5 Percentage of aligned reads using BWA (the rows show the percentage of alignments for samples s1 to s7 with one, two and three allowed mismatches respectively).

Mismatches	s1	s2	s3	s4	s5	s6	s7
1	77%	82%	83%	84%	82%	77%	76%
2	84%	86%	86%	87%	86%	82%	81%
3	88%	89%	88%	89%	88%	85%	85%

3 Comparison Study

We compared the performance of ELAND, Bowtie and BWA in terms of execution time, sensitivity and alignments. This required the implementation of different software utilities which can be found at: http://bios.ugr.es/comparisonEBB. Accordingly, the first section deals with the time required to complete the alignment process. The second section compares the total number of reads aligned. Finally, the third section compares the consistency of reads aligned with each tool. Note that additional comparisons between Bowtie and BWA can be found in scientific literature, such as [19 and 20]. However, these assessments are not comparable to our work because they use previous versions of the tools as well as different parameters of configuration and input data. In addition, these only compare quantity but not quality of the alignments.

With the purpose of comparing the **execution time of BWA vs. Bowtie**, both, BWA (version 0.5.8) and Bowtie (version 0.12.3) were executed using a Dell server with 4 quad-core Intel Xeon E7340 processors at 2.4 GHz and 64 GB of RAM. In order to launch the eight parallel search threads, the option "-p 8" was used in Bowtie and the option "-t 8" was used in BWA. The execution times of each run are shown in Table 6, for 1, 2 and 3 mismatches. The results show that Bowtie is usually faster that BWA, particularly as the number of errors increases. A reason is that Bowtie is designed to be very fast for small values in the parameter -m, and in our case, -m is established to 1. Alignment with ELAND was executed at the *Harvard Partner Center for Genetics and Genomics* and we do not have information about execution time.

Table 6 Execution time in minutes ("Bo" means Bowtie and "BW" means BWA).

	s1		s2		s3		s4		s5		s6		s7		Average	
Mismatches	*Bo*	*BW*	*Bo*	*BW*	*Bo*	*BW*	*Bo*	*BW*	*Bo*	*BW*	*Bo*	*BW*	*Bo*	*BW*	*Bo*	*BW*
1	3	20	4	27	5	27	4	24	3	22	4	24	3	21	3.71	23.57
2	7	34	9	44	9	44	7	38	7	35	8	40	8	36	7.85	38.71
3	12	138	13	168	13	167	11	137	11	129	13	165	12	144	12.14	149.71

In addition, we compared the **sensitivity of ELAND, Bowtie and BWA**. Since the reads for each sample are obtained from a concrete human genome, we think that it is reasonable to consider that the majority of the reads should have an alignment with the reference human genome. Thus, if we ignore errors arising from the sequencing process, it is expected that alignment "coverage" will be highly correlated with alignment sensitivity. Note that the simple existence of an alignment does not imply that it is correct, but we can assume that sensitivity will be proportional to the total amount of aligned sequences (especially when the number of allowed mismatches is so low).

Fig. 1 shows the sensitivity of each tool for each sample. In order to make the comparison with ELAND more precise, we use the percentage of alignment in Bowtie and BWA when three errors are permitted. In all samples, BWA is the most sensitive (third column in each sample) and Bowtie is the least sensitive (second column). Note that at most 3 errors were allowed in the alignments generated with Bowtie and BWA, while some of the alignments generated with ELAND included up to 11 mismatches (although, as explained above, the ELAND alignments with a high number of errors is very scaled down). In addition, as discussed above, in Bowtie and BWA, alignments that have multiple possibilities in best stratum are not reported, while in ELAND this circumstance is not considered (again this is a infrequent event).

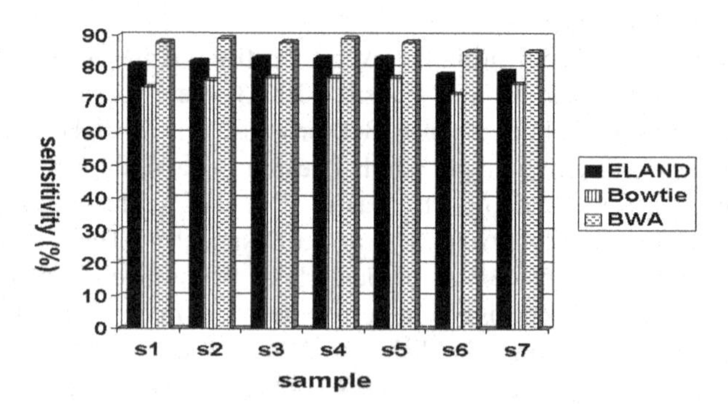

Fig. 1 Comparison of the percentage of reads aligned in ELAND, Bowtie and BWA (for 3 mismatches).

This initial analysis suggests that BWA aligns a larger number of reads than Bowtie. However, a new question arises: when the same read is aligned by both tools, is the read aligned to the same reference position? To answer this question, we compared the **alignment produced by Bowtie and BWA** to obtain: a) the number of reads that are aligned to the same genomic location by both tools, and b) the number of reads that are aligned to different genomic locations. To conduct this comparison, we took into account the fact that in Bowtie, the position is the 0-based offset into the forward reference strand where leftmost character of the alignment occurs, while in BWA is the 1-based leftmost position.

Table 7 Comparison of alignment quality in BWA *vs*. Bowtie (for sample 1).

Number of mismatches	1	2	3
Number of alignments in BWA	16,062,001 (77%)	17,532,464 (84%)	18,236,355 (88%)
Number of alignments in Bowtie	13,569,946 (66%)	14,632,541 (70%)	15,274,162 (74%)
Reads aligned in BWA, but not in Bowtie	2,492,055 (15'51%)	2,899,923 (16,54%)	2,962,193 (16,24%)
Reads aligned in Bowtie, but not in BWA	0	0	0
Identical alignments in both tools	13,569,946 (84,49%)	14,632,541 (83,46%)	15,274,162 (83,76%)

Table 7 summarizes the results for one sample and shows that reads aligned by Bowtie are aligned by BWA, but BWA aligns a larger number of reads (it may be because Bowtie does not report reads with several best alignments stratum while BWA does). In addition, the larger number of reads aligned by BWA relative to Bowtie increases as more errors are allowed. The results obtained in the other 6 samples also confirm that BWA consistently aligns more reads than Bowtie. For example, in sample 1, when 1 mismatch is allowed, the read HWI-EAS00184_0005_FC131:1:1:1055:8306#0 is aligned in the position 113,928,812 of the chromosome 4 in BWA while it is not aligned in Bowtie. In addition, the results of this comparison show that all reads aligned by both tools are aligned in the same way, that is, in the same positions in the reference genome (for 1, 2 and 3 mismatches). For example, for sample 1 with 1 allowed mismatch, the read HWI-EAS00184_0005_FC131:1:1:1033:6665#0 is aligned in the position 1,236,604 of the chromosome 1 by both tools. The complete comparison results for the sample 1 can be seen in Table 7.

4 Conclusions

High-throughput sequencing technologies are able to generate millions of short reads that have to be aligned to a reference genome before any further analysis can be conducted. A necessary and important task is the careful choice of the alignment algorithm and tool. Different proposals have been developed in the last few years and many more are emerging. To help investigators decide what tool is best to use in the initial phase of analysis of sequence data, we performed a comparison between ELAND, Bowtie and BWA that are among of the most popular tools for aligning short sequence reads. Both Bowtie and BWA are BWT-based aligners and require a similar memory space, which is independent of the number of reads.

Our comparison shows that Bowtie and BWA outperform ELAND in terms of sensitivity. In addition, Bowtie was always faster than BWA, but it aligned a smaller number of reads compared to BWA. Furthermore, the quality of reads aligned by both tools was comparable, so that BWA appears to increase sensitivity without sacrificing quality. The differences in execution time and genome coverage between the two tools can be partially explained by the fact that BWA performs gapped alignment and calculates a quality score of each match. In our experience, both tools are an excellent choice to align single-end short sequences.

If execution time is the first priority, Bowtie is a better choice than BWA. If sensitivity is the priority, BWA appears to be superior. Both tools support the SAM format to represent the performed alignments. However, the default Bowtie output (not SAM format) includes some information not included in the SAM output, such as the number of other instances where the same sequence aligned. Another important feature of BWA is that it permits configuration of more parameters such as the maximum number of gaps or deletions.

Our evaluation focused on ELAND, BWA and Bowtie provides additional results to a growing body of literature that evaluates aligners for next generation sequencing technology [21 and 22]. Future work will need to perform a more thorough comparison, including other parameter settings, additional aligners, additional elements of analysis such as splicing junctions and quality scores and combine real data with simulations.

Acknowledgements. This research is supported by the Andalusian Government R+D project P08-TIC-03717 and RC2HL101212/HL/NHLBI NIH HHS/United States.

References

1. Craig Venter, J., et al.: The sequence of the human genome. Science 291(5507), 1304–1351 (2001); doi:10.1126/science.1058040
2. Sinsheimer, R.L.: Sequencing the human genome: summary report of the Santa Fe workshop. Genomics 5(4), 954–956 (1989)
3. US Department of Health and Human Services and Department of Energy: Understanding our genetic inheritance. The U.S. human genome project: the first five years. US Dept. of Health and Human Services, Washington, DC (1990)
4. Strauss, E.C., Kobori, J.A., Siu, G., Hood, L.E.: Specific-primer-directed DNA sequencing. Anal. Biochem. 154(1), 353–360 (1986)
5. Yang, G., Ho, M.-H., Hubbell, E.: High-throughput microarray-based genotyping. In: IEEE Computational Systems Bioinformatics Conference, pp. 586–587 (2004)
6. Hall, N.: Advanced sequencing technologies and their wider impact in microbiology. The Journal of Experimental Biology 210(9), 1518–1525 (2007); doi:10.1242/jeb.001370
7. Pop, M., Salzberg, S., Shumway, M.: Genome sequence assembly: algorithms and issues. IEEE Computer 35, 47–54 (2002)
8. Mount, D.M.: Bioinformatics: sequence and genome analysis. Cold Spring Harbor Laboratory Press, Cold Spring Harbor,(2004); ISBN: 0-87969-608-7
9. Needleman, S.B., Wunsch, C.D.: A general method applicable to the search for similarities in the amino acid sequence of two proteins. Journal of Molecular Biology 48(3), 443–453 (1970)
10. Smith, T.F., Waterman, M.S.: Identification of common molecular subsequences. Journal of Molecular Biology 147(1), 195–197 (1981)
11. Drummond, A.J., Ashton, B., Buxton, S., Cheung, M., Cooper, A., Heled, J., Kearse, M., Moir, R., Stones-Havas, S., Sturrock, S., Thierer, T., Wilson, A.: Geneious v5.1 (2010), http://www.geneious.com
12. CLC Main Workbench: A comprehensive workbench for advanced DNA, RNA, and protein analyses, http://www.clcbio.com

13. Langmead, B., Trapnell, C., Pop, M., Salzberg, S.L.: Ultrafast and memory-efficient alignment of short DNA sequences to the human genome. Genome Biology 10(3), R25 (2009)
14. Li, H., Durbin, R.: Fast and accurate short read alignment with Burrows-Wheeler transform. Bioinformatics 25(19), 1754–1760 (2009)
15. Giardine, B., Riemer, C., Hardison, R.C., Burhans, R., Elnitski, L., Shah, P., Zhang, Y., Blankenberg, D., Albert, I., Miller, W., et al.: Galaxy: a platform for interactive large-scale genome analysis. Genome Research 15(10), 1451–1455 (2005)
16. Illumina: Illumina sequencing, `http://www.illumina.com`
17. Nelson, M.: Data compression with the Burrows-Wheeler transform. Dr. Dobb's Journal of Software Tools 21(9), 46–50 (1996)
18. Li, H., Handsaker, B., Wysoker, A., Fennell, T., Ruan, J., Homer, N., Marth, G., Abecasis, G., Durbin, R.: 1000 genome project data processing subgroup. The sequence alignment/map format and SAMtools. Bioinformatics 25(16), 2078–2079 (2009)
19. Li, H., Durbin, R.: Fast and accurate short read alignment with Burrows-Wheeler transform. Bioinformatics 25(14), 1754–1760 (2009)
20. Hoffmann, S., Otto, C., Kurtz, S., Sharma, C.M., Khaitovich, P., Vogel, J., Stadler, P.F., Hackermuller, J.: Fast mapping of short sequences with mismatches, insertions and deletions using index structures. PLoS Computational Biology 5(9), R1000502 (2009)
21. Ruffalo, M., Laframboise, T., Koyutürk, M.: Comparative analysis of algorithms for next-generation sequencing read alignment. Bioinformatics 27(20), 2790–2796 (2011)
22. Heng, L., Nils, H.: A survey of sequence alignment algorithms for next-generation sequencing. Briefings in Bioinformatics 11(5), 473–483 (2010)

SAMasGC: Sequencing Analysis with a Multiagent System and Grid Computing

Roberto González, Carolina Zato, Rocío Benito, María Hernández, Jesús M. Hernández, and Juan F. De Paz

Abstract. Advances in bioinformatics have contributed towards a significant increase in available information. Information analysis requires the use of distributed computing systems to best engage the process of data analysis. This study proposes a multiagent system that incorporates grid technology to facilitate distributed data analysis by dynamically incorporating the roles associated to each specific case study. The system was applied to genetic sequencing data to extract relevant information about insertions, deletions or polymorphisms.

Keywords: Multiagent system, Grid Computing, genetic sequencing, distributed computing, bioinformatics.

1 Introduction

Advances in genetic sequencing have made it possible to carry out massive sequencing and to perform studies on genetic sequencing in multiple patients [2] [3]. One area of medicine in its height of development and fundamental in the application of techniques that facilitate the automatic treatment of data and the extraction of knowledge is genomics. This increase in information has made it necessary to create systems that can perform a distributed analysis of information and be

Roberto González · Carolina Zato · Juan F. De Paz
Department of Computer Science, University of Salamanca
Plaza de la Merced, s/n, 37008, Salamanca, Spain
e-mail: {rgonzalezramos,carol_zato,fcofds}@usal.es

Rocío Benito · María Hernández · Jesús M. Hernández
IBMCC, Cancer Research Center, University of Salamanca-CSIC, Spain
e-mail: {beniroc,jhmr}@usal.es, mahesa2504@hotmail.com

Jesús M. Hernández
Servicio de Hematología, Hospital Universitario de Salamanca, Spain

M.P. Rocha et al. (Eds.): 6th International Conference on PACBB, AISC 154, pp. 209–216.
springerlink.com

adapted to different types of analysis such as with genetic sequencing. The development of distributed applications and parallel computing currently requires the use of complex software and libraries [7], while some, such as MPI and CUDA [8], even use combinations of libraries. It has become necessary to develop systems that facilitate the creation of distributed systems that allow the distributed implementation and execution of different types of analyses+ by applying technologies such as grid computing [6].

The rise in bioinformatics, primarily following the emergence of microarrays and sequencing in particular, has made it possible to generate great volumes of information [5]. With the appearance of expression arrays, specifically BAC arrays and more importantly Exon arrays [5], it became necessary to create systems that would allow the distributed analysis of information to improve the output of algorithms. The use of NGS (next generation sequencing) has noticeably increased the amount of information, which has in turn led to an improvement in output and a reduction in execution time. As a result, it has become necessary to create systems that facilitate the management of distributed systems. These systems must facilitate the creation of algorithms that are executed in a distributed way, which enables the dynamic generation of control flows.

This study proposes the use of multiagent systems [9] capable of distributed execution performed by the integration of grid technology [6]. The multiagent system integrates agents to manage the roles in the case study that is being carried out. Within the context of this study, the proposed system focuses on detecting relevant patterns and mutations, insertions, deletions or polymorphisms, within the sequence data taken from patient samples provided by the Cancer Institute of the University of Salamanca. The analysis of sequencing data requires various types of processes: i) assembly [4] ii) alignment [4] and iii) knowledge extraction [5] in order to analyze sequence data. The Cancer Institute of the University of Salamanca is striving to develop tools to automate the evaluation of data and to facilitate the analysis of information. This proposal is a step forward in this direction and the first step toward the development of a multiagent system.

This article is structured as follows: section 2 reviews the state of the art in genetic sequencing; section 3 presents the proposed architecture and adapts the architecture to the case study; section 4 presents the results and conclusions.

2 Massive Analysis and Sequencing

Sequencing began in the 60s, although it was not until the 80s and the Sanger method [4] that gene and genome sequencing emerged. The sequencing process was a laborious manual process; following the development of automated sequencing in the late 80s the volume of information increased dramatically. The process of separating DNA fragments with automated sequencing was initially performed with gel electrophoresis [1], subsequently replaced by capillary electrophoresis [1], after which pyrosequencing [1] was developed. There are currently various types of NGS with different capabilities in base pairs. Zhang et al. [4] describe the different manufacturers. The length of the fragments of the base pairs can vary according to the sequencing used, from 25 bp to the 500 bp used with sequencing by the Roche company, which can perform de-novo sequencing [4]. In the near future, the length of sequenced base pairs is expected to increase considerably; in fact, new research in techniques has developed SMRT (single molecule real time) sequencing, which can achieve 10,000 bp, facilitating the processes of new genome assembly and sequencing.

The human genome is estimated to be about 3,000 million base pairs long and contain around 25,000 genes [10]. Consequently, sequencing genome fragments of 500 bp at a time is costly and requires computational techniques that can join contig fragments to generate the complete genome. Sequencing is not usually applied to just any part of the genome; instead, specific exon sequences corresponding to the DNA code are selected. Exons are the part of the DNA that is represented in the messenger RNA. The regions that are transcribed in the messenger RNA can later be converted into proteins [11], hence the relevance of its analysis and the detection of variants.

3 Proposed SMASasGC Architecture

Bioinformatic data analysis requires different processes and algorithms that vary according to the data recovered. Nevertheless, these different types of analyses all share a common characteristic, namely the high computing cost involved. It has become necessary to develop systems that facilitate the efficient development of distributed systems. In order to analyze distributed data in an efficient manner, grid technology [6] and multiagent systems [9] are integrated to produce a high-performance hybrid architecture. The architecture created for this study contains three separate layers: the coordination layer contains agents assigned to maintain the algorithms specific to the case study; the control layer contains agents responsible for controlling the grid; and the specialization layer contains the agents and the processes specific to the case study.

Figure 1 displays the agents that correspond to the coordination and control layer. These layers are independent of the case study, allowing their functionality to be reused.

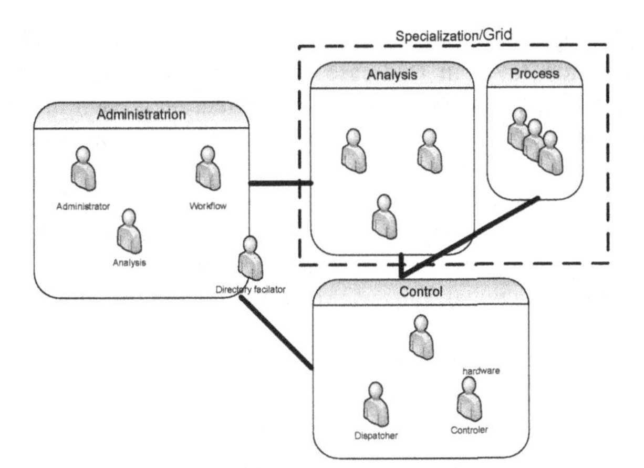

Fig. 1 Architecture for system coordination and control

The coordination layer includes the administrator, analysis, workflow and directory facilitator agents. The administrator agent is in charge of storing and controlling project data for their subsequent analysis. Each project contains information about the flow of analysis and the results obtained by the applied algorithms. Additionally administrator agent includes all associated roles including the status of the project process, launched tasks, and the type of data associated with the different case studies. The analysis agent is in charge of recovering previous flows of analysis and storing new flows of analysis that may be used to recommend flows of execution with similar data. The workflow agent is responsible for creating new flows of execution based on existing algorithms for different types of analyses. Finally, the directory facilitator (DF) stores and administers existing algorithms for each of the possible types of analysis; its functionality is similar to that of a web services DF.

The control layer includes the agents responsible for controlling the state of execution for the grid. The dispatcher, hardware and controller agents are available in this layer. The dispatcher agent is in charge of recovering and distributing tasks between the nodes. The hardware agent controls machine resources available on the grid. The controller agent controls the machine load level to control the state of execution for the machines containing grid.

The specialization layer is composed of agents and processes that are executed in the grid nodes. These processes and agents are specific to each case study and are responsible for defining the hardware needs for their execution, and breaking the tasks down into subtasks that are sent to the dispatcher.

3.1 Sequence Analysis

The process of sequence data analysis varies according to the results that one wants to obtain. It normally requires a process of assembly, alignment and knowledge extraction to automatically process the data. As the architecture must be specific to this end, agents and processes specialized in performing these tasks are required. The agents are responsible for establishing the restrictions and procedures for distributing tasks along the grid nodes according to available resources.

The specialization layer in this case study was composed of the following agents: assembly, alignment and knowledge extraction. Each agent defines the following roles for the purpose of carrying out the task for which they were added to the system: manage available algorithms to execute the task; manage the resources needed to apply each algorithm; determine the preconditions for executing tasks; manage the nodes required to execute the tasks; break the tasks down into subtasks that are subsequently queued in the dispatcher.

Each agent in this stage has various processes that are executed through grid in a distributed manner. The processes can vary according to the algorithm that is selected in the work flow of the project created in the analysis. Thus, the algorithms developed for each stage of the case study are as follows:

3.1.1 Assembly

The assembly process varies according to the size of each reading that is used. In this particular case, we chose to use the algorithm provided by the manufacturer of the sequencer being used. Different assembly algorithms can be seen in [4] and [16]. Roche provides the Newbler assembler [15], which was used in this study.

3.1.2 Alignment

The alignment process consists of establishing the fragment of the reference genome that is most similar to the fragment of the patient being treated. The alignment algorithms are applied to different fields in addition to bioinformatics.

While there are many different ways to carry out the alignment process, performance is ultimately the most important factor. The alignment algorithms used are local, since the sequence to be aligned, or the contigs, is smaller in size than the reference genome. Local alignments are based on the Smith-Waterman algorithm [12]. The alignments can be given in pairs or groups according to the number of fragments that must be analyzed simultaneously. There are currently many alignment algorithms, but the most commonly used are BLAST (Basic Local Alignment Search Tool) [13] and BLAT (BLAST-Like Alignment Tool) [14]. BLAT can perform an alignment faster than BLAST, but it cannot ensure that the final alignment is the best one possible, although performance is greatly improved. Additionally, there are many algorithms that can be found in different review articles such as [4] and [16].

3.1.3 Extraction of Knowledge

The classification algorithms can be divided in: decision trees, decision rules, probabilistic models, fuzzy models, based on functions, ensemble. During the extraction of knowledge phase, different analyses of the variations were performed to detect the following types of alterations: SNP, point mutation not SNP, insertions and deletions. The process of detecting SNP and other point mutations is simple since it only involves searching information in databases that contain the previously published information.

4 Results and Conclusions

Genetic sequencing was applied to a data set taken from patients with leukemia. Specific genes were sequenced from a total of 8 patients, each of whom had approximately 110,000 sequence fragments that corresponded to the regions relevant to this study. The sequenced fragments vary in length for the different patients, as shown in figure 2.

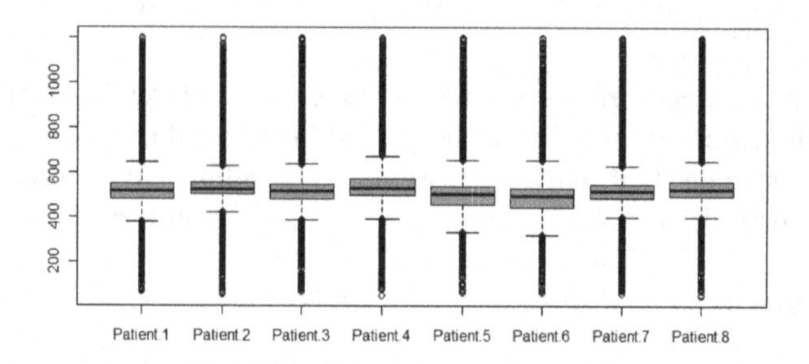

Fig. 2 Box diagram of the lengths of the fragmented segments.

The version of the reference genome used in this study corresponds to HG18; this is because the information used as a reference in selecting the sequence regions was obtained in previous studies using the same version.

The system was tested to validate the generic architecture proposed and to validate the system's capacity for analyzing the proposed case study. The main goal in validating the architecture was to determine the efficiency of the system and its ability to distribute tasks according to the existing nodes. To validate the increase in performance, we selected the Newbler, BLAT. Figure 3 demonstrates the size of the contigs that were assembled once the Newbler algorithm was applied and executed on the grid.

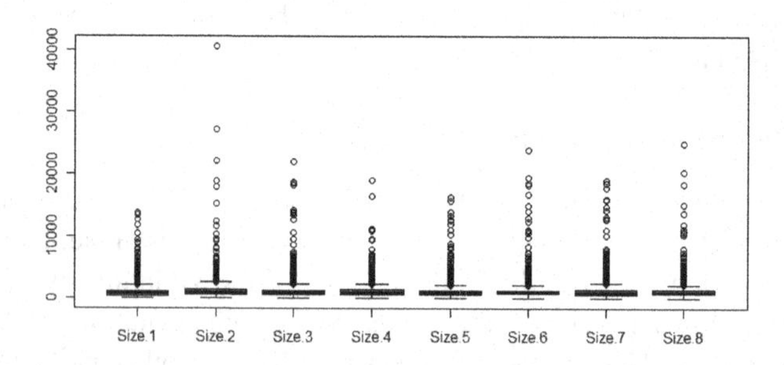

Fig. 3 Box diagram with the length of the assembled contigs

The average sizes for each of the aligned patients are as follows: 1023.134 1356.855 1292.022 1251.492 1239.192 1306.707 1302.321 1362.151.

After completing the assembly process, the fragments were aligned using the BLAT algorithm, which obtained PSL output files. The specific files used to predict the alterations are the following: matches, misMatches, repMatches, nCount, qNumInsert, qBaseInsert, tNumInsert, tBaseInsert, qSize, qStart, qEnd, tSize, tStart, tEnd, blockCount, blockSizes. The analyzed alterations are insertions, deletions and polymorphisms, in order to detect these alterations the system used the SNP130 table from UCSC. The final number of alterations in the patient 1 are shown in table 1, the number of unknown alterations (insertions and deletions) is high due to the pathology and the analyzed regions.

Table 1 Variants

Initial Contigs	Contigs with variants	SNPs	Unknown variants
1310	1100	6584	2021

The next step was to analyze the execution time in order to observe the system's scalability according to the number of nodes on which the processes were dropped. The execution time decrease linearly with the number of nodes in the system. The table 2 shows the execution times in seconds for several nodes.

Table 2 Time in seconds as the number of nodes

	1 Node	2 Nodes	3 Nodes
Segmentation	691s	405s	241s
Polymorphism	2340s	1145s	710s

The multiagent system has made it possible to integrate algorithms that can adapt to a specific case study, facilitating the distributed execution of work flows. The system facilitates the integration of algorithms for different case studies and reduces the execution time in an efficient manner, so long as it remains possible to improve performance by separating tasks for their more effective execution in grid technology.

Acknowledgments. This work has been supported by the MICINN TIN 2009-13839-C03-03.

References

[1] Mitchelson, K.R., Hawkes, D.B., Turakulov, R., Men, A.E.: Chapter Overview: Developments in DNA Sequencing. Perspectives in Bioanalysis 2, 3–44 (2007)
[2] Soo, R.A., Wang, L.Z., Ng, S.S., Chong, P.Y., Yong, W.P., Lee, S.C., Liu, J.J., Choo, T.B., Tham, L.S., Lee, H.S., Goh, B.C., Soong, R.: Distribution of gemcitabine pathway genotypes in ethnic Asians and their association with outcome in non-small cell lung cancer patients. Lung Cancer 63(1), 121–127 (2009)
[3] Esteban, F., Royo, J.L., González-Moles, M.A., Gonzalez-Perez, A., Redondo, M., Moreno-Luna, R., Rodríguez-Sola, M., Gonzalez, A., Real, L.M., Ruiz, A., Ramírez-Lorca, R.: CAPN10 alleles modify laryngeal cancer risk in the Spanish population. European Journal of Surgical Oncology (EJSO) 34(1), 94–99 (2008)
[4] Zhang, J., Chiodini, R., Badr, A., Zhang, G.: The impact of next-generation sequencing on genomics. Journal of Genetics and Genomics 38(3), 95–109 (2011)
[5] Corchado, J.M., De Paz, J.F., Rodríguez, S., Bajo, J.: Model of experts for decision support in the diagnosis of leukemia patients. Artificial Intelligence in Medicine 46(3), 179–200 (2009)
[6] Gregoretti, F., Laccetti, G., Murlib, A., Olivaa, G., Scafuri, U.: MGF: A grid-enabled MPI library. Future Generation Computer Systems 24(2), 158–165 (2008)

[7] Jin, H., Jespersen, D., Mehrotra, P., Biswas, R., Huang, L., Chapman, B.: High performance computing using MPI and OpenMP on multi-core parallel systems. Parallel Computing (in Press)

[8] Rakić, P.S., Milašinović, D.D., Živanov, Ž., Suvajdžin, Z., Nikolić, M., Hajduković, M.: MPI–CUDA parallelization of a finite-strip program for geometric nonlinear analysis: A hybrid approac. Advances in Engineering Software, 42(5), 273–285 (2011)

[9] Wooldridge, M.: Introduction to Mult iAgent Systems. John Wiley & Sons (2002)

[10] International Human Genome Sequencing Consortium, Finishing the euchromatic sequence of the human genome. Nature 431, 931–945 (2004)

[11] Sahua, S.S., Pand, G.: dentification of Protein-Coding Regions DNA Sequences Using A Time-Frequency Filtering Approach. Genomics, Proteomics & Bioinformatics 9(1-2), 45–55 (2011)

[12] Khajeh-Saeed, A., Poole, S., Perot, J.B.: Acceleration of the Smith–Waterman algorithm using single and multiple graphics processors. Journal of Computational Physics 229(11), 4247–4258 (2010)

[13] Altschul, S.F., Gish, W., Miller, W., Myers, E.W., Lipman, D.J.: Basic local alignment search tool. Journal of Molecular Biology 215(3), 403–410 (1990)

[14] Kent, W.J.: BLAT—the BLAST-like alignment tool. Genome. Res. 12, 656–664 (2002)

[15] Margulies, M., Egholm, M., Altman, W.E., Attiya, S., Bader, J.S., Bemben, L.A., Berka, J., Braverman, M.S., Chen, Y.J., Chen, Z., Dewell, S.B., Du, L., Fierro, J.M., Gomes, X.V., Godwin, B.C., He, W., Helgesen, S., Ho, C.H., Irzyk, G.P., Jando, S.C., Alenquer, M.L., Jarvie, T.P., Jirage, K.B., Kim, J.B., Knight, J.R., Lanza, J.R., Leamon, J.H., Lefkowitz, S.M., Lei, M., Li, J., Lohman, K.L., Lu, H., Makhijani, V.B., McDade, K.E., McKenna, M.P., Myers, E.W., Nickerson, E., Nobile, J.R., Plant, R., Puc, B.P., Ronan, M.T., Roth, G.T., Sarkis, G.J., Simons, J.F., Simpson, J.W., Srinivasan, M., Tartaro, K.R., Tomasz, A., Vogt, K.A., Volkmer, G.A., Wang, S.H., Wang, Y., Weiner, M.P., Yu, P., Begley, R.F., Rothberg, J.M.: Genome sequencing in microfabricated high-density picolitre reactors. Nature 437, 376–380 (2005)

[16] Miller, J.R., Koren, S., Sutton, G.: Assembly algorithms for next-generation sequencing data. Genomics 95(6), 315–327 (2010)

Exon: A Web-Based Software Toolkit for DNA Sequence Analysis

Diogo Pratas, Armando J. Pinho, and Sara P. Garcia

Abstract. Recent advances in DNA sequencing methodologies have caused an exponential growth of publicly available genomic sequence data. By consequence, many computational biologists have intensified studies in order to understand the content of these sequences and, in some cases, to search for association to disease. However, the lack of public available tools is an issue, specially when related to efficiency and usability. In this paper, we present Exon, a user-friendly solution containing tools for online analysis of DNA sequences through compression based profiles.

Keywords: web-based software toolkit, DNA sequence analysis, DNA compression.

1 Introduction

The construction and analysis of DNA complexity profiles has been an important topic of research, due to its applicability in the study of regulatory functions of DNA, comparative analysis of organisms, genomic evolution and others [7, 11]. For example, it has been observed that low complexity regions of DNA are often associated with important regulatory functions [6].

Several measures have also been proposed for evaluating the complexity of DNA sequences. Among those, we find the compression-based approaches the most promising and natural, because compression efficiency is clearly defined (it can be measured by the number of bits generated by the encoder) [4, 1, 5, 15].

One of the key advantages of DNA compression based on finite-context models [13, 9, 10, 8] is that the encoders are fast and have $\mathcal{O}(n)$ time complexity.

Diogo Pratas · Armando J. Pinho · Sara P. Garcia
Signal Processing Lab, IEETA / DETI, University of Aveiro,
3810–193 Aveiro, Portugal
e-mail: {pratas,ap,spgarcia}@ua.pt

M.P. Rocha et al. (Eds.): 6th International Conference on PACBB, AISC 154, pp. 217–224.
springerlink.com

Most of the effort spent by previous DNA compressors is in the task of finding exact or approximate repeats of sub-sequences or of their inverted complements. No doubt, this approach has proven to give good returns in terms of compression gains, but it may be disadvantageous in terms of time consuming to perform the compression. Although slow encoders could be tolerated for storage purposes (compression could be ran in batch mode), for interactive applications they are certainly not appropriate. For example, the currently best performing DNA compression technique, eXpert-Model (XM) [3], could take hours for compressing a single human chromosome. Compressing one of the largest human chromosomes with the techniques based on finite-context models (FCM) takes less than ten minutes. Along with this inconvenience, there is the need for a strong computational system to perform these operations, particularly with regard to memory (RAM). On the other hand, these tools require local installation, sometimes on particular operating systems, which makes them inconvenient to use. Moreover, there is the need of designing a graphical interface to make Exon attractive to biologists and not only informatics, instead of the prior command line approach.

In this paper, we provide solutions to the mentioned issues using Exon, a web-based user-friendly software toolkit that analyses DNA sequences using compression based approaches, such as finite-context modelling [13, 9, 8] and XM [3]. Since the software is web-based, it is available to any computer (with a web browser linked to the Internet) and without the need of any software installation. Moreover, Exon is hosted at a virtual web server composed by 8 cores of Intel(R) Xeon(R) CPU X5650 @ 2.67GHz, with at least 8 GB of RAM and with 2 TB of storage, running the CentOS linux distribution.

This paper is organized as follows. In Section 2, we describe the methods used, in particular the compression approaches. In Section 3, we provide some examples of the Exon usage. Finally, in Section 4, we draw some conclusions.

2 Materials and Methods

Using technologies and programming languages such as HTML, PHP, Java, Javascript, CSS, PostgresSQL, shell script and C, we were able to integrate several methods in the Exon toolkit. All these methods fall into one of three categories (pre-encoding, encoding and post-encoding).

The first category is pre-encoding (sequence edition before encoding). In this category, there are the following methods: reverse (a sequence), concatenate (two sequences) and generate (a sequence). The first two methods are self-explanatory, due to their names. Sequence generation is based on multiple competing finite-context models. In short, this generator allows to generate a synthetic sequence based on statistics collected from a template sequence, using a stochastic process [12].

The second category is encoding (sequence compression). In this category, there are two possible models: XM or FCM (finite-context modelling). The first method, XM [3], relies on a mixture of experts for providing symbol by symbol probability estimates, which are then used for driving an arithmetic encoder. The algorithm comprises three types of experts: (1) order-2 Markov models; (2) order-1 context Markov models, i.e., Markov models that use statistical information only of a recent past (typically, the 512 previous symbols); (3) the copy expert, that considers the next symbol as part of a copied region from a particular offset. The probability estimates provided by the set of experts are then combined using Bayesian averaging and sent to the arithmetic encoder. Although the XM approach is inappropriate for large DNA sequences, we have included it in the Exon toolkit, because it can be used in small sequences (normally below 2 MB) and also for comparison with other methods. The other method, FCM [8], is an approach based on multiple finite-context models of different orders that compete for encoding the data. Figure 1 gives an example of these multiple models. The competitive procedure implies that the best of the models is chosen for encoding each DNA block, i.e., the one that requires less bits is used for representing the current block.

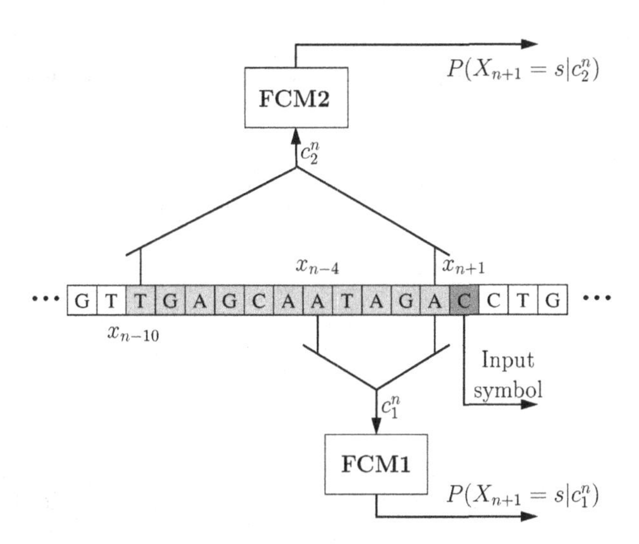

Fig. 1 Example of the use of multiple finite-context models for encoding DNA sequence data. In this case, two models are used, one with a depth-5 context and the other using an order-11 context.

The third (and last) category is the post-encoding (computation and manipulation of the information content received from category two). In this particular category, there is a module capable of processing the information content, by applying signal processing techniques, such as filtering, in order to improve the visual output of the complexity profiles.

The process of building a complexity profile for visualization can be seen in Figure 2.

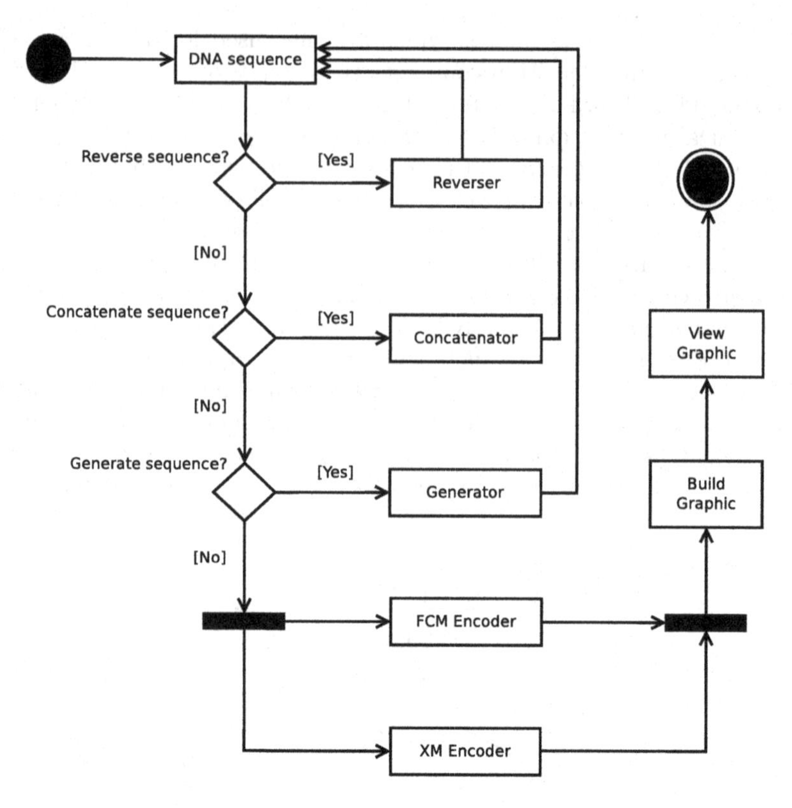

Fig. 2 Activity diagram of the process of building a complexity profile for visualization.

Exon has a pyramidal hierarchy structure of 3 levels (roles). The first and lowest is the guest user. Here, the user can perform the basic operations, namely those described in Figure 2, although, with certain restrictions such as limited number of FCM models to be chosen. In the middle, there is the operator level. This is a registered user that has the same privileges as the guest user, but with access to more features such as the uploading of sequences. The upper level is administrator. This is the unlimited user, that can also control logs, edit files, edit accounts, edit models, among other operations.

2.1 Software Availability

This web-based software toolkit is publicly available for non-commercial purposes at http://exon.ieeta.pt.

3 Some Examples

In this section, we provide some examples where we used Exon. For that purpose, we used the *Saccharomyces cerevisiae* genome (uid128) from the National Center for Biotechnology Information (NCBI).

Figure 3 shows an example of a complexity profile (corresponding to chromosome 2 of the *S. cerevisiae*), generated by a multiple finite-context model DNA encoder in Exon.

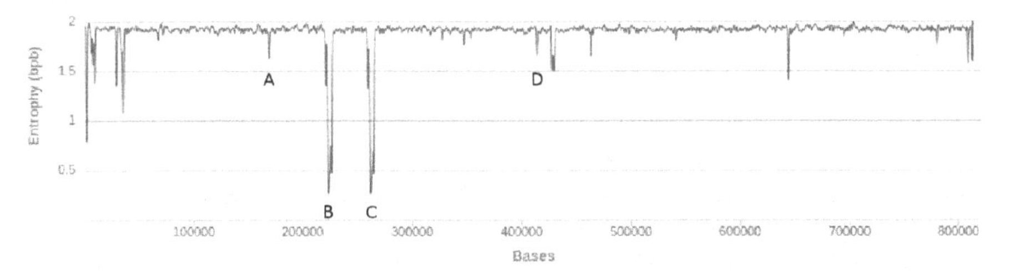

Fig. 3 Complexity profile of chromosome 2 from the *S. cerevisiae* genome. The information content was processed in both directions, combined using the minimum value of each direction, and low-pass filtered using a Blackman window of size 11.

We can observe several regions where the complexity is very small, meaning that a reduced number of bits was required for compressing those regions. Of particular interest are the two regions that are below 0.5 bpb (marked with letters B and C). These regions, which have been zoomed-in in Figure 4, correspond to transposons (sequences of DNA that can move or transpose themselves to new positions within the genome of a single cell). There are strong evidences that genetic diseases are caused by transposition events, for more information see [14].

The transposon marked with letter B (YBLWTy1-1) has 5,915 bases (from base 221,037 to base 226,952). The transposon marked with letter C (YBRWTy1-2) has 5,917 bases (from base 259,578 to base 265,494). Moreover, a Blast search [2] indicates that these sequences have ~99% of identity.

Similarly, the regions marked with letters A and D represent homologous genes. In particular, the region marked with letter A, with 954 bases (from base 168,423 to base 169,376), represents the gene RPL19B. On the other hand, the region marked with letter D, with 1,076 bases (from base 414,186 to base 415,261), represents the gene RPL19A. A Blast search indicates that these sequences have ~93% of identity.

Another application of Exon is the ability to perform inter-sequence analysis, i.e., using sequence concatenation it is possible to unveil low complexity zones which are usually associated to zones of potential biological

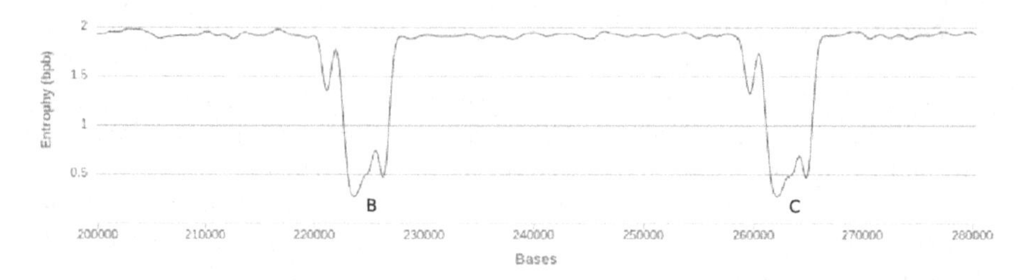

Fig. 4 Zoom of the complexity profile (from base 200,000 to base 280,000) of chromosome 2 from the *S. cerevisiae* genome. The information content was processed in both directions, combined using the minimum value of each direction, and low-pass filtered using a Blackman window of size 11.

interest. Thereby, we have concatenated chromosome 2 and 4 of of *S. cerevisiae* genome (showing just the results for the range of chromosome 2) and built the correspondent complexity profile (see Figure 5).

In Figure 5, it is possible to observe (second row) that some new low complexity zones have been unveiled, comparing with the complexity profile of chromosome 2 alone (first row). These regions have been marked with letters E and F. Relatively to region E, the complexity profile shows two locations, almost coincident, of tRNA (tRNA-Ile and tRNA-Gly), which are contained also in chromosome 4. Letter F marks a region containing 1,089 bases, corresponding to gene RPL4A (from base 300,166 to base 301,254). Thereafter,

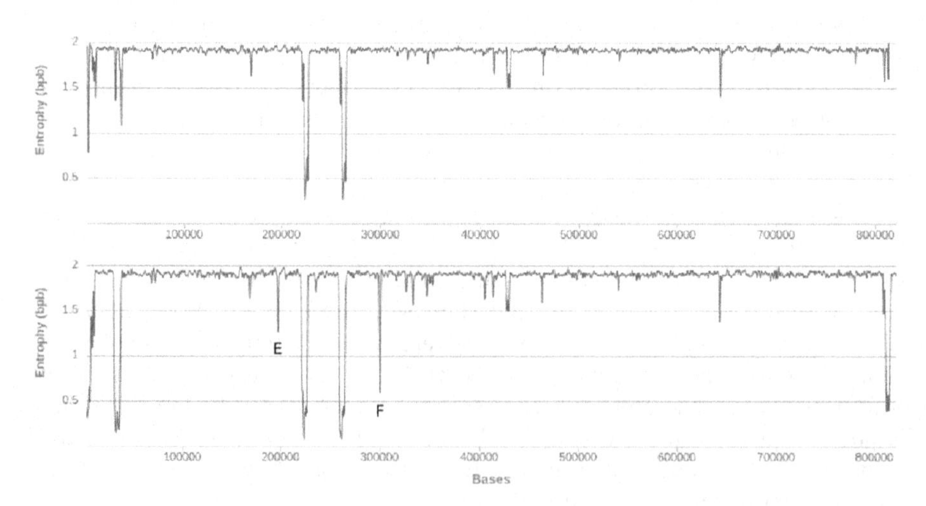

Fig. 5 Complexity profile comparison of chromosome 2 from the *S. cerevisiae* genome. The first row shows the information content of chromosome 2. The second row shows chromosome 2 with information added from chromosome 4. The information content was processed in both directions, combined using the minimum value of each direction, and low-pass filtered using a Blackman window of size 11.

we determined the corresponding source, leading to a region in chromosome 4 containing also 1,089 bases corresponding to gene RPL4B (from base 414,186 to base 415,261). Blast search indicates that these sequences have ~100% of identity, i.e., these sequences are equal. These very similar genes have homologous in other species, such as in the *Homo sapiens*.

4 Conclusion

In this paper, we have presented Exon, a user-friendly solution containing tools for online analysis of DNA sequences through compression based profiles.

The main aim was to bridge the field of biology and the field of informatics, allowing a biologist without expertise in computer science to create complexity profiles and analyse DNA sequences. To attain that, we have provided a graphical interface, avoiding the command line approach, together with the availability of a powerful web server, not requiring software installation (with common usage, accessible to any computer connected to the Internet).

Also, we have demonstrated the usefulness of Exon, building complexity profiles of a few sequences from the *Saccharomyces cerevisiae* genome. In particular, we have identified transposons, tRNA genes, similar and highly similar genes.

Acknowledgements. This work was partially funded by FEDER through the Operational Program Competitiveness Factors - COMPETE and by National Funds through FCT - Foundation for Science and Technology in the context of the project FCOMP-01-0124-FEDER-010099 (FCT reference PTDC/EIA-EIA/103099/2008). Sara P. Garcia acknowledges funding from the European Social Fund and the Portuguese Ministry of Education and Science.

References

1. Allison, L., Stern, L., Edgoose, T., Dix, T.I.: Sequence complexity for biological sequence analysis. Computers & Chemistry 24, 43–55 (2000)
2. Altschul, S.F., Gish, W., Miller, W., Myers, E.W., Lipman, D.J.: Basic local alignment search tool. Journal of Molecular Biology 215(3), 403–410 (1990); doi:10.1006/jmbi.1990.9999
3. Cao, M.D., Dix, T.I., Allison, L., Mears, C.: A simple statistical algorithm for biological sequence compression. In: Proc. of the Data Compression Conf., DCC 2007, Snowbird, Utah, pp. 43–52 (2007)
4. Crochemore, M., Vrin, R.: Zones of low entropy in genomic sequences. Computers & Chemistry, 275–282 (1999)
5. Dix, T.I., Powell, D.R., Allison, L., Bernal, J., Jaeger, S., Stern, L.: Comparative analysis of long DNA sequences by per element information content using different contexts. BMC Bioinformatics 8(suppl. 2), S10 (2007); doi:10.1186/1471-2105-8-S2-S10

6. Gusev, V.D., Nemytikova, L.A., Chuzhanova, N.A.: On the complexity measures of genetic sequences. Bioinformatics 15(12), 994–999 (1999)
7. Nan, F., Adjeroh, D.: On the complexity measures for biological sequences. In: Proc. of the IEEE Computational Systems Bioinformatics Conference, CSB 2004, Stanford, CA (2004)
8. Pinho, A.J., Ferreira, P.J.S.G., Neves, A.J.R., Bastos, C.A.C.: On the representability of complete genomes by multiple competing finite-context (Markov) models. PLoS ONE 6(6), e21, 588 (2011); doi:10.1371/journal.pone.0021588
9. Pinho, A.J., Pratas, D., Ferreira, P.J.S.G.: Bacteria DNA sequence compression using a mixture of finite-context models. In: Proc. of the IEEE Workshop on Statistical Signal Processing, Nice, France (2011)
10. Pinho, A.J., Pratas, D., Ferreira, P.J.S.G., Garcia, S.P.: Symbolic to numerical conversion of DNA sequences using finite-context models. In: Proc. of the 19th European Signal Processing Conf., EUSIPCO 2011, Barcelona, Spain (2011)
11. Pirhaji, L., Kargar, M., Sheari, A., Poormohammadi, H., Sadeghi, M., Pezeshk, H., Eslahchi, C.: The performances of the chi-square test and complexity measures for signal recognition in biological sequences. Journal of Theoretical Biology 251(2), 380–387 (2008)
12. Pratas, D., Bastos, C.A.C., Pinho, A.J., Neves, A.J.R., Matos, L.: DNA synthetic sequences generation using multiple competing Markov models. In: Proc. of the IEEE Workshop on Statistical Signal Processing, Nice, France (2011)
13. Pratas, D., Pinho, A.J.: Compressing the human genome using exclusively Markov models. In: Proc. of the 5th Int. Conf. on Practical Applications of Computational Biology & Bioinformatics, PACBB 2011. AISC, vol. 93, pp. 213–220 (2011)
14. Roy, A., Carroll, M., Kass, D., Nguyen, S., Salem, A., Batzer, M., Deininger, P.: Recently integrated human alu repeats: finding needles in the haystack. Genetica 107(1-3), 149–161 (1999)
15. Troyanskaya, O.G., Arbell, O., Koren, Y., Landau, G.M., Bolshoy, A.: Sequence complexity profiles of prokaryotic genomic sequences: a fast algorithm for calculating linguistic complexity. Bioinformatics 18(5), 679–688 (2002)

On the Development of a Pipeline for the Automatic Detection of Positively Selected Sites

David Reboiro-Jato, Miguel Reboiro-Jato, Florentino Fdez-Riverola, Nuno A. Fonseca, and Jorge Vieira

Abstract. In this paper we present the ADOPS (Automatic Detection Of Positively Selected Sites) software that is ideal for research projects involving the analysis of tens of genes. ADOPS is a novel software pipeline that is being implemented with the goal of providing an automatic and flexible tool for detecting positively selected sites given a set of unaligned nucleotide sequence data.

Keywords: Positively selected sites, Adaptation, Phylogenetics, Software integration.

1 Introduction

The understanding of the molecular basis of species adaptation is one of the main goals of Biology. Changes in gene expression levels and proteins may be adaptive and several approaches to infer deviations from neutrality are available [1, 2, 3]. When changes at a few amino acid sites are the target of selection, such positively selected amino acid sites may be detected using a phylogenetic approach [4].

Although several programs are available for detecting positively selected sites [4, 5], a most used approach, as evidenced by the over 700 citations (according to ISI web of knowledge database - `http://apps.webofknowledge.com/`),

David Reboiro-Jato · Miguel Reboiro-Jato · Florentino Fdez-Riverola
Departamento de Informática, Universidade de Vigo, Spain

Nuno A. Fonseca
CRACS-INESC Porto LA, Universidade do Porto, Portugal

EMBL-European Bioinformatics Institute, UK

Jorge Vieira
Instituto de Biologia Molecular e Celular,
Universidade do Porto, Portugal

M.P. Rocha et al. (Eds.): 6th International Conference on PACBB, AISC 154, pp. 225–229.
springerlink.com

is based on the method proposed by Yang and Nielsen [4]. The software developed by these authors requires as input files the aligned nucleotide coding sequences, a tree showing the inferred relationship of the sequences being used, and a control file. Since the programs commonly used to align nucleotide sequences and infer phylogenetic relationships use different input file formats this is a tedious and time consuming process. Moreover, multiple aspects of the process may interfere with the ability to infer positively selected amino acid sites or, even worse, erroneously infer positively selected amino acid sites, such as, for instance, the inclusion of highly divergent sequences, recombinant sequences, or the use of amino acid positions that are aligned with low confidence [6]. The coding sequences may also lead to erroneous conclusions since, for many commonly studied species, there is no transcript data to support the genome annotation [7]. In order to comprehensively address the possible impact of every one of these factors, the output from different programs must be analyzed, and the entire process ran several times, using each time, different input files making the whole process even more tedious and time consuming.

In this paper we address the problem of automatizing the whole process of detecting positively selected sites. A prototype, called ADOPS, is presented that effortlessly tries to detect positively selected amino acid sites from a set of unaligned sequences. The pipeline uses existing well established software to perform several steps involved in the process of detecting positively selected sites: i) alignment of the sequences (T-coffee [8]); ii) inference of the phylogenetic trees (MrBayes [9]); and iii) detection of positively selected amino acid sites (*codeml* [10]). The ADOPS has a graphical interface that provides an integrated view of all the results, including intermediate stages (such as the tree used and convergence statistics). Furthermore, ADOPS software can also be used in the command line.

The rest of the paper has the following structure. Section 2 describes the pipeline and Graphical User Interface. Conclusions and future work are presented in the last section of the paper.

2 System Prototype

ADOPS starts with a file with a set of unaligned DNA sequences (in FASTA format). T-coffee [8] is used to align and manipulate the sequences. T-coffee allows the user to select a variety of alignment methods and produces confidence statistics for each amino acid position in the alignment. The tree is generated by MrBayes [9]. MrBayes allows the implementation of a parameter-rich generalized time reversible (GTR) model of sequence evolution, among-site rate variation, and a proportion of invariable sites. By default, the *codeml* [10] random-sites models M0, M1, M2, M3, M7 and M8 are used. The tree and aligned sequences are given as input to *codeml* to detect positively selected amino acid sites.

The integration of the results produced by T-coffee and *codeml* is achieved by using a reference sequence and a multiple track system (as implemented

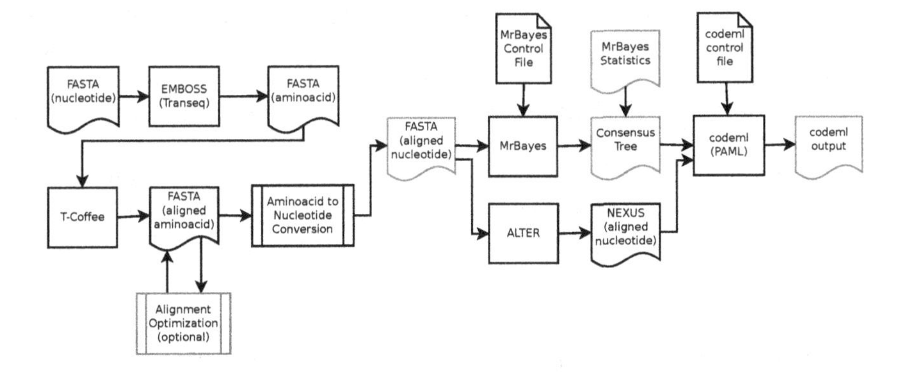

Fig. 1 Schematic ADOPS workflow.

in PileLine [11]). By default, amino acid sites identified as positively selected are shown as well as their probability of being positively selected and whether they were identified using Nave empirical Bayesian (NEB) or Bayes empirical Bayes (BEB) analyses [10]. The degree of confidence for that amino acid alignment position, computed using T-coffee, is also shown. The effect of the inclusion of a given sequence, or the use of a given alignment software can thus be tested without having to go through the hassle of preparing input files for the different applications.

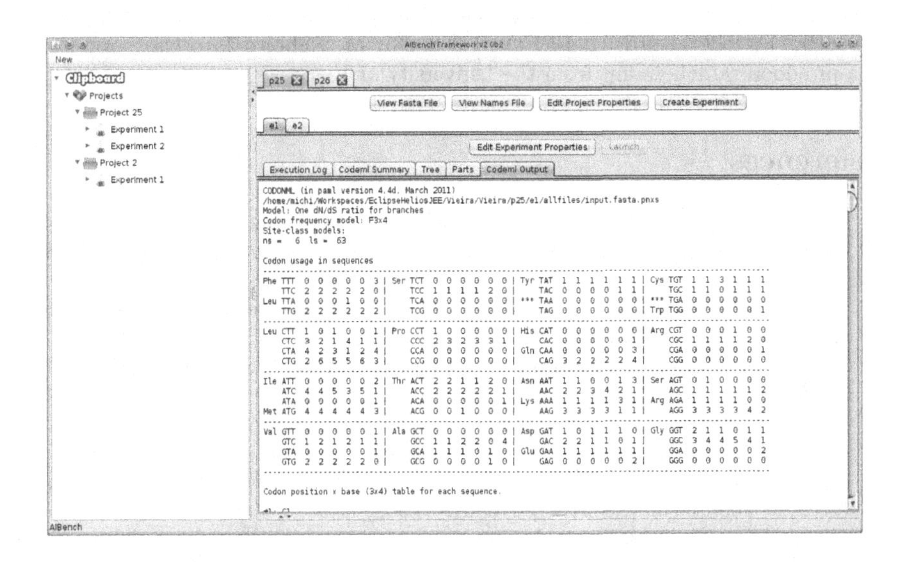

Fig. 2 ADOPS graphical user interface

In order to implement the final-user application for the workflow depicted in Figure 1 we have used our previous AIBench project [12], a Java application framework for scientific software development specially designed for giving support to the IPO (Input-Process-Output) model. Figure 2 shows a screenshot of the ADOPS software in which problem-specific *data-types*, supported *operations* and custom *views* are integrated for assisting the automatic detection of positively selected sites.

3 Conclusions

ADOPS is ideal for research projects involving the analysis of tens of genes. Although a good and easy to use graphical interface is provided, the entire pipeline can be run using a command line option as well thus being adequate to process hundreds or thousands of genes as well. ADOPS software can be freely downloaded from `http://sing.ei.uvigo.es/ADOPS/`.

Acknowledgements. This work has been partially supported by the following projects: the Spanish Ministry of Science and Innovation, the Plan E from the Spanish Government and the European Union from the ERDF (TIN2009-14057-C03-02 and AIB2010PT-00353), the projects PTDC/EIA-EIA/100897/2008, PTDC/BIA-BDE/66765/2006 and PTDC/BIA-BEC/099933/2008 comp-01-0124-FEDER-008916, funded by Programa Operacional para a Ciência e Inovação (POCI) 2010, co-funded by Fundo Europeu para o Desenvolvimento Regional (FEDER) funds and Programa Operacional para a Promoção da competividade (COMPETE; FCOMP-01-0124-FEDER-008923). M. Reboiro-Jato was supported by a pre-doctoral fellowship from the University of Vigo.

References

1. Fu, Y.X.: Statistical Tests of Neutrality of Mutations Against Population Growth, Hitchhiking and Background Selection. Genetics 147(2), 915–925 (1997)
2. Hudson, R.R., Kreitman, M., Aguade, M.: A Test of Neutral Molecular Evolution Based on Nucleotide Data. Genetics 116(1), 153–159 (1987)
3. Tajima, F.: Statistical Method for Testing the Neutral Mutation Hypothesis by DNA Polymorphism. Genetics 123(3), 585–595 (1989)
4. Yang, Z., Nielsen, R.: Synonymous and nonsynonymous rate variation in nuclear genes of mammals. Journal of Molecular Evolution 46, 409–418 (1998)
5. Pond, S.L.K., Frost, S.D.W., Muse, S.V.: HyPhy: hypothesis testing using phylogenies. Bioinformatics 21(5), 676–679 (2005)
6. Guirao-Rico, S., Aguad, M.: Molecular evolution of the ligands of the insulin-signaling pathway: dilp genes in the genus Drosophila. Mol. Biol. Evol. 28(5), 1557–1560 (2011)
7. Reis, M., Sousa-Guimares, S., Vieira, C., Sunkel, C., Vieira, J.: Drosophila genes that affect meiosis duration are among the meiosis related genes that are more often found duplicated. Plos One 10(3), e17512 (2011)

8. Notredame, C., Higgins, D.G., Heringa, J.: T-Coffee: A novel method for fast and accurate multiple sequence alignment. Journal of Molecular Biology 302(1), 205–217 (2000)
9. Huelsenbeck, J.P., Ronquist, F.: MRBAYES: bayesian inference of phylogenetic trees. Bioinformatics 17(8), 754–755 (2001)
10. Yang, Z.: PAML 4: Phylogenetic Analysis by Maximum Likelihood. Molecular Biology and Evolution 24 (2007)
11. Glez-Peña, D., Gómez-López, G., Reboiro-Jato, M., Fdez-Riverola, F., Pisano, D.G.: PileLine: a toolbox to handle genome position information in next-generation sequencing studies. BMC Bioinformatics 12(1) (2011)
12. Glez-Peña, D., Reboiro-Jato, M., Maia, P., Rocha, M., Díaz, F., Fdez-Riverola, F.: AIBench: A rapid application development framework for translational research in biomedicine. Computer Methods and Programs in Biomedicine 98, 191–203 (2010)

Compact Representation of Biological Sequences Using Set Decision Diagrams

José Ignacio Requeno and José Manuel Colom

Abstract. Nowadays, the exponential availability of biological sequences and the complexity of the computational methods that use them as input motivate the research of new compact representations. To this end, we propose an alternative method for storing sets of sequences based on set decision diagrams instead of classical compression techniques. The set decision diagrams are an extension of the reduced ordered binary decision diagrams, a graph data structure used as a symbolic compact representation of sets or relations between sets. Some experiments with genes of the mitochondrion DNA support the feasibility of our approach.

Keywords: biological sequence, symbolic representation, set decision diagrams.

1 Introduction

DNA and protein sequences, with up to thousands or millions of nucleotides and aminoacids per string, retain meaningful biological information. For example, they are used in different application domains such as protein folding prediction or phylogenetic analysis [14]. In fact, the complexity of many computational methods that use them as input depends on the alignment size. Currently, one of the major concerns in bioinformatics is how to store and handle efficiently huge amounts of those biological sequences [18, 19].

Several techniques try to tackle this problem by means of grammars, specialized compression algorithms or suffix trees [8, 13]. They mainly differ in data structures and usability (efficiency of read-write operations and compression ratio). In this paper, we present an alternative approach for a symbolic compact representation of

José Ignacio Requeno · José Manuel Colom
Department of Computer Science and Systems Engineering (DIIS)/
Aragon Institute of Engineering Research (I3A),
Universidad de Zaragoza, C/ María de Luna 1, 50018 Zaragoza, Spain
e-mail: {nrequeno,jm}@unizar.es

M.P. Rocha et al. (Eds.): 6th International Conference on PACBB, AISC 154, pp. 231–239.
springerlink.com

sets of biological sequences based on *set decision diagrams* (SDD) [5], an extension of the classical *reduced ordered binary decision diagrams* (ROBDD) [3] for multiple-valued variables.

As a prominent advantage of the ROBDDs we can say that they allow to manipulate sets in a symbolic way, making very adequate its usage in the context of formal verification like model checking techniques [16]. The ROBDD returns a boolean formula that is the canonical (minimal and unique) representation of a set of elements given an ordering function of its components. The characteristic function of a set allows an efficient set comparison as it only checks boolean formulas instead of the whole sets. Different ordering functions will be posed.

Alternatively, another benefit of using ROBDDs is the automatic collapse of common prefix and suffix paths in the graph structure, as well as identical subtrees do not exist. This feature is particularly helpful for mining frequent common substrings, allowing a potential greater compression rate than stand-alone prefix or suffix trees. It is specially interesting for the analysis of populations in a specific species, where the genotype differences between two individuals are really small. Moreover, the number of nodes of the graph structure promises to be a good metric of the evolutive distance between species in a set as closely related taxons with a high conservation degree will generate smaller decision diagrams.

Unlike other compression methods, the addition of new sequences to the data structure is incremental: it uses previous intermediate results for the recomputation of the entire ROBDD. The concatenation of ROBDDs representing substrings is also feasible, enabling the parallelization and reconstruction of longer sequences. The ROBDDs have no information loss and they are compatible with other compression techniques that can be applied in cascade.

In our case, the items of the set will consist of sequences of characters of a genomical alphabet (nucleotides or aminoacids), which motivates the use of SDDs for multiple-valued alphabets instead of binary variables.

This document is divided in 5 sections. After this introduction, Section 2 explains the advantages of using SDDs for storing sets of biological sequences in opposition to the ROBDDs. Secondly, Section 3 introduces the operations that convert a ROBDD into a SDD. Next, Section 4 draws some experimental results that support the feasibility of our approach. Finally, Section 5 briefs the conclusions.

2 Set Decision Diagrams in the Field of Sequences

The simplest way to represent a set of sequences with a binary ROBDD is by defining an absence (presence) bit per alphabet character in each sequence position, named *sequence BDD* (seqBDD) [10, 7]. However, this proposal leads to an extensive number of nodes, with up to $log|\Sigma|$ boolean variables per position in alphabets Σ with cardinality greater than two. An intuitive solution that alleviates the node explosion in genomical alphabets consists of the redefinition of decision diagrams with multiple-valued variables.

Multiple-valued decision diagrams (MDD) [12] and *data decision diagrams* (DDD) [4] are extensions of the ROBDDs for multiple-valued variables; that is,

Table 1 Example of an input alignment.

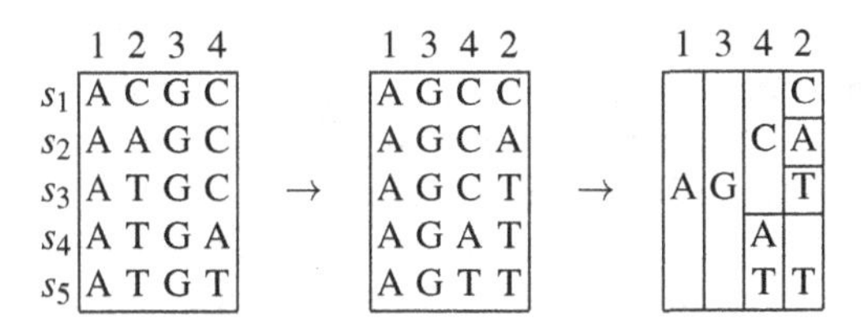

	1	2	3	4
s_1	A	C	G	C
s_2	A	A	G	C
s_3	A	T	G	C
s_4	A	T	G	A
s_5	A	T	G	T

→

1	3	4	2
A	G	C	C
A	G	C	A
A	G	C	T
A	G	A	T
A	G	T	T

→

1	3	4	2
			C
		C	A
A	G		T
		A	
		T	T

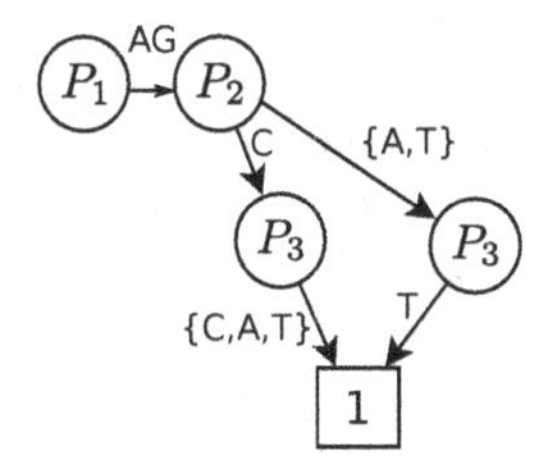

Fig. 1 Characteristic function of the SDD: $AG[C(C+A+T)+(A+T)T]$

the edges accept integer labels instead of bit values. This feature potentially reduces the number of nodes in the graph while enlarging the domain of values. The *set decision diagrams* (SDD) [5] are a special case of the previous ones that include edges labeled with sets of elements as well as simple tags. In our approach, we will label the edges with substrings, single characters or a set combination of both. The binary operations (like concatenation of SDDs) and the equivalence tests can be done in an analogous way to the ROBDDs [11]. We must take into account the peculiarity of the arcs labelled with sets of characters. For a formal definition of the SDDs and their properties the reader can see [15].

The Figure 1 illustrates the SDD of the alignment example (Table 1). In this case, we reordered the conserved columns and joined common substrings (*AG*) as detailed in Section 3. The nodes P_i represent the variable at the i-th position of the sequence and the edges represent the values that each node can take.

The size of the SDD is determined by the characteristic function being represented and the chosen ordering of the variables. Unfortunately, the search for the best ordering function that minimizes the characteristic function is NP-complete [2]. In this particular application, the size of the graph is constrained by the similarity of sequences and the position of conserved characters: a high conservation degree benefits our compactation method. Specially when the number of sequences in an alignment (m) is bigger than the size of the alphabet (n), the probability that any group of nodes collapse at a specific column increments with the number of sequences of the alignment.

In other cases, when the size of the alignment is small ($m \leq n$) and no character is conserved, we get a plain representation of the original alignment and no gain is obtained. Theoretically, the worst case involves an exponential number of nodes with respect to the length l of the alignment ($\mathcal{O}(n^{l-2}+n+1)$). That node explosion is due to the branching structure in the internal positions of the alignment: from the 2nd to the $(m-1)$-th column, the number of non-collapsed children is potentially multiplied by n, plus the root and n terminal characters at the m-th position (they are merged due to the SDD properties). In the opposite side, a set of completely equal items is stored in only one node of the structure.

To this end, we present some classical heuristic tricks that are adapted to strings in order to efficiently obtain a good approximation of the optimal solution. These heuristics are grouped in two different blocks. The first one consists of a sequence preprocessing where the characters are reordered with respect to the column conservation degree of the alignment in order to delay the appearance of the first branching point (dissimilar character) in the graph structure. The permutations can be stored so that we could recover the original sequences in the future.

As a huge number of nodes overloads the memory system with redundant pointers and long linear paths, our second approach proposes a bunch of reduction rules that minimizes the number of superfluous nodes and edges by merging trivial cases. They do not compress the alignment themselves because the labels of the edges are preserved, but they work as a fine optimization that simplifies the complexity of the structure and reduces the memory consumption. These operations can be applied iteratively in a predefined order until no new nodes are collapsed. They allow to convert a DDD into a SDD. Finally, the graph obtained promises to be lighter (or at least equal) than original one in terms of number of nodes.

3 Reduction Rules

The traditional left-to-right read of biological sequences is not habitually the best ordering function for coding the alignment because the initial character may differ in several sequences while bursts of common substrings can appear in forward positions of the alignment. This produces a fan out situation in the data structure that generates extra nodes which can be avoided with a previous reordering of columns. Reordering the alignment with respect to the highest conserved columns is one of the simplest operations that usually offers good optimization results. Conserved columns can be moved either to the head (delaying the divergence point) or the tail (merging equivalent leaves) of the alignment.

The nodes P_i of the graph are character variables that represent the i-th position of the sequence; and the edges represent the domain of values of the input alignment that those nodes can take in that place. From a low-level computational point of view, long linear paths (nodes with only one successor or predecessor) introduces a memory and time overhead with redundant pointers. Thus, deleting or merging nodes with one "parent" and/or "child" simplifies the size of the graph. To act in this way, we must expand the labels supported by the edges to deal with substrings, characters and sets. Implicitly, we offer a way of translating a DDD into a SDD.

Three simplification policies can be defined. Next reduction rules are inspired from those in [17]. From left to right (Figure 2), the first one swaps the nodes placed in a linear path. The second reduction collapses convergent paths and the resultant

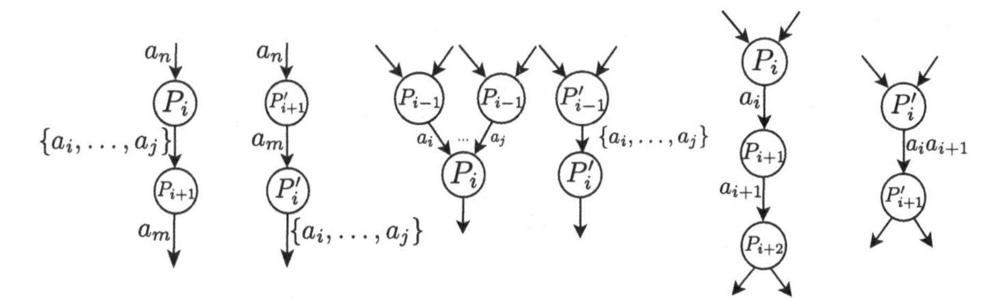

Fig. 2 Reduction rules.

edge is labeled with a set of characters or strings. Finally, the last reduction involves the concatenation of sequential nodes in a unique substring.

The reconstruction of the initial sequences is a direct process as the concatenation of substrings and sets preserves the order of the sequence. On the other hand, the i-th index in the node label P_i helps us to trace the permutation of nodes and the column reordering at the preprocessing phase. We recover the full alignment by traversing the graph from the root to the leaves through every path.

Lets recall that these reduction rules only delete unnecessary pointers (edges) and nodes, and return a smaller image of the graph; that is, they convert a DDD into a SDD. As the edge labels are conserved through all this process (labels are concatenated, swapped or stored in sets, but never deleted), there is no extra compression in the strict sense of the word.

4 Experimental Results

In this section, we outline some experiments with biological sequences that compare seqBDDs and SDDs and illustrate the feasibility of our approach. To this end, we use an Erlang implementation of the seqBDDs [6]. We also use the package libDDD version 1.7.0 [9], which is a LGPL library in C++ that implements DDD and SDD diagrams for integer variables. In order to obtain a simple communication with the libDDD library, the input alignment is encoded with the ASCII integer values of its characters as an intermediate step. We extended the libDDD library with a post-processing program that takes the computed DDD and applies the optimizations explained in Section 3 in order to obtain a SDD.

The performance evaluation of our system has been measured with human protein alignments that we have retrieved from GenBank [1] using the procedures supplied there to obtain our alignment of genes. In particular, we selected genes of respiratory complex I encoded in mitochondrial DNA (mtDNA). They are biologically interesting and varied in length, which makes them suitable for a complete performance analysis. Experimental results will draw out approximate upper bounds for work with mtDNA genes and help estimate costs elsewhere. All tests have been run on a scientific workstation (Intel Core 2 Duo E6750 @ 2.66 GHz, 8 GB RAM).

Table 2 Number of nodes of the seqBDD without column reordering.

		Set Size						
Seq. size	Gene	500	750	1000	1250	1500	1750	2000
98	ND4L	178	180	180	209	210	210	210
115	ND3	325	407	297	297	442	462	464
174	ND6	468	538	540	552	558	624	625
318	ND1	1844	2339	2457	2670	3179	3180	3327
347	ND2	1424	1621	2166	2378	2472	2702	2756
459	ND4	1538	1567	2018	1931	2277	3012	3014
603	ND5	4881	7436	8215	9626	10432	11253	11440

Table 3 Number of nodes of the seqBDD including column reordering.

		Set Size						
Seq. size	Gene	500	750	1000	1250	1500	1750	2000
98	ND4L	118	127	127	135	138	138	138
115	ND3	131	139	145	145	155	160	167
174	ND6	239	309	313	327	344	362	344
318	ND1	534	615	612	717	1130	1177	1179
347	ND2	475	503	563	577	662	713	777
459	ND4	620	644	709	701	747	801	802
603	ND5	965	1194	1297	1688	1895	1934	2015

Tables 2 and 4 outline the number of nodes in a seqBDD and DDD structure respectively. A bit (char byte) per node in the seqBDD (DDD) plus up to n pointers to the successor nodes ($|\Sigma| = n$) is a good approximation of the node size. The positive influence of the column reordering is captured in Tables 3 and 5. Althought the sizes of the DDDs outperform those of the seqBDDs, the results are not so striking. Probably, it is due to the extreme conservation of the data set, which constrains the complete genomical alphabet to a couple of biological viable values per column (similar to the SNPs). Conversely, a higher number of common nodes benefits the concatenation of common substrings in the SDD (Table 6), which results in a reduction rate of 8-10 times the number of nodes of the DDD.

Finally, the number of nodes of the data structure highly depends on the similarity of the input sequences of the set. Usually, longer sequences or distant species involve a greater number of dissimilar positions. In the other hand, if all the sequences come from the same species (conservation rate of 90% or more), the SDD seems to scale sublinearly with respect to the set size. This effect is encouraging for the application of our approach to large data bases of specific species.

Table 4 Number of nodes of the DDD without column reordering.

		Set Size						
Seq. size	Gene	500	750	1000	1250	1500	1750	2000
98	ND4L	172	172	172	197	197	197	197
115	ND3	319	399	289	289	430	448	448
174	ND6	453	515	524	524	524	587	587
318	ND1	1808	2294	2409	2615	3109	3109	3253
347	ND2	1395	1583	2122	2328	2413	2640	2691
459	ND4	1501	1519	1964	1875	2217	2950	2950
603	ND5	4827	7363	8133	9529	10324	11140	11325

Table 5 Number of nodes of the DDD including column reordering.

		Set Size						
Seq. size	Gene	500	750	1000	1250	1500	1750	2000
98	ND4L	111	118	118	123	125	125	125
115	ND3	126	132	137	137	143	146	151
174	ND6	225	288	290	301	312	327	308
318	ND1	500	567	561	660	1058	1104	1102
347	ND2	447	467	518	528	604	653	714
459	ND4	583	597	656	646	688	740	739
603	ND5	912	1128	1222	1598	1796	1828	1905

Table 6 Number of nodes of the SDD.

		Set Size						
Seq. size	Gene	500	750	1000	1250	1500	1750	2000
98	ND4L	12	14	14	18	20	20	20
115	ND3	11	14	15	15	21	25	29
174	ND6	27	41	45	49	59	64	66
318	ND1	65	84	89	98	123	126	130
347	ND2	49	62	76	87	102	107	114
459	ND4	68	83	93	95	104	109	111
603	ND5	99	129	143	171	190	204	212

5 Conclusions

In this paper, we presented the SDD as a suitable data structure for storing big sets of biological sequences. The advantages of using SDDs are fourfold. Firstly, the SDD outperforms the seqBDD, which is based on the ROBDDs, an historically efficient data structure for symbolic manipulation and storage of sets of elements. Secondly, they allow fast set comparisons because they return the characteristic (canonical) function of a set of sequences. Thirdly, the SDDs are incremental and they do not have to recompute the whole process in opposition to other compression techniques. Finally, the existence of generic public libraries enables the adaptation of the SDDs for biological sequences so that to obtain better compactation results.

We have also evaluated the performance of this data structure using genomic data. In particular, we have analyzed proteins coded by genes from the mtDNA genome, which are quite smaller than those from nuclear DNA. We have seen that the number of nodes of the data structure depends on the similarity of the input sequences of the set. Usually, longer sequences or distant species involve a greater number of dissimilar positions. Alternatively, it seems to scale sublinearly with respect to the set size if all the sequences come from the same species.

We can conclude that our first approximation to biological sequences using SDDs is innovative and encouraging in terms of efficiency. Our future work aims to scale up this technique for bigger sets and longer sequences. Furthermore, the reconstruction of phylogenetic trees using the number of nodes of the data structure as a distance metric shall be considered.

Acknowledgements. This work was supported by the Spanish Ministry of Science and Innovation (MICINN) [TIN2011-27479-C04-01] and the Government of Aragon [B117/10].

References

1. Benson, D.A., Karsch-Mizrachi, I., Lipman, D.J., Ostell, J., Sayers, E.W.: GenBank. Nucleic. Acids Res. 51, D46–D51 (2010)
2. Bollig, B., Wegener, I.: Improving the variable ordering of OBDDs Is NP-Complete. IEEE Trans. Comput. 45, 993–1002 (1996)
3. Bryant, R.E.: Symbolic manipulation of boolean functions using a graphical representation. In: Proc. of the 22nd ACM/IEEE Design Automation Conference, pp. 688–694. ACM (1985)
4. Couvreur, J.-M., Encrenaz, E., Paviot-Adet, E., Poitrenaud, D., Wacrenier, P.-A.: Data Decision Diagrams for Petri Net Analysis. In: Esparza, J., Lakos, C.A. (eds.) ICATPN 2002. LNCS, vol. 2360, pp. 101–120. Springer, Heidelberg (2002)
5. Couvreur, J.-M., Thierry-Mieg, Y.: Hierarchical decision diagrams to exploit model structure. In: Wang, F. (ed.) FORTE 2005. LNCS, vol. 3731, pp. 443–457. Springer, Heidelberg (2005)
6. Denzumi, S., Arimura, H., Minato, S.: Implementation of sequence BDDs in Erlang. In: Proceedings of the 10th ACM SIGPLAN W. on Erlang, pp. 90–91. ACM (2011)
7. Denzumi, S., Yoshinaka, R., Arimura, H., Minato, S.: Notes on Sequence Binary Decision Diagrams: Relationship to Acyclic Automata and Complexities of Binary Set Operations. In: Holub, J., et al. (eds.) Proc. of the Prague Stringology Conference 2011, pp. 147–161, Prague Stringology Club (2011)
8. Giancarlo, R., Scaturro, D., Utro, F.: Textual data compression in computational biology: a synopsis. Bioinformatics 25, 1575–1586 (2009)
9. LIP6/Move: the libDDD environment (2011), `http://ddd.lip6.fr`
10. Loekito, E., Bailey, J., Pei, J.: A binary decision diagram based approach for mining frequent subsequences. Knowl. Inf. Syst. 24, 235–268 (2010)
11. Meinel, C., Theobald, T.: Algorithms and data structures in VLSI design: OBDD - Foundations and applications. Springer, Heidelberg (1998)

12. Miller, D.M.: Multiple-valued logic design tools. In: Proceedings of the 23rd Int. Symposium on Multiple-Valued Logic, pp. 2–11. IEEE (1993)
13. Nalbantoglu, Ö.U., Russell, D.J., Sayood, K.: Data compression concepts and algorithms and their applications to bioinformatics. Entropy 12, 34–52 (2009)
14. Polanski, A., Kimmel, M.: Bioinformatics. Springer, Heidelberg (2007)
15. Requeno, J.I.: Análisis filogenético mediante lógica temporal y model checking. Master Thesis in Computer Science and Engineering, University of Zaragoza (2010)
16. Requeno, J.I., Blanco, R., de Miguel Casado, G., Colom, J.M.: Phylogenetic Analysis Using an SMV Tool. In: Rocha, M.P., Rodríguez, J.M.C., Fdez-Riverola, F., Valencia, A. (eds.) 5th International Conference on Practical Applications of Computational Biology & Bioinformatics (PACBB 2011). AISC, vol. 93, pp. 167–174. Springer, Heidelberg (2011)
17. Rudell, R.: Dynamic variable ordering for ordered binary decision diagrams. In: Proc. of the 1993 IEEE/ACM Int. Conf. on Computer-aided Design, pp. 42–47. IEEE (1993)
18. Strelets, V.B., Lim, H.A.: Compression of protein sequence databases. Comput. Appl. Biosci. 11, 557–561 (1995)
19. White, W.T., Hendy, M.: Compressing DNA sequence databases with coil. BMC Bioinformatics 9, 242–257 (2008)

Computational Tools for Strain Optimization by Adding Reactions

Sara Correia and Miguel Rocha

Abstract. This paper introduces a new plug-in for the OptFlux Metabolic Engineering platform, aimed at finding suitable sets of reactions to add to the genomes of microbes (wild type strain), as well as finding complementary sets of deletions, so that the mutant becomes able to overproduce compounds with industrial interest, while preserving their viability. The optimization methods used are Evolutionary Algorithms and Simulated Annealing. The usefulness of this plug-in is demonstrated by a case study, regarding the production of vanillin by the bacterium *E. coli*.

1 Introduction

An important challenge in Metabolic Engineering (ME) consists in the identification of genetic manipulations to be applied to an organism, with the aim of constructing a mutant strain able to produce compounds of industrial interest. Based on the knowledge about the biological system and, more specifically, its metabolic network, we can manipulate the environment in which it develops, or alter it genetically, to maximize the production of a given compound [6].

Recently, advances have been achieved concerning the available knowledge of some biological organisms, for instance from the sequencing of their genomes and also from various types of high-throughput experimental data (e.g. gene expression, proteomics). However, the lack of tools to perform the analysis and interpretation of biological data still limits the use and interconnection of that knowledge [2].

In this context arises the OptFlux (http://www.optflux.org) [9], an open-source and modular platform for ME, incorporating strain optimization tasks, using Evolutionary Algorithms (EAs) [1] and Simulated Annealing (SA) [5]. OptFlux also allows the use of stoichiometric metabolic models for phenotype simulation of both

Sara Correia and Miguel Rocha
CCTC, University of Minho, Campus de Gualtar,
Braga, Portugal
e-mail: sarag.correia@gmail.com, mrocha@di.uminho.pt

M.P. Rocha et al. (Eds.): 6th International Conference on PACBB, AISC 154, pp. 241–250.
springerlink.com

wild-type and mutant organisms, *Metabolic Flux Analysis* and pathway analysis using Elementary Flux Modes, among other features.

When performing strain optimization, some limitations arise from the metabolic models are incomplete [11] or the desired product can not be produced. In both cases, it will be necessary to find reactions to add to a metabolic model. In this paper, we present a new plug-in for OptFlux that allows to incorporate a set of reactions from an external database into an existing metabolic model performing phenotype simulation using those added reactions. Also, optimization methods will be put forward to allow the selection of the best set of reactions to add to the model. according to a given objective function (e.g. maximizing the production of a compound or filling gaps in the model).

2 Methods for Phenotype Simulation and Strain Optimization

The simulation process allows the prediction of the organism phenotype, using methods based on fundamental restrictions to the biological system. One of these methods is Flux Balance Analysis (FBA), that calculates the flux distribution making it possible to predict the growth rate of an organism or the rate of production of a metabolite, based on stoichiometric, reversibility and fluxes constraints [4]. FBA assumes that metabolic networks will reach a steady state constrained by the stoichiometry.

Predicting the metabolic state of an organism after a genetic manipulation (e.g. gene knockout) is a challenging task, because mutants are generally not subjected to the same evolutionary pressure that shaped the wild type. In these cases, other methods such *Minimization of Metabolic Adjustment* (MOMA) [12] and *Regulatory On/Off Minimization of metabolic fluxes* (ROOM) [13] are proposed to find a flux distribution for mutant strains.

Based on these methods, a question arises: how to find the ideal set of genes to be deleted to reach the desired phenotype? To try answer this question, the *OptGene* algorithm proposed by Patil et al [7] and its extensions made by Rocha et al [10] were proposed. In this last work, the authors' research group proposed a set-based representation that considered variable-sized solutions, allowing for solutions with different numbers of knockouts during the optimization process. Two optimization algorithms were developed: SA and Set-based EAs (SEAs). Both search for the optimum set size in parallel with the search for the optimum set of gene deletions.

This work aims to enlarge the set of possible genetic modifications by addressing gene additions. In this case, using SEAs or SA approaches, the optimization process finds a set of new reactions to be added to the model. Optionally, a complementary set of reactions to remove can also be optimized. Optimization methods are the same that were used previously. The main difference lies in the representation of the solutions. Although still using a representation based on sets, it is necessary to integrate information regarding the reactions to be added. Thus, a new way of representing solutions including two independent sets (knockouts and added reactions) was created. In Figure 1, the representation of one solution is depicted.

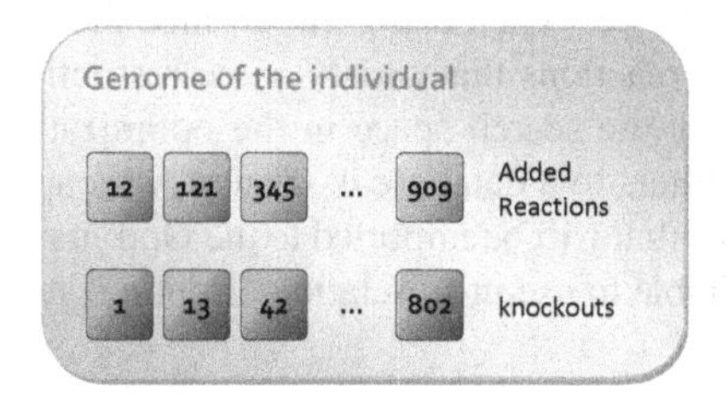

Fig. 1 Representation of the genome of an individual. Green squares represent reactions that will be added to the model (numbers are the reactions indexes in the external database). The knockouts are represented by red squares (numbers are indexes of reactions in the model).

3 OptFlux Plug-In for Adding Reactions

A new plug-in was developed for OptFlux to allow the addition of external reactions to a metabolic model. The addition of new reactions can be made for phenotype simulation or conducting a strain optimization process. Methods to import, filter and visualize the external database of reactions are also available. The new functionalities can be accessed by the "Plugins/ Add Reactions" menu.

3.1 Import Database of Reactions

Importing an external database of reactions into OptFlux can be made using the same methods used for creating metabolic models (SBML [3] and flat text files). Also, a new format of text files is defined (details are in the site documentation) to

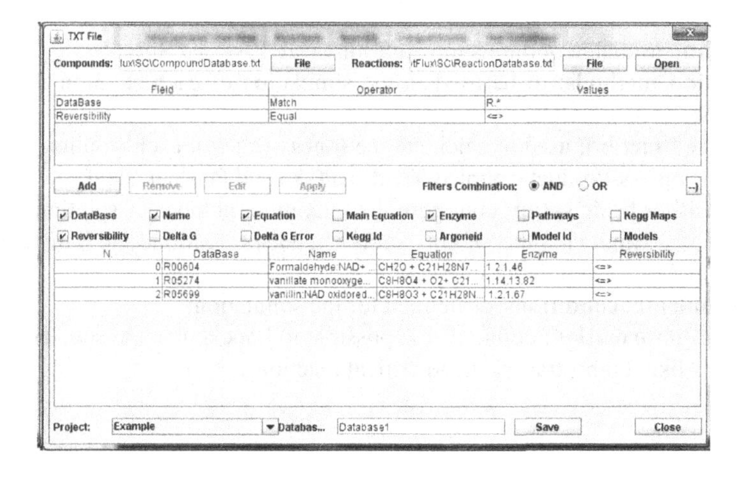

Fig. 2 Interface for selecting reactions and importing them to OptFlux. In this example, the user chooses only the reactions where ids start with "R" and that are reversible.

allow a more flexible scheme. When using this format, the user can filter the input data files to select only reactions that satisfy some restrictions. This is useful for readability and to reduce the search space in the optimization tasks. In Figure 2, the application of two filters to a database is shown. After applying filters, the user obtains a set of reactions that will be imported to the OptFlux platform. The reaction database becomes available to use in simulation or optimization processes.

3.2 Mutant Simulation by Adding Reactions

The phenotype simulation functionality allows mutant simulation by adding new reactions and optionally removing others from the model. After selecting the model to use, a previously loaded database is selected and the set of reactions to be added is chosen. Also, a set of knockouts can be selected. In Figure 3, the simulation interface is presented. During the configuration process, the user selects the simulation methods (FBA, MOMA or ROOM), the environmental conditions (the rates at which external metabolites can be consumed/ produced), and the objective function (e.g. the maximization/ minimization of a selected flux).

The result of mutant simulation can be observed in a specific interface (Figure 4), where the user can check the main results of the simulation: the list of added reactions, list of knockouts and values for all fluxes in the model.

3.3 Strain Optimization by Adding Reactions

The strain optimization process tries to find a set of reactions to be added to the model to improve a given objective function (e.g. the production of specific product). The search can be for only a set of reactions to be added or the combination of added reactions and knockouts. In the interface (Figure 5), the user selects:

- **algorithm**: available optimization algorithms are EAs and SA;
- **simulation methods**: to be used in the simulation of each solution evaluated (FBA, MOMA or ROOM);
- **objective function**: used to calculate the fitness value of each solution; options are the Biomass-Product Coupled Yield (BPCY) and Product Yield;
- **optimization basic setup**: configure the maximum number of solution evaluations, the maximum number of knockouts and added reactions and if the genome size should be fixed or have a variable size;
- **environmental conditions**: as defined for the simulation;
- **essential information**: define if it is possible to knockout some special type of reactions like drains, transport and critical reactions.

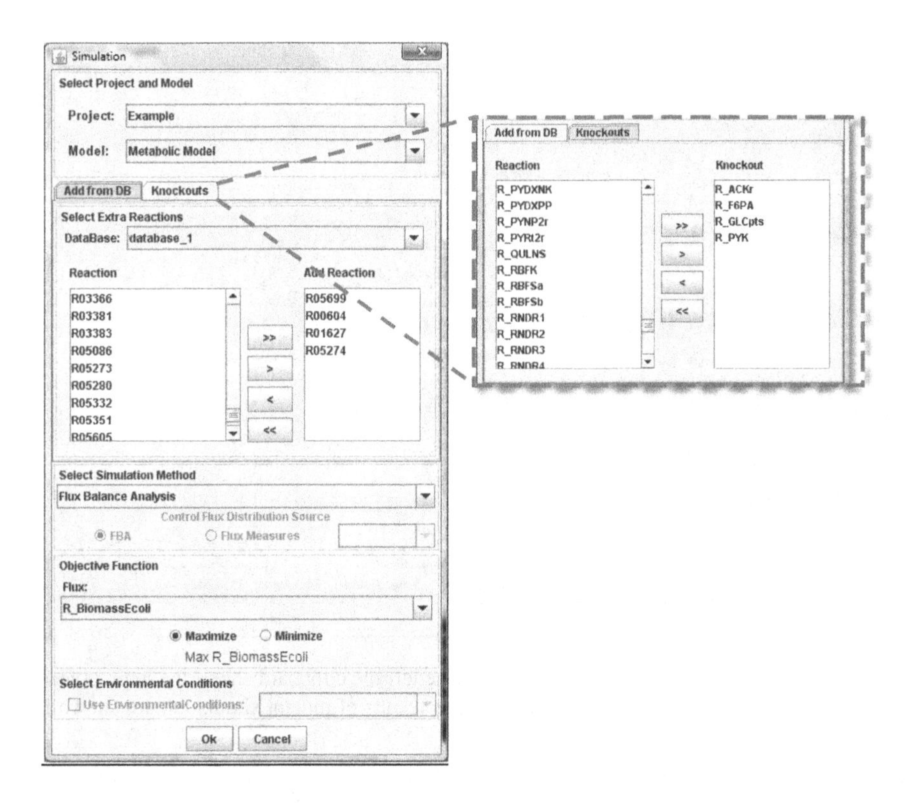

Fig. 3 Interface for mutant simulation. The case study of vanillin production is shown here (see below). In the example, 4 knockouts and 4 added reactions are selected.

4 Results

4.1 Rebuilding Gaps in the Metabolic Model

In this case study, used for validation purposes, OptFlux simplification methods were used to identify reactions constrained to a flux value of zero in the *E.coli* model. The model is reduced eliminating those reactions and a database is created with the removed reactions (407). In each run, three randomly selected reactions are further removed from the new reduced model and inserted into the database. The optimization methods must find these reactions and re-integrate them in the model to maximize biomass production. This process was repeated 10 times for SA and EA. The number of evaluations needed to find the solution in each run are given in Table 1.

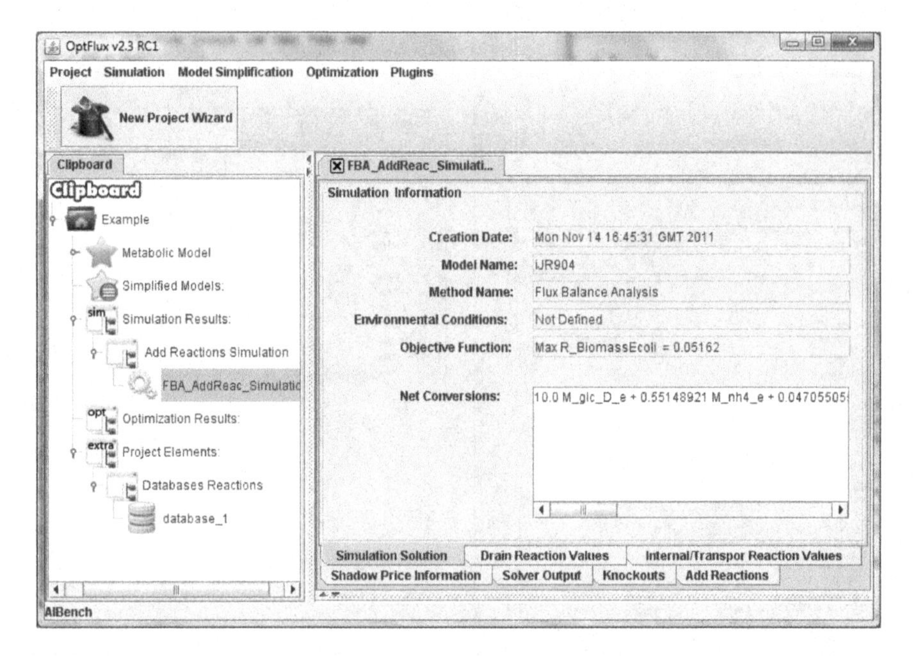

Fig. 4 Interface showing simulation results: in the left the clipboard shows the main objects and in the right side the visualization of the main results of mutant simulation are shown in distinct tabs.

Table 1 Number of function evaluations to find the optimal solution using SA and EA.

Test reactions	EA	SA
TPI,TKT1 e TKT2	500	300
IGPS, IDOND e ENO	2060	1120
MDH, ICDHyr e CBMK	2700	2930
IPPS, HSST e GSNK	9550	1735
PANTS, P5CR e ORPT	8240	4270
ADCL, IMPD e PSERT	6750	11103
RPI, TALA e ACLS	1215	1680
ACOTA, DDPA e PFL	2020	998
PRPPS, SPMS e TRDR	1035	5302
A5PISO, RPI e TYRTA	9065	7650

4.2 Vanillin Case Study

This case study aims to identify new pathways for the production of vanillin from glucose in *E. coli* and validate the implemented simulation method. To demonstrate the validity of the simulation process, we used the previous study with the OptStrain framework [8]. To proceed with the test it was required to build a database of

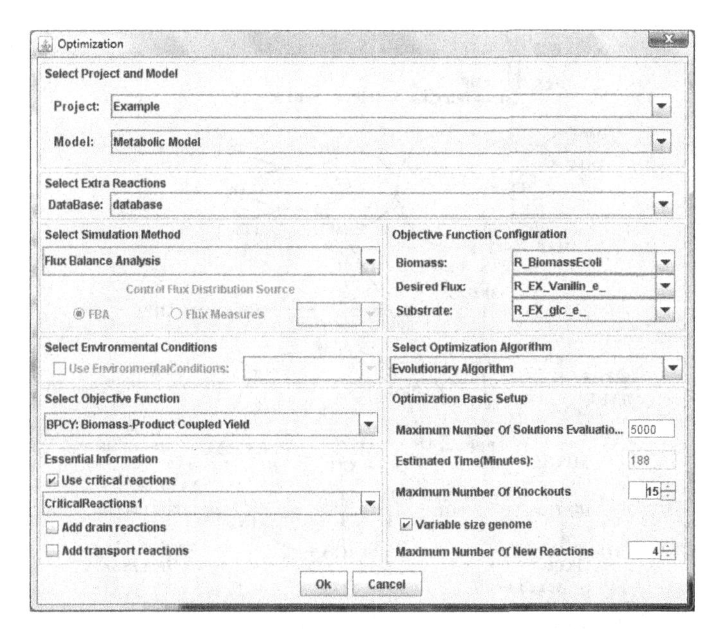

Fig. 5 Interface for strain optimization processes. In this example, an EA is configures, the simulation method is FBA, the objective function is BPCY, essential information uses the critical reactions, a maximum of 15 knockouts and 4 new reactions are permitted in variable sized sets.

reactions to add to the metabolic model. The added reactions to the metabolic model can be observed in Figure 6.

The simulation was performed for each of the three sets of knockouts in the paper, considering the substrate flux of 10 $mmol/gDWh^{-1}$ and the objective function the maximization of biomass. FBA was used in simulation process. The obtained results agree with the one from the previous work [8], thus validating our implementation.

The next step was to run the strain optimization process to find a set of added reactions and knockouts, that maximizes vanillin production coupled with the organism growth. The process was run 30 times for each EA and SA using as objective function the Biomass-Product Coupled Yield (BPCY). Previously, it was necessary change the metabolite ids of metabolic model for those used in database.

Table 2 shows the 95% confidence interval of results obtained in the optimization process, considering the best solution from each run.

Comparing these results with the ones obtained in the previous study [8], we see that the BPCY value of their solution was 0.035 ($BPCY = (6.787 \times 0.052)/10 = 0.035$). Although the vanillin production is lower in our case, the BPCY value increased significantly given that the biomass is much higher, which mean that our strain has a larger growth rate.

Afterwards, we focused in increasing the production of vanillin, without considering the biomass formation as a priority. Considering this, the tests were repeated

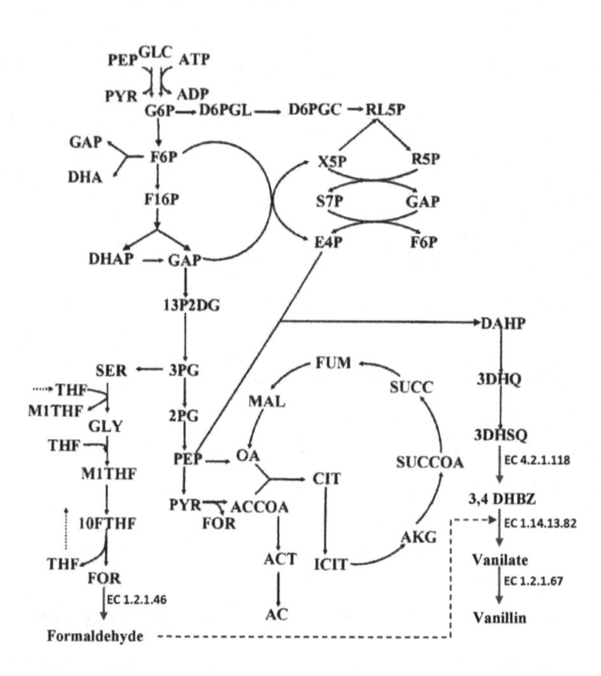

Fig. 6 The added pathway for the vanillin production.

Table 2 The 95% confidence interval of results obtained in the optimization process.

	EA	SA
Fitness (BPCY)	[0.17;0.181]	[0.177;0.189]
Biomass	[0.309;0.351]	[0.323;0.437]
Product	[5.264;5.478]	[4.822;5.407]
Number of knockouts	[8.78;11.421]	[11.134;14.332]
Number of added reactions	[8.741;9.259]	[8.409;9.058]

Table 3 Best results of strain optimization for vanillin production using the Yield objective function for each algorithm (EA and SA).

	Product	Biomass	No. added reactions	No. Knockouts
	6.948	0.022	20	4
EA	6.945	0.022	19	7
	6.944	0.023	20	4
	6.948	0.022	17	6
SA	6.948	0.022	18	4
	6.948	0.022	19	5

with a new objective function, by maximizing the flux of the product, ensuring a minimum limit of biomass production (5% of the wild type value). The results shown in Table 3 contain solutions considering new pathways, where the production of vanillin is higher than the obtained in [8].

The smaller set of added reactions suggested by the optimization process include the reactions with KEGG (http://www.genome.jp/kegg) ids: R01216, R01627, R05273, R05274. A supplementary file containing the full results obtained in the experiments summarized here is given in http://darwin.di.uminho.pt/pacbb2012/.

5 Conclusion

This paper presents methods for the simulation of strains by adding external reactions to the metabolic model, aiming to produce a desired product or to fill gaps. In this approach, information is added to the stoichiometry model regarding new reactions, thus making an extension to the initial model. Methods for strain optimization were developed, using EAs and SA, to find a sets of external reactions to be added and the necessary knockouts to maximize an objective function, typically related to the production of a compound of interest.

To provide these features to the scientific community, a plug-in has been developed for the OptFlux ME platform that allows simple and intuitive phenotype simulation and strain optimization with the addition of external reactions to the metabolic model. Thus, the tool set available for ME experts has been enlarged with useful techniques.

Future work will be devoted to the validation of these methods with other real world case studies.

Acknowledgements. This work is supported by project PTDC/EIA-EIA/115176/2009, funded by Portuguese FCT and Programa COMPETE.

References

1. Bäck, T.: Evolutionary algorithms in theory and practice: evolution strategies, evolutionary programming, genetic algorithms. Oxford University Press, Dortmund (1996)
2. Edwards, J.S., Covert, M., Palsson, B.: Metabolic modelling of microbes: the flux-balance approach. Environ. Microbiol. 4(3), 133–140 (2002)
3. Hucka, M., Finney, A., Sauro, H.M., Bolouri, H., et al.: The systems biology markup language (sbml): a medium for representation and exchange of biochemical network models. Bioinformatics 19(4), 524–531 (2003)
4. Kauffman, K.J., Prakash, P., Edwards, J.S.: Advances in flux balance analysis. Curr. Opin. Biotechnol. 14(5), 491–496 (2003)
5. Kirkpatrick, S., Gelatt, C.D., Vecchi, M.P.: Optimization by Simulated Annealing. Science 220(4598), 671–680 (1983)
6. Nielsen, J.: Metabolic engineering. Applied Microbiology and Biotechnology 55(3), 263–283 (2001)
7. Patil, K.R., Rocha, I., Förster, J., Nielsen, J.: Evolutionary programming as a platform for in silico metabolic engineering. BMC Bioinformatics 6, 308 (2005)

8. Pharkya, P., Burgard, A.P., Maranas, C.D.: Optstrain: a computational framework for redesign of microbial production systems. Genome Res. 14(11), 2367–2376 (2004)
9. Rocha, I., Maia, P., Evangelista, P., Vilaca, P., Soares, S., Pinto, J.P., Nielsen, J., Patil, K.R., Ferreira, E.C., Rocha, M.: OptFlux: an open-source software platform for in silico metabolic engineering. BMC Syst. Biol. 4, 45 (2010)
10. Rocha, M., Maia, P., Mendes, R., Pinto, J.P., Ferreira, E.C., Nielsen, J., Patil, K.R., Rocha, I.: Natural computation meta-heuristics for the in silico optimization of microbial strains. BMC Bioinformatics 9, 499 (2008)
11. Satish Kumar, V., Dasika, M.S., Maranas, C.D.: Optimization based automated curation of metabolic reconstructions. BMC Bioinformatics 8, 212 (2007)
12. Segrè, D., Vitkup, D., Church, G.M.: Analysis of optimality in natural and perturbed metabolic networks. Proc. Natl. Acad. Sci. USA 99(23), 15112–15117 (2002)
13. Shlomi, T., Berkman, O., Ruppin, E.: Regulatory on/off minimization of metabolic flux changes after genetic perturbations. Proc. Natl. Acad. Sci. USA 102(21), 7695–7700 (2005)

Computational Tools for Strain Optimization by Tuning the Optimal Level of Gene Expression

Emanuel Gonçalves, Isabel Rocha, and Miguel Rocha

Abstract. In this work, a plug-in for the OptFlux Metabolic Engineering platform is presented, implementing methods that allow the identification of sets of genes to over/under express, relatively to their wild type levels. The optimization methods used are Simulated Annealing and Evolutionary Algorithms, working with a novel representation and operators. This overcomes the limitations of previous approaches based solely on gene knockouts, bringing new avenues for Biotechnology, fostering the discovery of genetic manipulations able to increase the production of certain compounds using a host microbe. The plug-in is made freely available together with appropriate documentation.

1 Introduction

To reach biological ways of production for compounds of industrial interest, the metabolism of the selected host microbe usually needs to be modified. Metabolic Engineering (ME) is an essential field addressing these issues [1], being one of its main tasks to identify gene manipulations that increase production yields. To attain this goal, there is the need to develop accurate and efficient computational methods at three different levels: reconstruction of genome-scale metabolic models (GSMs), phenotype simulation and strain optimization. Currently, several GSMs containing information regarding the microbes' metabolism and the transcriptional information are available, for several microbes (e.g. *Escherichia coli* [2]).

One popular method to simulate the phenotype of cells, assuming a pseudo-steady state, is Flux Balance Analysis (FBA) [3]. FBA aims to maximize cell

Emanuel Gonçalves · Miguel Rocha
CCTC, University of Minho, Campus de Gualtar, 4710-057 Braga, Portugal
e-mail: {pg17598,mrocha}@di.uminho.pt

Isabel Rocha
CEB / IBB, University of Minho, Campus de Gualtar, 4710-057 Braga, Portugal
e-mail: irocha@deb.uminho.pt

M.P. Rocha et al. (Eds.): 6th International Conference on PACBB, AISC 154, pp. 251–258.
springerlink.com

growth, by solving a linear programming problem, subject to stoichiometric and reversibility constraints. Other approaches have been suggested, such as parsimonious enzyme usage FBA [4].

A bi-level strain optimization problem can be formulated assuming an inner-layer, the phenotype simulation method, and an outer-layer, optimization algorithms that search for the best set of genetic modifications to apply to a microbe. Several methods based on this structure were suggested, such as OptKnock [5] which aims to identify a set of reactions to be deleted from the metabolic model. Evolutionary Algorithms (EA) have been proposed as an alternative, providing methods such as OptGene [6], identifying near optimal solutions for large problems in a reasonable amount of time. Recent work by the authors' research group proposed alternatives based on Simulated Annealing (SA) and a variable size set-based solution representation enabling the identification of the optimal number of reaction deletions [7]. Also, they proposed methods that use the transcriptional/translational information available in some GSMs, allowing the prediction of a set of gene knockouts, instead of reaction deletions [8].

Another approach was suggested in OptReg [9] that aims to overcome one of the major limitations of previous methods, which restrict modifications to gene knockouts. OptReg works with reaction under/overexpression, where a reaction is over (under) expressed if it has a flux considerably higher (lower) when compared to the wild type fluxes. OptKnock and OptReg have some limitations, mainly due to the computational burden imposed by the mixed integer linear programming formulation, as well as the restriction that imposes the linearity of the objective functions. EAs and SA have proven to address well these issues in previous work [7].

In this work, the focus is in the development of computational tools for ME that allow to work with sets of over/under expressed genes. These tools provide the implementation of *(i)* methods for phenotype simulation of under and overexpression mutant strains and *(ii)* strain optimization algorithms based on SA and EA allowing to reach sets of genes to over/underexpress optimizing the production of a specific target compound. These tools are integrated as a plug-in within the OptFlux platform [10], being validated with a simple model for consistency check and a real case with *E. coli* as a host for the production of succinic and lactic acid.

2 Methods

2.1 Phenotype Simulation with over/under Expressed Genes

Flux Balance Analysis (FBA) [3] depicts the mass balance of all internal metabolites as linear equations. It assumes that the sum of all consumptions and productions is null. Therefore, for M metabolites and N reactions we have the following constraints: $\sum_{j=1}^{N} S_{ij} v_j = 0, i = 1, \ldots, M$. Other restrictions (reversibility, thermodynamics, capacity constraints) are represented as inequalities: $\alpha_j \leq v_j \leq \beta_j, j = 1, \ldots, N$. Using linear programming, FBA maximizes a specific optimization function, usually the biomass formation, subject to the previous restrictions.

In this work, we use the parsimonious enzyme consumption FBA (pFBA) method [4]. The pFBA firstly runs a FBA simulation optimizing the biomass flux value. Afterwards, it adds a constraint which forces the biomass flux value to be equal to the optimal value found in the previous step. Finally, a new optimization problem is formulated minimizing the sum of all fluxes and it is solved. This method tends to predict better the phenotype of the microbe [4].

The simulation of the mutant has as input a list of genes and associated expression values, relating the gene expression to its level in the wild type. If the expression value is 0, it encodes a gene deletion, if it is less than 1, it deems a gene underexpression; a value higher than 1 is an overexpression. Since these are gene expression values, they need to be propagated to reaction flux values. The simulation process is made in two steps: *(i)* using the transcriptional/translational information, the gene expression values are decoded into constraints over reactions fluxes (the process of decoding is depicted in Figure 1b); *(ii)* these constraints are then added to the pFBA optimization problem and solved as usual (Figure 1c).

Regarding *(i)*, transcriptional/translational information is given in the form of gene-reaction rules represented as Boolean expressions. Since we are working with numerical values of expression, a strategy to convert the Boolean operators to numerical values is needed. The AND operator is transformed into the minimum function and the OR operator is the average. To generate the constraints to add to the linear programming problem in pFBA, a reference set of flux values is provided by the simulation of the wild type strain (also with pFBA). If the relative value of a given reaction is equal to one, no constraints are added. On the other hand, if the value is larger (smaller) than 1, a constraint is added forcing the flux to be higher (lower) than the wild type reference value (see Figure 1 c).

2.2 Strain Optimization

In the representation adopted, each solution is composed by a list of pairs, each made by a gene index and the respective expression level (two integers). The expression level encodes a series of possible expression values. A discrete representation is used, greatly reducing the solution space and, therefore, the complexity of the optimization task.

To fairly distribute the discrete values of expression between under and overexpression and not disregarding knockouts, we defined a logarithmic scale containing the values: $0, 2^{-n}, 2^{-n+1}, \ldots, 2^{-1}, 2, 2^{2}, \ldots, 2^{n-1}, 2^{n}$. The parameter n will define the maximum (minimum) range of overexpression (underexpression). As depicted in Figure 1, after decoding the gene expression values to reaction flux levels and applying the constraints to the model, a pFBA simulation is run and returns the fluxes of the reactions in the model. To measure the quality of the solution, we use an objective function termed as Biomass-Product Coupled Yield (BPCY) [6].

In this work, EAs and SA were used as the optimization methods. These are similar to the ones previously proposed by the authors research group for knockout optimization [7]. The major change in these algorithms was the novel representation presented above. Six reproduction operators were used: *(i)* an operator based

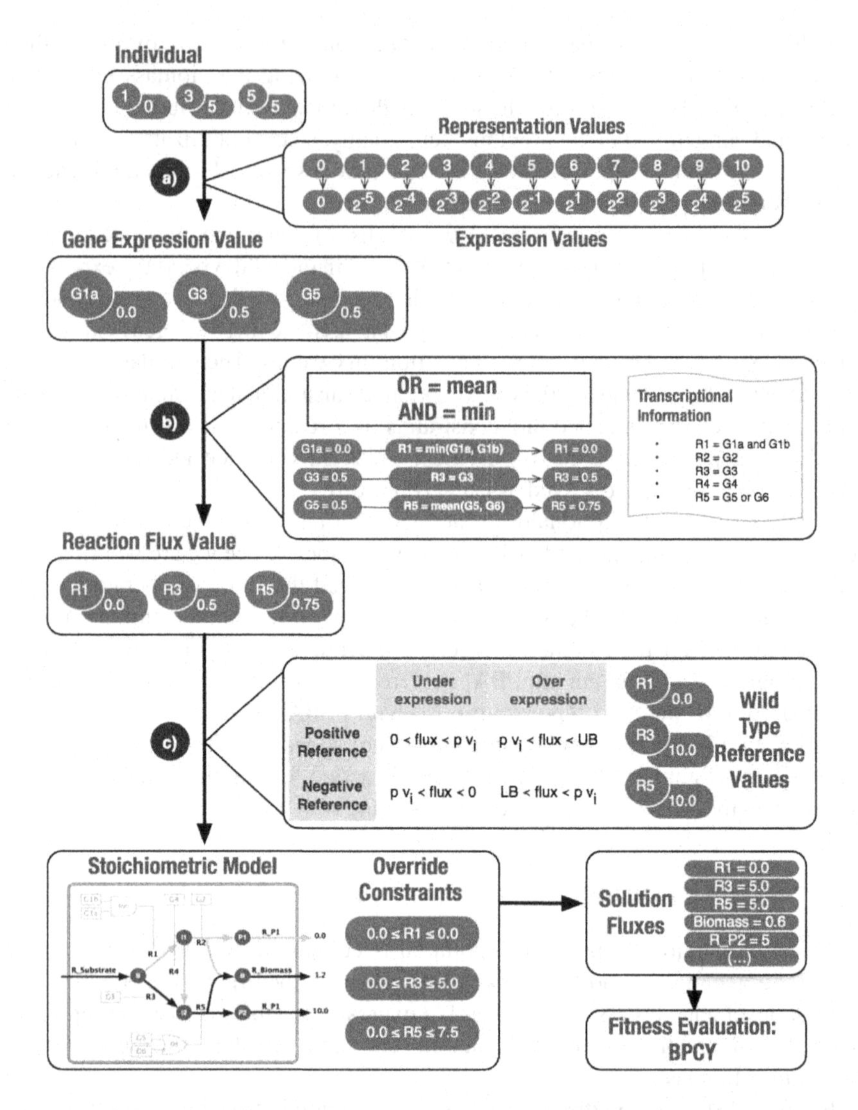

Fig. 1 Global scheme of the solution's decoding and evaluation: a) shows the decoding of the expression index to real values; b) shows the transformation of gene expression values to reaction flux values, using the transcriptional information available in the model; c) the under and overexpression constraints are calculated based on the reference values, and then applied as override constraints to the model; the simulation of the mutant is run and the reaction fluxes are sent to the BPCY evaluation function to measure the quality of the solution.

on the uniform crossover, which works as follows: the genes regulated on both parent genomes are sent to one of the offspring; the others are randomly sent to the offspring; *(ii)* two random mutation operators that randomly modify the gene and the expression value indexes; *(iii)* a mutation operator which changes the expression value index of a pair by the next or previous value; *(iv)* to generate solutions with different sizes two mutations operators: the shrink mutation operator removes a selected pair and the grow mutation operator adds a new pair.

3 Implementation

The framework proposed in this work was developed in JAVA within the OptFlux platform (`http://www.optflux.org`) [10]. OptFlux is an open source ME platform with an user-friendly interface. It has a modular structure, enabling the possibility to easily develop plug-ins adding new features to the application. OptFlux includes operations to run *in silico* phenotype simulations of the wild type or mutant strains using several methods. Users can also perform strain optimization selecting gene knockouts or reaction deletions. The software supports importing/exporting GSMs in several file formats, such as flat files and standard SBML.

OptFlux interface is built on top of AIBench (`http://www.aibench.org`) [11], using it to help producing GUI components such as the clipboard, the panels and views. AIBench is a Java framework that eases the development of applications based in the Input-Processing-Ouput paradigm, being developed according to the MVC (Model-View-Controller) strategy.

The plug-in proposed enables OptFlux users to perform operations involving mutants with a set of genes that are over/underexpressed. The user can perform the phenotype simulation of mutant strains specifying a set of genes and their relative expression levels or a set of reactions with relative flux levels. The methods available for the simulation are the same provided for knockouts, including the FBA and pFBA described above, among others.

Users have the possibility to perform strain optimization to find both sets of genes or reactions and their relative levels of expression. Both EA and SA are available to be selected as optimization methods and the user can configure the objective function, the termination criterion (e.g. maximum number of solutions evaluated) and other parameters. Also, the target compound and the biomass reaction need to be defined.

The plug-in is available for download through the OptFlux web site. Documentation is also available regarding the plug-in, with a detailed How-to where the main steps to configure and perform the operations are depicted. Figure 2 shows the configuration panels of the simulation and optimization operations, providing also an example of the views used to present the output solutions.

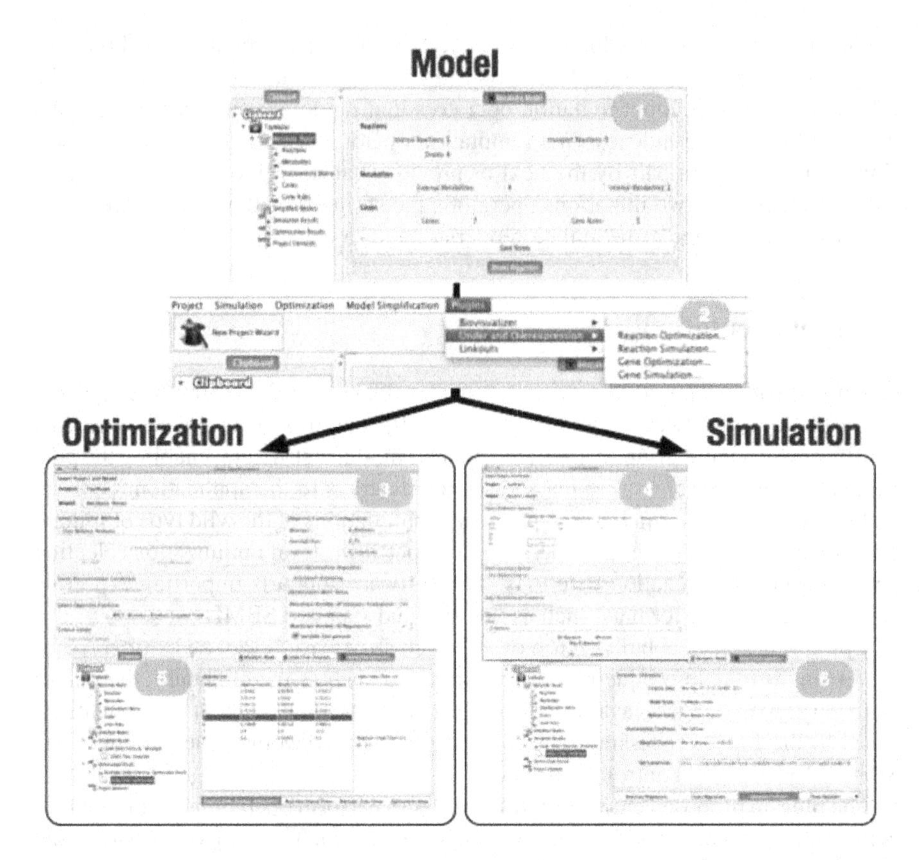

Fig. 2 A summary of the panels and menus of the proposed plug-in. Panel 1 shows the model loaded on OptFlux; through the menu bar in 2 the user can chose to simulate the mutant phenotype or to run strain optimization; 3 depicts the menu for configuration of the optimization, while 5 shows the results; simulation is made through the menu in 4, where the user can choose each gene and the expression level (the framework previews the constraints to be applied); the simulation results appear in a panel shown in 6.

4 Case Studies

The validation process for this software, in a first step, has been done using a small model containing transcriptional/translational information [12]. The model is portrayed in Figure 3 a), showing the fluxes of the wild type with an FBA simulation, where it produces P2 and does not produce P1. Therefore, in order to validate our framework we defined as objective function the maximization of the production of P1. Regarding this problem, it is easily perceptible that two solutions may be found: the knockout of G3 and G4 and therefore the deletion of R3 and R4; or the deletion of G5 and G6 that leads to the deletion of R5 (both gene deletions are needed since they regulate R5 by an OR operator). The drain reactions (R_P1, R_P2, R_Biomass and R_Substrate) were set to be critical and by that means impossible to be regulated. We ran the framework for this problem with a number of function evaluations enough to assure that a good solution was found. Easily, the framework found a solution producing the maximum level of P1, as desired, suggesting the first solution mentioned: the deletion of G3 and G4 (Figure 3 b).

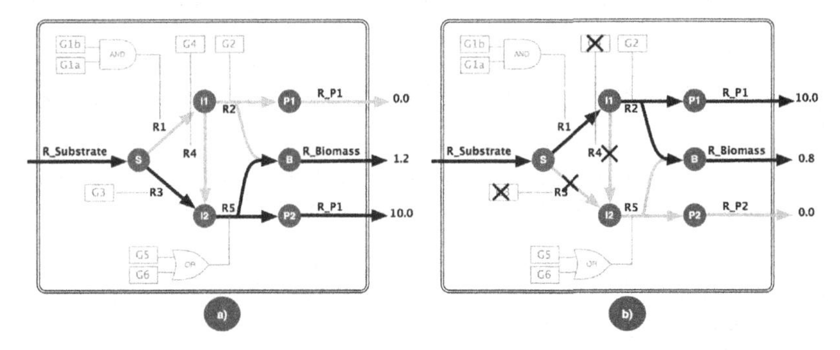

Fig. 3 Toy model from [12] showing the wild type flux distribution (flux of P2 = 10 and biomass flux = 1.2), a). The second picture, b), shows the mutant simulation with the best solution suggested by the framework optimizing the production of P1. R1 and R2 are now active leading to the maximum production of P1 (10) and to a flux 0.8 in biomass.

This framework was tested with a larger case study, involving the production of succinic and lactic acid in *Escherichia coli*. The model used is from [2] and it contains 1075 reactions, 761 metabolites and 904 genes. The obtained results are presented elsewhere [13] and are considerably better when compared to the strain optimization based on gene knockouts, being the difference substantially larger with lactic acid production. Another important result is the fact of the under and overexpression solutions are, in general, quite small, increasing the feasibility of the genetic manipulations suggested.

5 Conclusion

Strain optimization based on gene knockouts is widely used in Biotechnology, since it can greatly reduce production costs. Here, a framework is proposed based on gene/reaction under/overexpression that overcomes the knockout optimization limitations. Thus, this work expands the available set of computational tools available for ME experts, being well integrated within a broad platform such as OptFlux.

Future work will focus on further validation of the framework with other case studies. An improvement to the handling of constraints regarding reactions with zero flux in the wild type will be considered. Also, existing tools to analyse the solutions *in silico* will be improved.

Acknowledgements. Support of FCT and Programa COMPETE (ref. PTDC/EIA-EIA/115176/2009).

References

1. Stephanopoulos, G., Aristidou, A., Nielsen, J.: Metabolic Engineering. Acad. Press (1998)
2. Reed, J., Vo, T., Schilling, C., Palsson, B.: An expanded genome-scale model of Escherichia coli K-12 (iJR904 GSM/GPR). Genome Biology 4, R54 (2003)
3. Edwards, J., Covert, M.: Minireview Metabolic modelling of microbes: the flux-balance approach. Environmental Microbiology 4, 133–140 (2002)
4. Lewis, N., Hixson, K., Conrad, T., et al.: Omic data from evolved E. coli are consistent with computed optimal growth from genome-scale models. Molec. Syst. Biol. 6(390) (2010)
5. Burgard, A., Pharkya, P., Maranas, C.: Optknock: a bilevel programming framework for identifying gene knockout strategies for microbial strain optimization. Biotechnology and Bioengineering 84, 647–657 (2003)
6. Patil, K., Rocha, I., Förster, J., Nielsen, J.: Evolutionary programming as a platform for in silico metabolic engineering. BMC Bioinformatics 6(308) (2005)
7. Rocha, M., Maia, P., Mendes, R., et al.: Natural computation meta-heuristics for the in silico optimization of microbial strains. BMC Bioinformatics 9(499) (2008)
8. Vilaça, P., Maia, P., Rocha, I., Rocha, M.: Metaheuristics for Strain Optimization Using Transcriptional Information Enriched Metabolic Models. In: Pizzuti, C., Ritchie, M.D., Giacobini, M. (eds.) EvoBIO 2010. LNCS, vol. 6023, pp. 205–216. Springer, Heidelberg (2010)
9. Pharkya, P., Maranas, C.: An optimization framework for identifying reaction activation/inhibition or elimination candidates for overproduction in microbial systems. Metabolic Engineering 8, 1–13 (2006)
10. Rocha, I., Maia, P., Evangelista, P., Vilaça, P., Soares, S., et al.: OptFlux: an open-source software platform for in silico metabolic engineering. BMC Systems Biology (2010)
11. Glez-Peña, D., Reboiro-Jato, M., Maia, P., Rocha, M., Dìaz, F., Fdez-Riverola, F.: AIBench: a rapid application development framework for translational research in biomedicine. Computer Methods and Programs in Biomedicine 98, 191–203 (2010)
12. Kim, J., Reed, J.: OptORF: Optimal metabolic and regulatory perturbations for metabolic engineering of microbial strains. BMC Systems Biology 4, 53 (2010)
13. Gonçalves, E., Pereira, R., Rocha, I., et al.: Optimization approaches for the in silico discovery of optimal targets for gene over/underexpression. J. Computational Biology (in press)

Efficient Verification for Logical Models of Regulatory Networks

Pedro T. Monteiro and Claudine Chaouiya

Abstract. The logical framework allows for the qualitative analysis of complex regulatory networks that control cellular processes. However, the study of large models is still hampered by the combinatorial explosion of the number of states. In this manuscript we present our work to analyse logical models of regulatory networks using model checking techniques. We propose a symbolic encoding (using NuSMV) of logical regulatory graphs, also considering priority classes. To achieve a reduction of the state space, we further label the transitions with the values of the input components. This encoding has been implemented in the form of an export facility in GINsim, a software dedicated to logical models. The potential of our symbolic encoding is illustrated through the analysis of the segment-polarity module of the *Drosophila* embryo segmentation.

Keywords: Regulatory networks, Logical modelling, Model-checking.

1 Introduction

The control of essential cellular processes is driven by a great amount of molecular actors interacting through a variety of regulatory mechanisms. Large and complex interaction networks are thus delineated, calling for dedicated modelling methods to understand (and predict) their behaviours. Here, we focus on the logical formalism, which proved useful to study a variety of regulatory processes [10]. We need to cope with challenging issues due to the complexity of ever larger models, for which manual verifications quickly become intractable. This justifies the use of suitable automated techniques to handle model analyses.

Recently, formal verification techniques based on model checking have been successfully employed in systems biology, within different modelling frameworks (*e.g.*,

Pedro T. Monteiro · Claudine Chaouiya
Instituto Gulbenkian de Ciência, Rua da Quinta Grande 6, P-2780-156 Oeiras, PT
e-mail: {ptgm, chaouiya}@igc.gulbenkian.pt

M.P. Rocha et al. (Eds.): 6th International Conference on PACBB, AISC 154, pp. 259–267.
springerlink.com

[1, 2]). Observed biological behaviours are specified as statements in temporal logic, and model checking algorithms are used to automatically verify if these statements are satisfied by a given model.

Here, we focus on the input language of NuSMV, a model checking tool for the specification of qualitative systems [3]. Our contribution lies on a symbolic encoding for the efficient analysis of regulatory networks. This encoding has been implemented in the form of a new model export functionality of GINsim, a software dedicated to the definition and analysis of logical models [6].

Section 2 provides the basics of the logical framework. Our symbolic encoding is presented in Section 3 and illustrated in Section 4 through the analysis of the segment-polarity module involved in the *Drosophila* embryo segmentation. The paper ends with conclusions in Section 5.

2 Logical Modelling of Regulatory Networks

Logical regulatory graphs. A logical model of a regulatory network is defined as follows.

Definition 1. *A logical regulatory graph (LRG)* is a directed multigraph $\mathscr{R} = (\mathscr{G}, \mathscr{I}, \mathscr{K})$ where,

- $\mathscr{G} = \{g_i, \ldots, g_m\}$ is the set of *internal regulatory components* and $\mathscr{I} = \{g_{m+1}, \ldots, g_n\}$ is the set of *input regulatory components* ($\mathscr{G} \cap \mathscr{I} = \emptyset$). For each component g_i, a discrete variable v_i denotes its *level*: $v_i \in \mathscr{D}_i = \{0, \ldots, M_i\}$. Then $v = (v_i)_{g_i \in \mathscr{G} \cup \mathscr{I}}$ is a *state* and $\mathscr{S} = \prod_{g_i \in \mathscr{G} \cup \mathscr{I}} \mathscr{D}_i$ is the *state space*.
- $\mathscr{K} = (\mathscr{K}_1, \ldots, \mathscr{K}_m)$ are the *regulatory functions* for each $g_i \in \mathscr{G} \cup \mathscr{I}$; $\mathscr{K}_i$ is a multi-valued function specifying the target value $\mathscr{K}_i(v)$ of g_i, given the current state v: $\mathscr{K}_i : \mathscr{S} \to D_i$. Input components are set to constant values in their domains: $\forall g_i \in \mathscr{I}, \forall v \in \mathscr{S}, \mathscr{K}_i(v) = constant \in D_i$.

The set of regulatory arcs (interactions) can be deduced from $\mathscr{K}$; there exists an interaction from g_i to g_j, that occurs when the level of its source g_i is greater than θ_{ij} (with $0 < \theta_{ij} \leq M_i$) if and only if: $\exists v, v' \in \mathscr{S}$ s.t. $v_i + 1 = \theta_{ij} = v'_i$ and $\mathscr{K}_j(v) \neq \mathscr{K}_j(v')$. The sign of $\mathscr{K}_j(v') - \mathscr{K}_j(v)$ indicates if the interaction is an activation or an inhibition.

State transition graphs. The dynamics of LRGs are represented by State Transition Graphs (STGs), where nodes represent states, and arrows denote transitions towards successor states.

Definition 2. Given a LRG $\mathscr{R} = (\mathscr{G}, \mathscr{I}, \mathscr{K})$, its (full) *State Transition Graph* (*STG*) $E = (\mathscr{S}, \mathscr{T})$ is defined by the nodes which are the states in $\mathscr{S}$, the arcs $(v, w) \in \mathscr{T} \subset \mathscr{S}^2$ which denote transitions between states, and:

(i) in the *synchronous policy*: w is the successor of v iff $w \neq v$ and

$$\begin{cases} \forall g_i \in \mathscr{G} \text{ s.t. } K_i(v) \neq v_i, & w_i = v_i + \frac{|K_i(v) - v_i|}{K_i(v) - v_i} \\ \forall g_j \in \mathscr{G} \text{ s.t. } K_j(v) = v_j, & w_j = v_j. \end{cases}$$

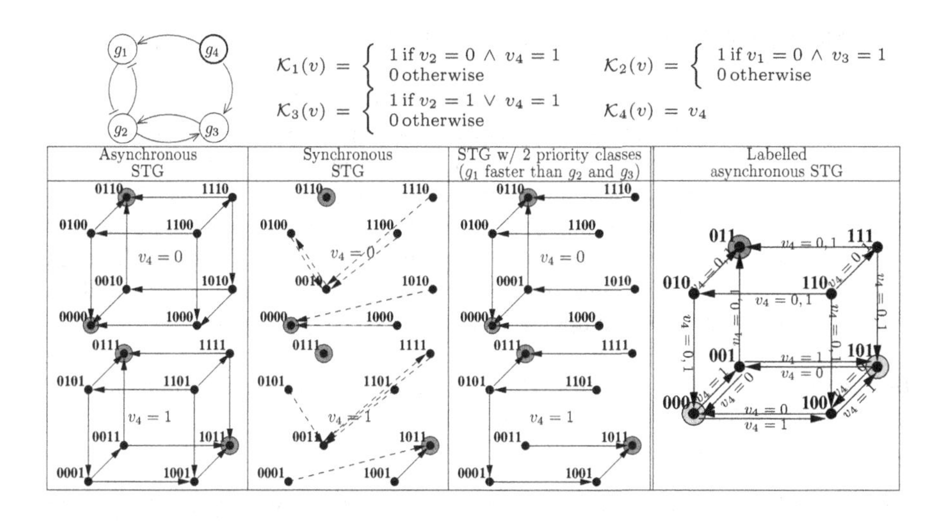

Fig. 1 A simple LRG and its dynamics. **Top:** the LRG and the associated logical functions. **Bottom:** the STGs obtained for different policies, the input component g_4 being constant in the first 3 columns. The third column considers two priority classes: g_1 in the fastest and g_2 and g_3 in the slowest. The fourth column displays the labelled STG (see Definition 3). States are defined as the vectors (v_1, v_2, v_3, v_4). States in cyan are stable states (strong in the labelled STG), weak stable states are denoted in pink.

(ii) in the *asynchronous policy*: w is a successor of v iff

$$\exists g_i \in \mathscr{G} \text{ s.t. } K_i(v) \neq v_i,\, w_i = v_i + \tfrac{|K_i(v)-v_i|}{K_i(v)-v_i} \quad \text{and} \quad \forall g_j \neq g_i,\, w_j = v_j.$$

In the synchronous policy, a state has at most one successor that accounts for all (simultaneous) updates, whereas in the asynchronous policy, a state has as many successors as the number of components called to update. Note that: variables increase or decrease by 1, both policies exclude self-loops and input components are constant. By specifying initial condition(s), a sub-graph of the full STG is explored. Terminal strongly connected components indicate stable asymptotic behaviours (or attractors), *i.e.* stable states or stable oscillations.

Priority Classes. The specification of priority classes, based on biologically founded assumptions, allows to reduce STGs sizes by discarding some spurious trajectories. A set of priority classes $\mathscr{C} = \{C_1, ..., C_k\}$ is defined as a k partition of $\mathscr{G}$. Each class C_i is associated with a rank $r(C_i)$ (with $1 \leq r(C_i) \leq k$, 1 being the fastest rank), as well as with an updating policy (synchronous or asynchronous). When classes have the same rank, concurrent updates of their components are triggered asynchronously. Whenever a component is called to change its value, this update is performed if and only if there is no component called to update and belonging to a class with a higher rank (see [4] for details).

3 Symbolic Encoding of Logical Models

GINsim has been extended with an export of logical models in the NuSMV language [3], according to two encodings: one where input components are kept constant, the other where input components can freely change their values.

We specify, in NuSMV, an implicit version of the model, yielding the same STG as the ones generated by GINsim. To account for the updating policies, we devised a generic encoding for priority classes, where the asynchronous and synchronous policies are defined as particular instances of priority classes. Hereafter, variables associated to the regulatory components are called *model variables*.

Declaration of Model Variables. They are declared under the environment VAR (see NuSMV Manual) as follows:

```
vi : { 0, ..., max_i};
```

Since NuSMV internally constructs a binary decision diagram (BDD) following the specification order of the variables, we speed up the verification procedure using as an heuristic the ordering of the model variables according to the number of the regulatory dependencies of the associated components. To this aim, we apply a topological sort algorithm.

Declaration of Priority Classes. A variable PCs is declared, whose possible values are the defined priority classes. For the *k*th class, a variable Ck_vars takes its values in the set of model variables belonging to that specific class. These *class variables* are declared under the environment IVAR (see NuSMV Manual), which are not considered as model variables, their valuations being entirely controlled by NuSMV.

To specify a synchronous policy, a single class variable is declared that contains all the model variables, whereas for an asynchronous policy, one class variable is declared for each model variable:

```
-- Asynchronous           -- Synchronous                 -- 2 priority classes
-- [v1] ... [vk]          -- [v1 ... vk]                 -- [v1 v2] [v3]
PCs: { C1, ..., Ck };     PCs: { C1 };                   PCs: { C1, C2 };
C1_vars: { C1_v1 };       C1_vars: { C1_v1_..._vk };     C1_vars: { C1_v1, C1_v2 };
...                                                      C2_vars: { C2_v3 };
Ck_vars: { Ck_vk };
```

Definition of Priority Ranks. A set of ranks is declared as possible valuations of the Rank variable. Each value of Rank is associated to a given (set of) class(es). For instance, in the case of the 2 priority classes mentioned above, the rank would be defined as follows:

```
Rank : { rank_1, rank_2 };
Rank := case
  (!v1_std | !v2_std ) : rank_1; // all vi in classes with rank 1
  (!v3_std ) : rank_2;
esac;
```

where vi_std refers to a Boolean variable that is true, whenever the model variable vi is stable (*i.e.* no call for update).

Computation of State Successors. NuSMV attributes all possible valuations to the variables under the `IVAR` environment, proceeding with the application of conditional rules for the model variables. Each of these rules contains a condition `update_vi_OK` enabling the update of the model variable `vi`. In the case of the 2 priority classes mentioned above, `update_v2_OK` would be defined as follows:

```
update_v2_OK := (PCs = C1) & (C1_vars = C1_v2) & (Rank = rank_1);
next(v2) := case
  update_v2_OK & (vi_inc) : v2 + 1;
  update_v2_OK & (vi_dec) : v2 - 1;
  TRUE : v2;
esac;
```

Selection of Valid Transitions. We must prevent any transition involving changes only over `IVAR` variables, unless these transitions correspond to a self-loop (*i.e.*, a stable state). We thus define a Boolean condition `TRANS` specifying that a transition involves at least one model variable update:

```
stableState := v1_std & ... & vi_std;
TRANS next(v1) != v1 | ... | next(vi) != vi | stableState;
```

Reduction of the State Space over Input Components. Regulatory functions of internal components define their behaviours, depending on the current state of their regulators. An input component has no regulator and its function is thus assumed to be constant. For a LRG with constant input values, its STG is composed by a set of disconnect graphs, one for each combination of input values (see Figure 1). The advantage of this representation is that, given an initial value of all the input variables, the behaviour is restricted to a sub-graph, easing the analysis of large systems. However, in concrete biological systems, input variables vary, often being externally controlled (e.g. light availability, presence of nutrients, heat shock, etc.).

When input components freely vary (i.e. under no specific control), the resulting STG encompasses transitions between all the states having the same values of the internal components, denoting the sole changes of the input components. Considering that states are characterised by model variables, the whole behaviour can thus be represented by a STG, with transitions labelled by the values of the input variables, yielding a compacted, labelled STG (see Figure 1). Below, we define such a projection over the input variables.

Definition 3. Given $\mathscr{R} = (\mathscr{G}, \mathscr{I}, \mathscr{K})$, a LRG with internal (in $\mathscr{G}$) and input components (in $\mathscr{I}$), and $E = (\mathscr{S}, \mathscr{T})$ its STG. The corresponding *labelled STG* $E^{|\mathscr{I}} = (\mathscr{S}^{|\mathscr{I}}, \mathscr{T}^{|\mathscr{I}})$ is defined as follows:

- $\mathscr{S}^{|\mathscr{I}} = \prod_{g_i \in \mathscr{G}} D_i$,
- $\forall v^{|\mathscr{I}}, w^{|\mathscr{I}} \in \mathscr{S}^{|\mathscr{I}}$, $(v^{|\mathscr{I}}, L, w^{|\mathscr{I}}) \in \mathscr{T}^{|\mathscr{I}}$ if $\exists v, w \in \mathscr{S}$ with $(v, w) \in \mathscr{T}$ such that $\forall g_i \in \mathscr{G}, v_i^{|\mathscr{I}} = v_i$ and $w_i^{|\mathscr{I}} = w_i$. Then, the label L of this transition is defined as the set of all the valuations of the input variables for which this transition is observed in E: $L = \left\{ u \in \Pi_{g_i \in \mathscr{I}} D_i \text{ s.t. } \forall g_i \in \mathscr{I}, v_i = u_i (= w_i) \right\}$.

When analysing real world models, this representation presents a true gain in the number of states ($\Pi_{g_i \in \mathscr{G}} |D_i|$ instead of $\Pi_{g_i \in \mathscr{G} \cup \mathscr{I}} |D_i|$), still keeping all the information regarding the *input variables*. Formal verification community already uses such a graph structure combining labels on both states and transitions [5]. To obtain this labelled STG, we declare the input variables in the IVAR environment (NuSMV freely attributes their valuations):

```
IVAR inputVar_i : { 0, ..., max_i};
```

In a labelled STG, one must distinguish between states that are stable for all the valuations of the input variables, from those that are stable only for a subset of valuations of the input variables. The definition below formalises this distinction. Note that the specification of valid transitions must change accordingly (not detailed in this manuscript).

Definition 4. Given a labelled STG $E^{|\mathcal{I}} = (S^{|\mathcal{I}}, T^{|\mathcal{I}})$, a state $v \in S^{|\mathcal{I}}$ is:

a *strong stable state* iff $\forall w \in S^{|\mathcal{I}}, \forall L \in \Pi_{g_i \in \mathcal{I}} D_i, w \neq v \Rightarrow (v, L, w) \notin T^{|\mathcal{I}}$,
a *weak stable state* iff $\forall w \in S^{|\mathcal{I}}, \exists L \in \Pi_{g_i \in \mathcal{I}} D_i, w \neq v \Rightarrow (v, L, w) \notin T^{|\mathcal{I}}$.

4 Application and Results: Segment-Polarity Module

To assess the efficiency of our symbolic encoding, we consider the logical model of the segment-polarity (SP) module defined in [8]. This module is involved in the formation of segment boundaries during early *Drosophila* embryogenesis. Briefly, to account for the intercellular interactions involved in the boundary formation, we consider two neighbouring cells (along the anterior-posterior axis) connected through Wingless (Wg) and Hedgehog (Hh) signals (Figure 2 top). This model has 5 cellular stable states: Trivial (T), Ci_Ciact (C), Nkd (N), En (E) and Wg (W), each characterised by an expression pattern of the internal components [8].

We first analyse the model composed by two interconnected cells, with no input signal (or, equivalently, setting the 4 input variables to 0). This model gives rise to 3 stable patterns: WE accounting for the wild-type pattern, with a wingless-expressing cell anterior to an engrailed-expressing cell; TT, with two cells expressing neither Wg nor En; the third stable state EW corresponds to the inverted wild-type. We analyse their reachability, considering an initial state (denoted PR) that accounts for the expression pattern specified by the preceding activity of the pair-rule module (see [8, 9] for further detail). The results are given in Figure 2 (middle), where it appears that: 1) the synchronous policy prevents the reachability of the WE pattern; 2) the asynchronous policy, much more time consuming, generates trajectories leading to the 3 stable patterns; 3) well-chosen priority classes (as defined in [8]) significantly reduce the verification time while preserving the reachability properties.

When considering the four input components, 17 stable patterns are obtained. We check the cross-reachability of these patterns through direct trajectories (*i.e.* not passing through a another stable pattern), considering the input variables as being either constant or varying (Figure 2 bottom).

Interestingly, the WE (wild type pattern) and EW are strong stable states. Note that, for some models, a notion of robust stable patterns might be appropriate, when variations of input values lead to outgoing transitions, but never to exit the basins of attraction. In contrast, all stable patterns are directly reachable from the TT stable

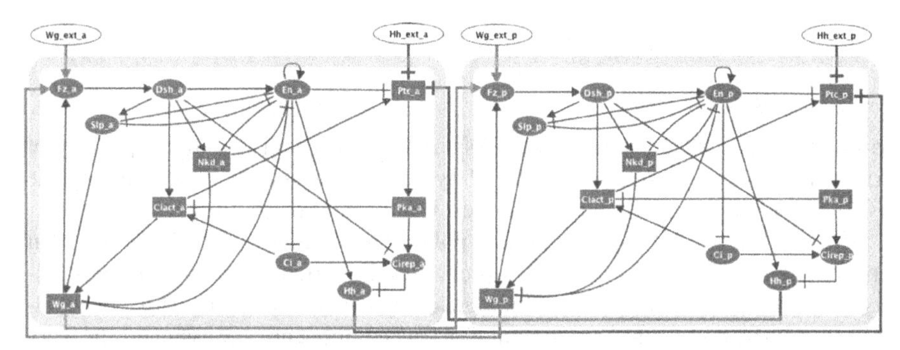

Stable patterns direct reachability: 0 inputs	Synchronous		Asynchronous		2 priority classes	
	Verdict	Time	Verdict	Time	Verdict	Time
Pair-rule → TT	TRUE	0.043s	TRUE	15m24s	TRUE	1.206s
Pair-rule → WE	FALSE	0.049s	TRUE	14m34s	TRUE	1.212s
Pair-rule → EW	FALSE	0.049s	TRUE	15m52s	TRUE	1.218s

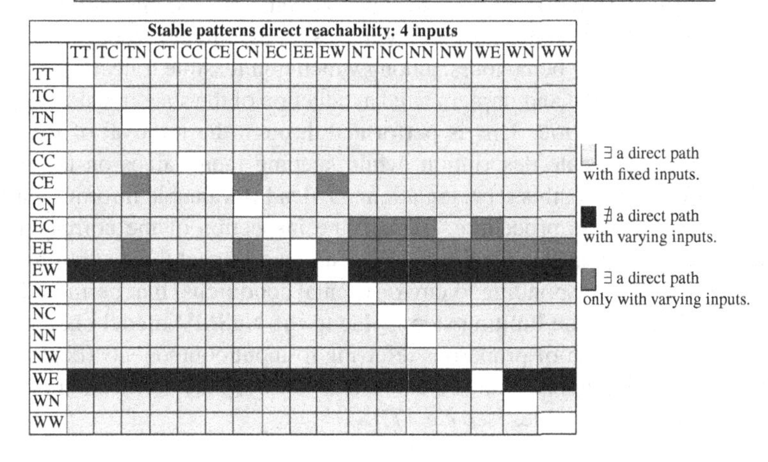

Fig. 2 Segment-polarity model. **Top:** two intra-cellular networks interconnected through the Hh and Wg signalling pathways, denoting the cells flanking the segmental border. External Hh and Wg account for signals from other neighbouring cells (bold interactions). Normal (resp. blunt) arrows depict activations (resp. inhibitions), and oval (resp. rectangular) nodes depict Boolean (resp. multi-valued) components. **Middle:** Direct reachability results for the 3 stable patterns starting from the pair-rule state, considering the two cells with no inputs, for different updating policies. Two priority classes are defined: the fastest contains Fz, Dsh, Ciact, Cirep and Pka; the slower contains the remaining variables. **Bottom:** Direct reachability results between all the 17 stable patterns, considering the two cells with 4 inputs, for the asynchronous policy. Reachability is tested both with (all combinations of) fixed inputs and varying inputs (see legend).

pattern, for at least one appropriate (fixed) combination of input values. But, to achieve the direct reachability of some patterns, input variables must vary along the path (see *e.g*. from EE to WE).

5 Conclusions

In this work, we have extended the GINsim modelling tool with the capability of converting logical models into NuSMV symbolic descriptions, thus enabling the use of model-checking techniques for their formal analysis and verification.

Through the analysis of the segment-polarity module, we show the impact that updating policies have on reachability properties. Importantly, the synchronous policy may prevent the emergence of biologically relevant behaviours. This policy, although relying on an unlikely assumption (all components are assigned the very same delay to update their states) is frequently favoured by modellers because it results in a deterministic behaviour, easier to analyse. Moreover, in some models, it happens to generate plausible behaviours (see *e.g*. [4]). Still, one should cautiously consider this policy since it can lead to spurious behaviours (in particular in what concerns cyclical attractors) and miss relevant trajectories, as demonstrated in this work. In contrast, the asynchronous policy makes no assumption about delays, but often generates too many behaviours, among which non feasible trajectories.

We have also proposed and implemented a reduction of the state transition graph, without loss of information. This is performed through the removal of the input components from the state description, while keeping their values on the transitions labels. Furthermore, this representation may lead to valuable information provided by the verification procedure, through the inspection of the corresponding witness/counterexample. This means that one can then inspect the evolution of the input components (corresponding to environmental conditions) that permitted such a behaviour. Currently, the limiting step is due to the NuSMV model-checker that impedes the specification of properties referring to input components (declared as `IVAR`). We thus contemplate NuSMV extensions allowing the verification of properties involving input variables (see *e.g*. [7]).

Acknowledgement. We acknowledge the support from the Portuguese FCT (post-doctoral fellowship SFRH/BPD/75124/2010 and project grant PTDC/EIACCO/099229/2008).

References

1. Batt, G., Ropers, D., de Jong, H., Geiselmann, J., Mateescu, R., Page, M., Schneider, D.: Validation of qualitative models of genetic regulatory networks by model checking: Analysis of the nutritional stress response in Escherichia coli. Bioinformatics 21 (suppl.), i19–i28 (2005)
2. Bernot, G., Comet, J., Richard, A., Guespin, J.: Application of formal methods to biological regulatory networks: Extending Thomas' asynchronous logical approach with temporal logic. J. Theor. Biol. 229(3), 339–348 (2004)

3. Cimatti, A., Clarke, E., Giunchiglia, E., Giunchiglia, F., Pistore, M., Roveri, M., Sebastiani, R., Tacchella, A.: NuSMV 2: An OpenSource Tool for Symbolic Model Checking. In: Brinksma, E., Larsen, K.G. (eds.) CAV 2002. LNCS, vol. 2404, pp. 359–364. Springer, Heidelberg (2002)
4. Fauré, A., Naldi, A., Chaouiya, C., Thieffry, D.: Dynamical analysis of a generic boolean model for the control of the mammalian cell cycle. Bioinformatics 22(14), 124–131 (2006)
5. Müller-Olm, M., Schmidt, D., Steffen, B.: Model-checking: A Tutorial Introduction. In: Cortesi, A., Filé, G. (eds.) SAS 1999. LNCS, vol. 1694, pp. 330–354. Springer, Heidelberg (1999)
6. Naldi, A., Berenguier, D., Fauré, A., Lopez, F., Thieffry, D., Chaouiya, C.: Logical modelling of regulatory networks with GINsim 2.3. Biosystems 97(2), 134–139 (2009)
7. Pecheur, C., Raimondi, F.: Symbolic Model Checking of Logics with Actions. In: Edelkamp, S., Lomuscio, A. (eds.) MoChArt IV. LNCS (LNAI), vol. 4428, pp. 113–128. Springer, Heidelberg (2007)
8. Sánchez, L., Chaouiya, C., Thieffry, D.: Segmenting the fly embryo: a logical analysis of the segment-polarity cross-regulatory module. Int. J. Dev. Biol. 52(8), 1059–1075 (2008)
9. Sánchez, L., Thieffry, D.: Segmenting the fly embryo: a logical analysis of the pair-rule cross-regulatory module. J. Theor. Biol. 224(4), 517–537 (2003)
10. Thomas, R., D'Ari, R.: Biological Feedback. CRC Press (1990)

Tackling Misleading Peptide Regulation Fold Changes in Quantitative Proteomics

Christoph Gernert, Evelin Berger, Frank Klawonn, and Lothar Jänsch

Abstract. Relative quantification in proteomics is a common strategy to analyze differences in biological samples and time series experiments. However, the resulting fold changes can give a wrong picture of the peptide amounts contained in the compared samples.

Fold changes hide the actual amounts of peptides. In addition posttranslational modifications can redistribute over multiple peptides, covering the same protein sequence, detected by mass spectrometry.

To circumvent these effects, a method was established to estimate the involved peptide amounts. The estimation of the theoretical peptide amount is based on the behavior of the peptide fold changes, in which lower peptide amounts are more susceptible to quantitative changes in a given sequence segment.

This method was successfully applied to a time-resolved analysis of growth receptor signaling in human prostate cancer cells. The theoretical peptide amounts show that high peptide fold changes can easily be nullified by the effects stated above.

1 Introduction

Bottom-up proteomics is a common way to identify posttranslational modifications, that are essential for signal transduction in cells. Proteins are digested into many shorter amino acid sequences, also called peptides. The peptides identified by mass spectrometry are reassigned to proteins by search engines for protein identification, e.g. Mascot or SEQUEST [6, 9].

Christoph Gernert · Evelin Berger · Lothar Jänsch
Helmholtz Centre for Infection Research
e-mail: Christoph.Gernert@Helmholtz-HZI.de

Frank Klawonn
Ostfalia University of Applied Science,
Helmholtz Centre for Infection Research

M.P. Rocha et al. (Eds.): 6th International Conference on PACBB, AISC 154, pp. 269–276.
springerlink.com

In quantitative proteomics peptides from various samples are differentially labeled by isotope-coded tags like iTRAQ and SILAC [4,5]. With iTRAQ, the relative ratio of the samples can be derived from the peaks in the scheduled reporter region of the peptide mass spectrum, 114 to 121 m/z. In complex protein samples absolute quantification is technically difficult. Relative quantification is often the only way to analyze differences in various biological samples, e.g. time series experiments, where cells are stimulated and observed at specific time points [12].

In any case, the relative amounts of the peptides in the samples, represented by the ratio of the reporter intensities, leads to the notorious regulation fold changes, often represented by heat maps or bar diagrams [11]. However, relative quantification of proteomics experiments aided by MS can produce many unwanted effects that can lead to flawed assumptions. Biologists look for highly positive or negative regulated peptides, to identify key events in the cell's signal transduction network. Problems with the peptide ionization in the MS device, noisy mass spectra [8], missed or wrongly associated peptides are only a few problems that can occur and distort the fold change on which the experiment analysis is based.

Despite technically related errors, misinterpretation can happen by not taking into account the total amount of the peptides. In addition, redistribution of phosphorylations between sites on peptides can produce conspicuous fold changes. These two problems are tackled in this work by organizing the peptides into groups and calculate their theoretical amount, which leads to the relative modified amount of each phosphorylation site. Based on this estimation also the theoretical peptide amount of the reporter reference can be determined.

In Section 2 the most common possibilities for peptide regulation distortion are listed. Utilized techniques for the estimation of the theoretical peptide amount and the fundamental quantification are given in section 3 *Methods*. In section 4 *Results* the mentioned methods are applied to a dataset from a signaling study. Possible problems that can appear by using this approach are considered in section 5 *Discussion*.

2 Known Problems of Relative Quantification in MS

Ionization of peptides. The detection of the peptides through MS depends on the ionisation. In most MS devices, peptides are ionized and accelerated by a magnetic or electric field before they can be detected. Because different peptides show differences in their molecular structure, some peptides can be ionized easier than others and even accumulate multiple charges [7]. If the abundance of a hard to ionize peptide is very low, the chance is high that it will not be detected.

Peptide Modification. Despite important posttranslational modifications like phosphorylation, which are important for the signaling of the cell, there can be other modifications. Only a limited number of posttranslational modifications can be set in search engines like Mascot. The protein search engine will not find peptides with modifications, other than the stated ones.

Arginine, lysine, and proline can oxidize to carbonyl derivatives [14]. Oxidation occurs in the aging process of peptides and it is a very common modification. On the other hand, modified peptides with low abundance tend to be not detected.

Miscleavages. In some cases, the digestion of the proteins is not complete. Trypsin digestion should cut the protein sequences after every lysine and arginine [13]. In some cases, trypsin can miss a lysine or arginine and the sequence will not be cut. This effect is called miscleavage [2]. Again, miscleaved peptides with low abundance tent to be not detected in MS. This can lead to very drastic fold changes.

Noisy Mass Spectra. Noisy spectra can influence the quantification of peptides, especially if the intensity of the reporter peaks is low. If the noise forms a big part of the peak, the outcome of the regulation can vary a lot. Previously work was done to estimate the variation of the peptide regulation caused by noise [8].

Sample Complexity. Ideally, the whole protein sequence should be covered by the digested peptides. Often this is not the case. Peptides can be missed by the reasons described above or through high sample complexety. In this case the MS is unable to single out specific peptides from the mixture [10].

In a complex protein mixture there is also a high chance that one peptide sequence can be associated with more than one protein. Often reliable estimation of the peptide regulation only can be done based on a unique protein association.

Results of Missed Peptides. The most dramatic error happens if two samples are compared, in which a certain peptide was detected in the first sample but not in the second one. The peptide regulation in this case is theoretically infinite.

This regulation can be reasonable, e.g. if a disease prevents phosphorylation of a certain protein and the modified version of the peptide is not present in the sample. It is very hard to tell if this effect is of biological or technical nature.

3 Methods

3.1 Quantification

The Quantification of the iTRAQ reporter intensities, which is based on the mascotdatfile Java library[1], calculates $\log_2$ fold changes for each peptide.

iTRAQ Signal Detection. The Kolmogorov–Smirnov test is used for automatic determination of the used iTRAQ reporters in the experiment. Hereby iTRAQ 4-plex and 8-plex kits are supported with arbitrary reporter allocation. For each possible reporter base, a result file is calculated.

Implicit Noise Recognition. As MS device an LTQ Orbitrap Velos with ESI ion source is used [1]. The MS device can detect the same peptide in multiple mass spectra. Each spectrum provides a different noise level, depending on the peak intensities. Since these spectra represent the same entity, iTRAQ reporter intensities can

[1] `http://code.google.com/p/mascotdatfile`

be aggregated for the calculation of one peptide's fold change. Spectra with lower intensities are more likely influenced by noise. Therefore, probable noisy spectra will only represent a small share of the total peptide intensity, which is used for the calculation of the fold change.

3.2 *Peptide Groups*

A certain peptide group consists of several modified versions of the same peptide sequence, which can be assigned uniquely to one protein. Optionally a completely unmodified peptide can be present in the group. However, this is not necessary for the further calculations. Peptide groups also can contain miscleaved peptides.

3.3 *Fold Changes*

Changes in quantity are stated as $\log_2$ fold changes in the result files of the basic quantification. Results from quantitative proteomics experiments are often published as $\log_2$ fold changes, if distinctions in very high positive or negative quantitative changes are negligible. For internal calculations also fold changes are used, where values less than one are denoted as the negative of its inverse, Eq. (2). The disadvantage of the method is that the fold changes "1" and "-1" represent the same value. In further calculations, this gap between 1 and -1 should be considered.

3.4 *Theoretical Peptide Amount*

Fold changes in one peptide group are distributed around the fold change of the whole protein, which is calculated from all unique assigned peptides.

An up regulation of one peptide automatically leads to the down regulation of one or more other peptides in a group, if the protein fold change itself is unchanged. Since the peptide fold changes are balanced around the protein fold change, it is assumed that in one peptide group the distance from the peptide to protein fold is inversely proportional to the peptide amount. That means that high positive or negative fold changes of small peptide amounts in one group are equalized by low fold changes of peptides with higher amounts.

3.5 *Calculation*

γ is the $\log_2$ fold change of a peptide i in a peptide group, normalized by the protein fold change, Eq. (1). $\log_2$ fold changes are given by the previous quantification of the reporter ion intensities for the purpose of symmetry.

The peptide regulation is projected to a fold change value where less than one is stated as the negative of its inverse $FC'_{pep\,i}$, Eq. (2). The properties of the logarithmic function are no longer valuable, since linear behaviour of the fold changes is needed for the calculation of the theoretical peptide amount.

The theoretical peptide amount for i in the iTRAQ reporter reference $q_{pep\,i}$ is the inverse modulus of peptide fold change $FC'_{pep\,i}$ divided by the protein fold change, Eq. (3). This amount multiplied by the fold change of the peptide i results the theoretical peptide amount of the reporter values $q'_{pep\,i}$, Eq. (4).

$$\gamma = \log_2 (FC_{pep\,i})_{norm} = \log_2 (FC_{prot}) - \log_2 (FC_{pep\,i}) \tag{1}$$

$$FC'_{pep\,i} = \begin{cases} 2^{\gamma} & \text{for } \gamma \geq 1 \\ -\dfrac{1}{2^{\gamma}} & \text{else} \end{cases} \tag{2}$$

$$q_{pep\,i} = \frac{1}{FC_{prot} \cdot \left| FC'_{pep\,i} \right|} \tag{3}$$

$$q'_{pep\,i} = q_{pep\,i} \cdot FC_{pep\,i} \tag{4}$$

3.6 *Site-Wise Representation*

Peptides with more than one modification site in their sequence will be detected in various combinations by MS. Due to the redistribution of the modified sites over the different peptides, these combinations can show differences in their regulation, but the amount of the modifications stays the same. The amount of a modification will stay balanced, if e.g. one peptide, containing a specific modification site, is up regulated and another peptide containing the same and another site is down regulated.

If the amount or in this case the theoretical amount of the peptides is known, the peptide amounts of a specific modification site $pep \in M$ can be summed up and set against the peptide amounts in which this site is not present $pep \notin M$; the unmodified fraction, Eq. (5).

$$r_{mod} = \frac{\sum\limits_{pep \in M} q_{pep}}{\sum\limits_{pep \notin M} q_{pep} + \sum\limits_{pep \in M} q_{pep}} \tag{5}$$

4 Results

The concepts stated in section 3 *Methods* were applied to a time series experiment for the estimation of the peptide amounts for each time point. For a time-resolved analysis of growth receptor signaling, human prostate cancer cells were stimulated for 3, 6, and 20 minutes and compared to a non-stimulated sample. Proteins were isolated, digested and phosphopeptides were enriched as shown elsewhere [3, 15]. Peptides of the four samples were quantitatively labeled with iTRAQ according to the manufacturer's protocol (Applied Biosystems), purified for MS and separated via nano-LC-MS/MS. Subsequently, the generated peptide spectra were queried against the UniProtKB/Swiss-Prot database using Mascot Daemon.

In this signaling experiment the AHNK_HUMAN protein shows high peptide fold changes, especially in peptide group 19. The peptide with the sequence LP*p*SGSGAA*p*SPTGSAVDIR, providing phosphorylation sites S210 and S216, shows the most interesting fold changes (Table 1).

Table 1 $\log_2$ fold changes for peptide group 19 of AHNK_HUMAN. In the peptide sequence, phosphorylations are marked as *p* followed by the modified amino acid. The site numbers and fold changes are stated in the following columns. The protein mean is calculated from all observed peptides assigned to the AHNK_HUMAN protein.

Peptide sequence	Phos. sites	3 Min	6 Min	20 Min
LPSGSGAASPTGSAVDIR	none	-0.1957	-0.2257	-0.1385
LP*p*SGSGAA*p*SPTGSAVDIR	S210, S216	0.3346	*0.8422*	*1.5323*
LPSG*p*SGAASPTGSAVDIR	S212	0.0951	0.5494	0.4464
LPSGSGAA*p*SPTG*p*SAVDIR	S216, S220	-0.0041	0.1630	0.0008
LPSG*p*SGAA*p*SPTGSAVDIR	S212, S216	0.5320	0.6256	0.1437
LP*p*SGSGAASPTGSAVDIR	S210	-0.2160	0.4247	0.6734
LPSGSGAA*p*SPTGSAVDIR	S216	0.2040	0.6073	0.2766
protein mean	all discovered	0.0222	0.1878	0.2037

As the theoretical peptide amount of the reporter reference was calculated based on each available reporter ratio, the protein fold change was reliably estimated by the mean [16] of all 161 peptides assigned to the AHNK_HUMAN protein, Eq. (3). The peptide that carries phosphorylations at the sites S212 and S216, and the peptide that carries phosphorylations at the sites S210 and S216 show relatively high variations in their amount (Fig 1a).

As they share the same phosphorylation site S216, a redistribution of the phosphorylation over the peptide can be assumed. Therefore site-wise representation was applied based on the theoretical peptide amounts, Eq. (5). This produces a consistent result of the phosphorylation site shares in the reporter reference (Fig 1b).

On basis of the theoretical peptide amount of the reference, the amount of the other reporter values were calculated Eq. (4) and site-wise representation was applied (Fig 1c). For an overview of all experimental time-points, the mean of the reference peptide amount (Fig 1b) was subjoined.

The high fold changes of the peptide, which carries the phosphorylation sites S210 and S216, was abrogated by a relatively low peptide amount and the redistribution of its phosphorylation sites on other peptides.

5 Discussion

We have demonstrated that our proposed method is able to tackle some problems that occur in relative quantification. However, other problems described in section 2 still remain unsolved, primarily the problem of the missed peptides in MS.

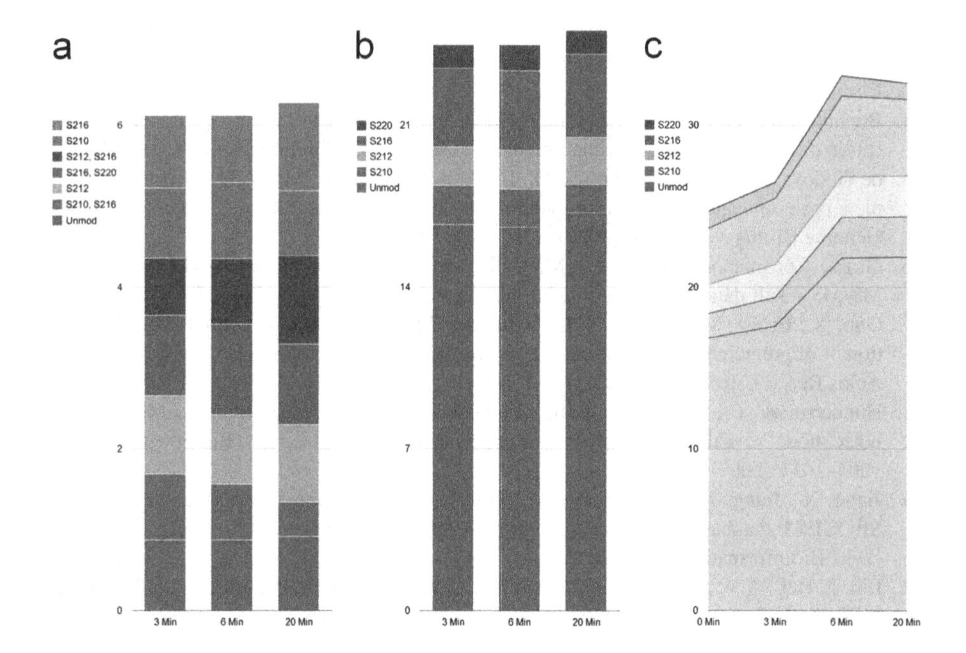

Fig. 1 a) The theoretical peptide amount in the reporter references calculated by the three reporter ratios. b) The site-wise representation of the calculated data. c) The combined site-wise representation of the theoretical peptide amount of the reporter reference and the other reporter values.

Missed peptides will distort the balance in a peptide group and lead to false results. In this case, the amount of the detected peptides is overrated in each instance. In addition, this method strongly relies on the correct protein fold change. The total estimated peptide amount can shift to a great degree, if the protein fold change is estimated incorrectly.

Distortions between the reporter ratios can be observed in the site-wise representation of the reference (Fig. 1b). Conspicuous variations at this point indicate that the estimation of the theoretical peptide amount is probability wrong. In the future work site-wise representation of the reference will be used to validate datasets and further analyze the cause of the distortions in the representation.

References

1. Adachi, J., Kumar, C., Zhang, Y., Olsen, J.V., Mann, M.: The human urinary proteome contains more than 1500 proteins, including a large proportion of membrane proteins. Genome Biol. 7, R80 (2006)
2. Alves, G., Ogurtsov, A.Y., Kwok, S., Wu, W.W., Wang, G., Shen, R.F., Yu, Y.K.: Detection of co-eluted peptides using database search methods. Biol. Direct 3, 27 (2008)

3. Bodenmiller, B., Aebersold, R.: Quantitative Analysis of Protein Phosphorylation on a System-Wide Scale by Mass Spectrometry-Based Proteomics, vol. 470, pp. 317–334. Elsevier (2010)
4. Boehm, A.M., Pütz, S., Altenhöfer, D., Sickmann, A., Falk, M.: Precise protein quantification based on peptide quantification using iTRAQ. BMC Bioinformatics 8, 214 (2007)
5. de Godoy, L., Olsen, J., de Souza, G., Li, G., Mortensen, P., Mann, M.: Status of complete proteome analysis by mass spectrometry: SILAC labeled yeast as a model system. Genome Biology 7(6), R50 (2006)
6. Grosse-Coosmann, F., Boehm, A.M., Sickmann, A.: Efficient analysis and extraction of MS/MS result data from mascot result files. BMC bioinformatics 6 (2005)
7. Guo, X., Bruist, M.F., Davis, D.L., Bentzley, C.M.: Secondary structural characterization of oligonucleotide strands using electrospray ionization mass spectrometry. Nucleic Acids Res. 33, 3659–3666 (2005)
8. Hundertmark, C., Fischer, R., Reinl, T., May, S., Klawonn, F., Jänsch, L.: Ms-specific noise model reveals the potential of itraq in quantitative proteomics. Bioinformatics 25, 1004–1011 (2009)
9. Jiang, X., Jiang, X., Han, G., Ye, M., Zou, H.: Optimization of filtering criterion for SEQUEST database searching to improve proteome coverage in shotgun proteomics. BMC Bioinformatics 8, 323 (2007)
10. Liu, J., Bell, A.W., Bergeron, J.J., Yanofsky, C.M., Carrillo, B., Beaudrie, C.E., Kearney, R.E.: Methods for peptide identification by spectral comparison. Proteome Science 5(1), 3 (2007)
11. Milano, A., Pendergrass, S.A., Sargent, J.L., George, L.K., McCalmont, T.H., Connolly, M.K., Whitfield, M.L.: Molecular subsets in the gene expression signatures of scleroderma skin. PLoS ONE 3(7), e2696 (2008)
12. Reinl, T., Nimtz, M., Hundertmark, C., Johl, T., Kéri, G., Wehland, J., Daub, H., Jänsch, L.: Quantitative phosphokinome analysis of the met pathway activated by the invasin internalin b from listeria monocytogenes. Mol. Cell Proteomics 8, 2778–2795 (2009)
13. Stigter, E.C., de Jong, G.J., van Bennekom, W.P.: Development of an open-tubular trypsin reactor for on-line digestion of proteins. Anal. Bioanal. Chem. 389, 1967–1977 (2007)
14. Tiku, M.L., Narla, H., Jain, M., Yalamanchili, P.: Glucosamine prevents in vitro collagen degradation in chondrocytes by inhibiting advanced lipoxidation reactions and protein oxidation. Arthritis Res. Ther. 9, R76 (2007)
15. Villén, J., Gygi, S.P.: The SCX/IMAC enrichment approach for global phosphorylation analysis by mass spectrometry. Nature Protocols 3(10), 1630–1638 (2008)
16. Webb-Robertson, B.J.M., Matzke, M.M., Jacobs, J.M., Pounds, J.G., Waters, K.M.: A statistical selection strategy for normalization procedures in lc-ms proteomics experiments through dataset-dependent ranking of normalization scaling factors. PROTEOMICS (2011)

Coffee Transcriptome Visualization Based on Functional Relationships among Gene Annotations

Luis F. Castillo, Oscar Gómez-Ramírez, Narmer Galeano-Vanegas, Luis Bertel-Paternina, Gustavo Isaza, and Álvaro Gaitán-Bustamante

Abstract. Simplified visualization and conformation of gene networks is one of the current bioinformatics challenges when thousands of gene models are being described in an organism genome. Bioinformatics tools such as BLAST and Interproscan build connections between sequences and potential biological functions through the search, alignment and annotation based on heuristic comparisons that make use of previous knowledge obtained from other sequences. This work describes the search procedure for functional relationships among a set of selected annotations, chosen by the quality of the sequence comparison as defined by the coverage, the identity and the length of the query, when coffee transcriptome sequences were compared against the reference databases UNIREF 100, Interpro, PDB and PFAM. Term descriptors for molecular biology and biochemistry were used along the wordnet dictionary in order to construct a Resource Description Framework (RDF) that enabled the finding of associations between annotations.

Luis F. Castillo · Gustavo Isaza
Universidad de Caldas. Manizales (Colombia),
Departamento de Sistemas e Informática
e-mail: {luis.castillo,gustavo.isaza}@ucaldas.edu.co

Luis F. Castillo
Universidad Nacional de Colombia-Manizales,
Departamento Ing. Industrial

Narmer Galeano-Vanegas · Álvaro Gaitán-Bustamante
Centro Nacional de Investigación del Café- CENICAFÉ
e-mail: {narmer.galeano,alvaro.gaitan}@cafedecolombia.com

Oscar Gómez-Ramírez · Luis Bertel-Paternina
Universidad Autónoma de Manizales, Grupo Investigación Ing. Del Software
e-mail: lbertel@umanizales.edu.co

M.P. Rocha et al. (Eds.): 6th International Conference on PACBB, AISC 154, pp. 277–283.
springerlink.com

Sequence-annotation relationships were graphically represented through a total of 6845 oriented vectors. A large gene network connecting transcripts by way of relational concepts was created with over 700 non-redundant annotations, that remain to be validated with biological activity data such as microarrays and RNA-seq. This tool development facilitates the visualization of complex and abundant transcripotome data, opens the possibility to complement genomic information for data mining purposes and generates new knowledge in metabolic pathways analysis.

Keywords. Gene ontology, vector visualization, metadata relationship, transcription network.

1 Introduction

Currently, the structural description of gene models in genomics research must be complemented with putative functional data and, on a larger scale, with associations to gene networks. Widely used bioinformatics tools such as BLAST and Interproscan are the first step towards the annotation of cDNA and protein sequences that belong to the transcriptome of species that have been scarcely studied at the experimental level [1][2][3]. Among these species is coffee, a crop circumscribed to tropical regions, and the second most important world commodity after oil. Bioinformatics developments applied to coffee genome studies are key for the exploitation of the biodiversity present in germplasm collections and the generation of new plant varieties that will enable coffee producers to face the challenges posed by economical, commercial and environmental demands. Networks are used ubiquitously throughout biology to represent the relationships between genes and gene products, and several tools are available for complex networks visualization [9]. These applications allow the display of existing relationships in a data set, helping in the interpretation and understanding in a simple way the connections in a wide data set. Among the representation utilities for bioinformatics and semantics are well-know examples such as Cytoscape, Zentity, VisANT, Pathway Studio and Patika, that are extensible in functionality through plugins. Citoscape is widely used in projects of data representation in biological, genomic and proteomic systems. Zentity and Pathway Studio are tools developed by Microsoft; Zentity is for visualize the existing relations in management process of research documents, and Pathway Studio is an application that combines an extensive set of features with a polished graphic user interface. It includes, most notably, customizable network display styles for assigning visual attributes such as node color, size and shape, multi-user support, and subcellular localization. Both tools provide development utilities for extend their functionalities. However, there is a lack of pipelines to perform data mining in genomic databases using semantics, in order to generate connections among the gene entities that have been annotated, and that could be visualized later on. This document presents a procedure to find explicit and implicit relationships among gene models based on their corresponding annotations, and applies the method to a coffee genomics database.

2 Methods

The experimental dataset was built with annotations from selected transcriptome sequences (contigs) from the Coffea species C. arabica, C. canephora, C. kapakata and C. liberica, publicly available in the Genbank and annotated at the National Coffee Research Center (Cenicafe). Out of 58,329 coffee transcriptome entries, those containing as the only description "no hits found" where eliminated in a first filtering round, followed by those annotated as "whole genome shotgun" [4] (Table 1). For the remaining 15,756 sequences, a quality threshold was determined based on the BLAST results which involved the percentage coverage of the reference sequence, the length of the query sequence involved in the match and the percentage similarity between the two local alignments [5]. In addition, a threshold was also determined for the Interproscan annotations, considering the e-value and the length of the match. Remaining redundancy after the two filters was eliminated to keep a single representative for each contig.

Table 1 Annotation status summary of the Coffea sp. transcripts at Cenicafe´s database by January 2011.

Species	Annotated contigs			Contigs with No hits	Total per species
	Annotated as "whole genome shotgun"	Annotation different from "whole genome shotgun"	Subtotal		
C canephora	5892	3336	9228	3929	13157
C arabica	16488	8250	24738	6735	31473
C kapakata	884	655	1539	508	2047
C liberica	6374	3515	9889	1763	11652
Total	29638	15756	45394	12935	58329

The procedure to find associations among selected transcripts was supported on a corpus created with the term descriptors listed in the Oxford Dictionary for Molecular Biology and Biochemistry [6], and enriched with the wordnet [7] set. Concepts present in the annotations were related to transcripts using oriented vectors and visualized in two dimensions. Annotations of the experimental dataset were encoded as Resource Description Framework (RDF) triplets, that served as entries for semi-automatic ontological analysis with Methontology [8]. Relationships among concepts and transcripts were then visualized using a graph visualization library using web workers and jQuery from http://arborjs.org.

3 Results and Discussion

An experimental set of sequences with biologically relevant annotations was selected from the available coffee transcripts. For the four coffee species included, the distribution of the transcripts according to the threshold parameters defined was the same (Figure 1, a to d), indicating that less than 8% of sequences (2,687) met simultaneously the three requirements of at least 40% similarity, 40% coverage and 400 bp in length, that defined significant confidence in the annotation. A comparison with the transcript set of the model species Arabidopsis thaliana showed that 35,5% (9,024) of the transcripts passed the same filter (data not shown). This reflects the current situation in gene annotation among plant species, where many of the genes found lack a significant putative annotation.

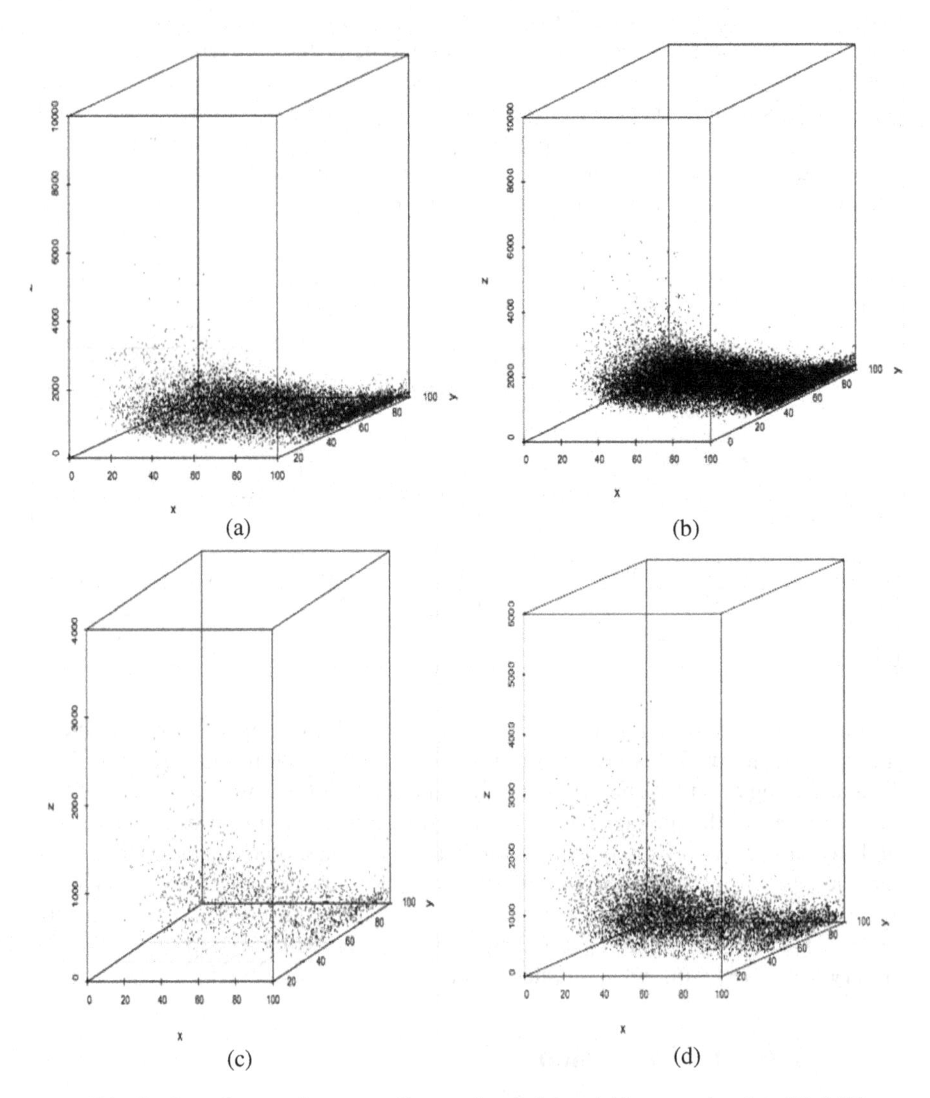

Fig. 1 Distribution of transcripts according to threshold variables associated to BLAST results. (a). C. canephora; (b). C. arabica; (c). C. kapakata; (d). C. liberica; x-axis: percentage coverage, y-axis: percentage identity, z-axis: contig length.

To widen the dataset, sequences displaying e-values smaller than e-20 and at least 200 bp of matching length in Interproscan (Figure 2) were added, completing the experimental set to 6,845 sequences. The oriented vector visualization (Figure 3) indicates that for some concepts, there are numerous transcripts associated, while in infrequent cases there were concepts unique to one or a few transcript annotations.

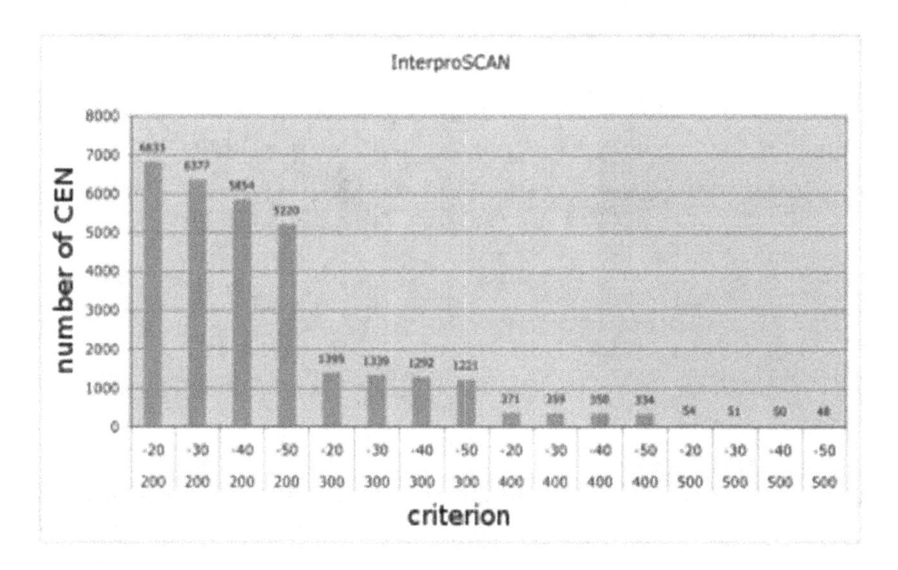

Fig. 2 Distribution of transcripts according to threshold variables associated to Interproscan results.

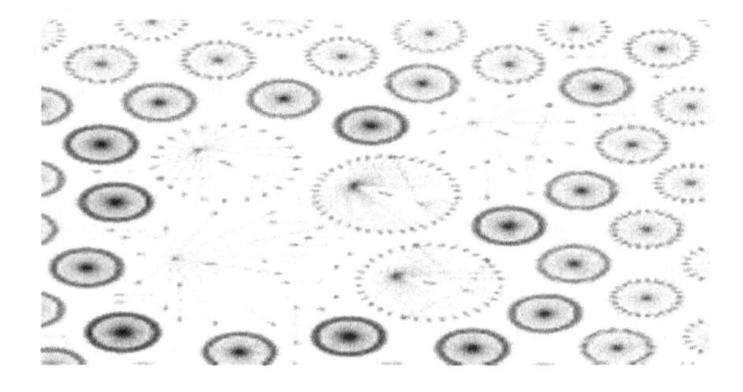

Fig. 3 Oriented vectors visualized in two dimensions. The base of the arrow corresponds to the annotation, while the tip represents the contig identification

Finally, a complex network of transcripts can be constructed based on the method applied (Figure 4), that still requires a manual examination to verify the biological correspondence of the connections.

Fig. 4 Complex network of coffee transcripts constructed on RDF associations from transcript annotations. Green: Contig annotation; Yellow: Relational Concept; Red: Annotation concep

4 Conclusions and Future Work

A semantic tool immersed in a biological environment was constructed to enable the coding and interchange of annotation data in the set of transcripts of an organism. Relationships found with this tools must resemble metabolic pathways already described in the scientific literature, but also new and uncovered associations that must be confirmed with wet-lab data such as microarrays and RNAseq expression experiments. The structured data can be reused to be mined with other metadata and complement the annotations on a genomic project

Acknowledgments. This work was supported for Cenicafé, Universidad Autónoma de Manizales and Universidad de Caldas.

References

[1] Li, W., Feng, J., Jiang, T.: 'Workshop: Transcriptome assembly from RNA-Seq data: Objectives, algorithms and challenges. In: 2011 IEEE 1st International Conference on Computational Advances in Bio and Medical Sciences (ICCABS), February 3-5, p. 271 (2011), doi:10.1109/ICCABS.2011.5729925

[2] Jing, L., Ng, M.K., Liu, Y.: Construction of Gene Networks With Hybrid Approach From Expression Profile and Gene Ontology. IEEE Transactions on Information Technology in Biomedicine 14(1), 107–118 (2010), doi:10.1109/TITB.2009.2033056
[3] Zheng, H., Azuaje, F., Wang, H.: seGOsa: Software environment for gene ontology-driven similarity assessment. In: 2010 IEEE International Conference on Bioinformatics and Biomedicine (BIBM), December 18-21, pp. 539–542 (2010), doi:10.1109/BIBM.2010.5706624
[4] Zhi, D., Keich, U., Pevzner, P., Heber, S., Tang, H.: Correcting Base-Assignment Errors in Repeat Regions of Shotgun Assembly. IEEE/ACM Transactions on Computational Biology and Bioinformatics 4(1), 54–64 (2007), doi:10.1109/TCBB.2007.1005
[5] Essinger, S.D., Rosen, G.L.: The Effect of Sequence Error and Partial Training Data on BLAST Accuracy. In: 2010 IEEE International Conference on BioInformatics and BioEngineering (BIBE), May 31-June 3, pp. 257–262 (2010), doi:10.1109/BIBE.2010.49
[6] Langari, Z., Tompa, F.W.: Subject classification in the Oxford English Dictionary. In: Proceedings IEEE International Conference on Data Mining, ICDM 2001, pp. 329–336 (2001), doi:10.1109/ICDM.2001.989536
[7] Ngo, V.M., Cao, T.H., Le, T.M.V.: Combining Named Entities with WordNet and Using Query-Oriented Spreading Activation for Semantic Text Search. In: 2010 IEEE RIVF International Conference on Computing and Communication Technologies, Research, Innovation, and Vision for the Future (RIVF), November 1-4, pp. 1–6 (2010), doi:10.1109/RIVF.2010.5633401
[8] Abulaish, M.: Relation Characterization Using Ontological Concepts. In: 2011 Eighth International Conference on Information Technology:New Generations (ITNG), April 11-13, pp. 585–590 (2011), doi:10.1109/ITNG.2011.107
[9] Matthew, S., Michael, H.: Tools for visually exploring biological networks. Bioinformatics Review 23(20), 2651–2659 (2007), doi:10.1093/bioinformatics/btm401

Author Index

Abáigar, María 121
Álvarez-Jarreta, Jorge 105
Alves, Joãd 129
Alves, Marco 129
Audic, Yann 33
Azevedo, Nuno F. 113

Baldwin, C.T. 197
Benito, Rocío 121, 209
Berger, Evelin 269
Bertel-Paternina, Luis 277
Bisbal, Jesus 63
Blanco, Roberto 95
Blanzieri, Enrico 33
Borrajo, L. 87
Broka, A. 197

Calvo-Dmgz, D. 53
Camacho, Rui 129
Campo, Livia 79
Carbonell, Jaime 43
Carreiro, André V. 11
Carvalhal, Carlos 73
Castaño, Andrés P. 157
Castillo, Andrés 157
Castillo, Luis F. 277
Chaouiya, Claudine 259
Colom, José Manuel 95, 231
Corchado, Juan M. 79
Correia, Sara 241
Cristóvão, Filipe 21

de Miguel Casado, Gregorio 95
De Paz, Juan F. 79, 121, 209

Deusdado, Leonel 73
Deusdado, Sérgio 73

Engelbrecht, Gerhard 63

Fdez-Riverola, Florentino 53, 189, 225
Fernandez, J.J. 137
Ferreira, Andreia 113
Ferreira, Artur J. 11
Figueiredo, Mário A.T. 11
Fonseca, Nuno A. 225
Frangi, Alejandro F. 63

Gaitán-Bustamante, Álvaro 277
Galeano-Vanegas, Narmer 277
Galvez, J.F. 53
Garcia, Enrique 79
Garcia, I. 137
Garcia, Sara P. 217
Gaspar, Paulo 43
Gernert, Christoph 269
Glez-Peña, Daniel 53, 189
Gómez-López, Gonzalo 189
Gómez-Ramírez, Oscar 277
Gonçalves, Emanuel 251
González, Roberto 209
Guivarch, Ronan 1

Heidenreich, Elvio 157
Hernández, Jesús M. 121, 209
Hernández, María 209

Iglesias, E.L. 87
Isaza, Gustavo 277

Jänsch, Lothar 269

Klawonn, Frank 269
Klings, E.S. 197

Lacey, S. 197
Lin, H. 197
Livi, Carmen Maria 33
Lopes, Pedro 173
López-Fernández, Hugo 189
Lourenço, Anália 113, 165, 181

Madeira, Sara C. 11, 21
Maia, Salomé 165
Martinez-Sanchez, A. 137
Mayordomo, Elvira 105
Medina-Medina, N. 197
Mendonça, Rafael 173
Monteagudo, Ángel 147
Monteiro, Pedro T. 259
Mouysset, Sandrine 1

Noailles, Joseph 1

Oliveira, Jorge 173
Oliveira, José Luís 43, 173

Paillard, Luc 33
Pereira, Luísa 129
Pereira, Maria Olívia 181
Pereira, Maria Olivia 113
Pinho, Armando J. 217
Pisano, David G. 189
Pratas, Diogo 217

Ramoa, Augusto 165
Reboiro-Jato, David 225
Reboiro-Jato, Miguel 189, 225
Requeno, José Ignacio 95, 231
Rocha, Hugo 173
Rocha, Isabel 251
Rocha, Miguel 241, 251
Rodríguez-Vicente, Ana 121
Romero, R. 87
Ruiz, Daniel 1
Ruiz-Pesini, Eduardo 105
Ruiz-Villa, Carlos A. 157

Santos, José 147
Santos, Rosário 173
Sebastiani, P. 197
Soares, Pedro 129
Sous, Ana Margarida 181
Steinberg, M.H. 197

Vera, Vicente 79
Vieira, Jorge 225
Vilarinho, Laura 173

Zato, Carolina 121, 209